Smart Innovation, Systems and Technologies

Volume 89

The Smart Innovation, Systems and Technologies book series encompasses the topics of knowledge, intelligence, innovation and sustainability. The aim of the series is to make available a platform for the publication of books on all aspects of single and multi-disciplinary research on these themes in order to make the latest results available in a readily-accessible form. Volumes on interdisciplinary research combining two or more of these areas is particularly sought.

The series covers systems and paradigms that employ knowledge and intelligence in a broad sense. Its scope is systems having embedded knowledge and intelligence, which may be applied to the solution of world problems in industry, the environment and the community. It also focusses on the knowledge-transfer methodologies and innovation strategies employed to make this happen effectively. The combination of intelligent systems tools and a broad range of applications introduces a need for a synergy of disciplines from science, technology, business and the humanities. The series will include conference proceedings, edited collections, monographs, handbooks, reference books, and other relevant types of book in areas of science and technology where smart systems and technologies can offer innovative solutions.

High quality content is an essential feature for all book proposals accepted for the series. It is expected that editors of all accepted volumes will ensure that contributions are subjected to an appropriate level of reviewing process and adhere to KES quality principles.

More information about this series at http://www.springer.com/series/8767

Valeriu Ungureanu

Pareto-Nash-Stackelberg Game and Control Theory

Intelligent Paradigms and Applications

Springer

Valeriu Ungureanu
Faculty of Mathematics
and Computer Science
Moldova State University
Chişinău
Moldova

ISSN 2190-3018 ISSN 2190-3026 (electronic)
Smart Innovation, Systems and Technologies
ISBN 978-3-030-09159-0 ISBN 978-3-319-75151-1 (eBook)
https://doi.org/10.1007/978-3-319-75151-1

Printed on acid-free paper

This Springer imprint is published by Springer Nature
The registered company is Springer International Publishing AG
The registered company address is: Gewerbestrasse 11, 6330 Cham, Switzerland

To

Valentina, Dorin, Mihai and Zina
Mihail, and Vladimir

Preface

This monograph is a result of prolonged work in the domain of game theory for nearly twenty years. It is dedicated to non-cooperative or strategic form games [1], simultaneous and sequential games [2], their mixtures and control [3]. Considered models are appropriate to games in extensive form [4–7]. Nevertheless, main interests remain essentially in the area of strategic form games.

The mathematical background for the book is somewhat more advanced than that for postgraduate students of the applied mathematics departments. It needs basic knowledge and skills from game theory, optimization methods, multi-criteria optimization, optimal control theory and fundamental mathematical disciplines as linear algebra, geometry, calculus and probability theory. Additionally, the book needs some knowledge of computer science foundations and the Wolfram language.

It must be mentioned that selected topics from all three parts of the book have been taught as an advanced specialization course "*Socio-Economical Problems and Game Theory*" for master's degree students of the applied mathematics specialization at Faculty of Mathematics and Computer Science at Moldova State University. Selected topics from this book were taught in the "*Game Theory*" and "*Operational Research*" courses for students of the same faculty.

The monograph consists of fifteen chapters divided into three parts that are dedicated respectively to non-cooperative games, mixtures of simultaneous and sequential multi-objective games, and to multi-agent control of Pareto-Nash-Stackelberg type. The Introduction chapter presents an overview. The book contains also the Bibliography, an Index, and a List of Symbols.

The book title may be seen as a compromise taken with the aim to have a short monograph's name which will reflect simply its content. It is an approximate synonym for the longer names "*theory of multi-objective multi-agent simultaneous and sequential games and optimal control mixtures*" or/and "*theory of multi-objective multi-leader multi-follower games and multi-agent control*". Sure, such names are seemed to be less acceptable. So, monograph's title was selected in order to be short and clear by associating it with the names of personalities who initiated well known branches of mathematics:

- **Pareto**—multi-objective/multi-criteria optimization,
- **Nash**—strategic/normal form simultaneous games,
- **Stackelberg**—strategic/normal form sequential games,
- **control of Pareto-Nash-Stackelberg type**—multi-objective multi-agent control taken as a mixture of simultaneous and sequential decision process in order to control states of a system.

The formal language used to expose *Pareto-Nash-Stackelberg game and control theory* is generally common for the enumerated above domains of mathematics. Nevertheless, its distinct features consist of being at the same time descriptive, constructive and normative [8]. More the more, the theory has its distinct and specific topics, models, concepts, problems, methods, results and large areas of investigations, extensions, and implementations. The purposes of the present work consist mainly and essentially of highlighting the mathematical aspects of the theory.

The monograph was prepared by the author himself in LaTeX. He is really conscious that it may admit some imperfections. So, suggestions, comments and observations are welcomed in order to continue efficiently investigations in a large spectrum of theoretical and practical unsolved problems. Undoubtedly, they will be treated very seriously, with great care and gratitude.

The book is addressed to researchers and advanced students both in mathematical and applied game theory, as well as multi-agent optimal control. Exposed theoretical results may have direct implementations in economic theory and different areas of human activity where strategic behaviour is underlying.

Chişinău, Moldova
November 2017

Valeriu Ungureanu

References

1. Alós-Ferrer, C., and K. Ritzberger. 2016. *The theory of extensive form games* (XVI+239 pp). Berlin: Springer.
2. Kuhn, H.W. 1953. Extensive games and the problem of information. In *Contributions to the theory of games*, Vol. II. *Annals of Mathematics Study*, ed. H. Kuhn and A. Tucker, Vol. 28, 217–243. Princeton: Princeton University Press.
3. Nash, J. 1951. Noncooperative games. *Annals of Mathematics* 54(2): 280–295.
4. Pontryagin, L.S., V.G. Boltyanskii, R.V. Gamkrelidze, and E.F. Mishchenko. 1961. *Mathematical theory of optimal processes* (393 pp) Moscow: Nauka (in Russian).
5. Von Neumann, J. 1928. Zur Theorie der Gesellschaftsspiele. *Mathematische Annalen* 100: 295–320 (in German).
6. Von Neumann, J., and O. Morgenstern. 1944. *Theory of games and economic behavior* (674 pp). Princeton, New Jersey: Annals Princeton University Press. 2nd edition, 1947.

7. Von Stackelberg, H. 1934. *Marktform und gleichgewicht* (*Market structure and equilibrium*) (XIV+134 pp). Vienna: Springer (in German).
8. Vorob'ev, N.N. 1984. *Foundations of game theory: Noncooperative games* (497 pp). Moscow: Nauka (in Russian); *Translated by Boas R.P.*, Basel-Boston: Birkhäuser, 1994.

Acknowledgements

I acknowledge Prof. D. Zambiţchi for introducing me in the wonderful world of mathematical research many years ago, Prof. V. Bălaş, for her support in launching this monograph, and all Springer staff who contributed to the realising of this project.

Contents

Symbols and Abbreviations

$\mathbb{N}$	Set of natural numbers
$\mathbb{N}^*$	$\mathbb{N}^* = \mathbb{N}\backslash\{0\}$
$\mathbb{N}^n$	Cartesian product of n sets $\mathbb{N}$
$\mathbb{Z}$	Set of integer numbers
$\mathbb{Q}$	Set of rational numbers
$\mathbb{R}$	Set of real numbers; real line
$\alpha, \beta, \ldots$	Notation for real numbers
$\mathbb{R}^n$	n-dimensional Euclidean space
N	$N = \{1, 2, \ldots, n\}$ is a set of players $1, 2, \ldots, n$
A^T	The transpose of a matrix A
$\mathbf{x} = x$	$\mathbf{x} = x = (x_1, x_2, \ldots, x_n)^T$— notation for $x \in \mathbb{R}^n$
$x \geq 0$	Sign constraints on x
$x > 0$	The components of vector x are positive
$x \geq y$	At least one component of x is strictly greater than the correspondent component of y
$x \geqq y$	Components of x are greater than components of y or they may be all equal
$x_i \gtreqless 0$	Greater than 0, equal to 0 or less than 0
$\alpha \gg \beta$	α is much greater than β
X, Y	Notation for subset of $\mathbb{R}^n$
$X \subseteq Y$	X is a subset of Y
$X \subset Y$	X is a proper subset of Y
$x \in X$	x is an element of X
$X \cap Y$	Intersection of X and Y
$X \cup Y$	Union of X and Y
$\|X\|, \#X$	Cardinality (power) of X
$\dim(X)$	Dimension of X
$f : X \rightarrow R$	Function f
$f(x),\ x \in X$	Alternative function notation
$f : X \multimap Y$	Multivalued mapping f

$\mathscr{P}(X)$	Power set of X
$\mathrm{gr}f(x)$	Graph of $f(x)$
$\mathrm{epi}f(x)$	Epigraph of $f(x)$
$\mathrm{hyp}f(x)$	Hypo-graph of $f(x)$
Γ	Normal form game $\Gamma = \langle N, \{X_p\}_{p\in N}, \{f_p(x)\}_{p\in N}\rangle$
Gr_p	Graph of pth player best response mapping
$NES(A,B)$	Nash Equilibrium Set function of A and B
γ^A, γ^B	Knowledge vectors
$\Gamma_{\gamma^A\gamma^B}$	Game with knowledge vectors γ^A *and* γ^B
$\hat{x}^{-p}$	$\hat{x}^{-p} = (\hat{x}^1, \hat{x}^2, \ldots, \hat{x}^{p-1}, \hat{x}^{p+1}, \ldots, \hat{x}^n)$
X_{-p}	$X_{-p} = X_1 \times X_2 \times \ldots \times X_{p-1} \times X_{p+1} \times \ldots \times X_n$
$\hat{x}^p \Vert \hat{x}^{-p}$	$\hat{x}^p \Vert \hat{x}^{-p} = (\hat{x}^p, \hat{x}^{-p}) = (\hat{x}^1, \hat{x}^2, \ldots, \hat{x}^{p-1}, \hat{x}^p, \hat{x}^{p+1}, \ldots, \hat{x}^n) = \hat{x}$
$f'(x)$	Gradient of $f(x)$
$O(\alpha)$	Infinitesimal of higher order than α
$L(x; u_0, u)$	Lagrangian function
$x^T y$	Scalar product of x and y
$\langle x, y\rangle$	Scalar product of x and y
$A_{m\times n}$	Matrix $A[m \times n]$ with m rows and n columns
a_{ij}	Element of a matrix A
AB	Matrix product
A^{-1}	Inverse of A
E	Identity matrix
$\mathrm{rank}(A)$	Rank of A
$\det(A)$, $\|A\|$	Determinant of A
$\Vert x \Vert$	$\Vert x \Vert = \sqrt{x_1^2 + \ldots + x_n^2} = \sqrt{x^T x}$ is the Euclidean norm of $x \in \mathbb{R}^n$
$\rho(x,y)$	$\rho(x,y) = \Vert x - y \Vert$ is the distance between x and y
$V_\varepsilon(x^*)$	$V_\varepsilon(x^*) = \{x \in X \mid \Vert x - x^* \Vert < \varepsilon\}$ is the ε-neighbourhood of x^*
$\mathrm{int}(X)$	Interior of a set X
$\bar{X}$	Closer of a set X
$\lfloor x \rfloor$	Integer part of $x \in R$; Maximal integer not greater than $x \in R$
$\lceil x \rceil$	Minimal integer not smaller than $x \in R$
$\{x\} = x - \lfloor x \rfloor$	Fractional part of $x \in R$
$\exists x$	Existential quantifier: "exists at least one" x
$\forall x$	Universal quantifier: "for all" x
$\|$, :	With properties
$\square$	End of a proof
$:=$, $=$	Assignment operators
$1^\circ, 2^\circ, \ldots$	Numbering steps of an algorithm
MNT	Maximin-Nash taxon
MT	Maximin taxon
NES	Nash Equilibrium Set
NSE	Set of Nash-Stackelberg equilibria
NT	Nash taxon
ONT	Optimum-Nash taxon

OST	Optimum-Stackelberg taxon
OT	Optimum taxon
PE	Set of pseudo-equilibria
PNES	Pareto-Nash equilibrium set
PPAD	Polynomial Parity Argument for Directed graphs
SNSE	Set of safe Nash-Stackelberg equilibria
SSES	Safe Stackelberg equilibrium set
ST	Stackelberg taxon
TSP	Traveling salesman problem
TSPT	Traveling salesman problem with transportation and fixed additional payments
USES	Unsafe Stackelberg equilibrium set

Chapter 1
Introduction

Abstract The essence of Game Theory springs from the cores of conflict and cooperation notions. Life existence resides inevitably in conflict and cooperation arising. What are the means of these concepts? Why they are so important? How and where conflict and cooperation appear? Who are the main actors in the situations of conflict and cooperation? What are the roles of the actors? What problems must they solve? We can continue the sequence of questions, but it's more appropriate to ask and respond to them at the right moments. So, let us highlight briefly the answers to these questions and let us give a general introductory description of game theory.

1.1 Basic Preliminary Terminology

Traditionally, a rigorous theory exposition starts from some undefinable but explainable terms used as a foundation for all the others. In the case of game theory, the first term of that kind is the notion of conflict.

1.1.1 Conflict and Cooperation Notions

What does the conflict notion mean? The answer seems to be simple because everyone has intuitive understanding of it. Nevertheless, for our purposes we need an explicit strong exposition.

To be alive, every being has needs in some resources, not obligatory of material nature only. If beings need to divide limited resources, they must agree on their division or must fight for them. So, the source of cooperation and conflict appears. Generally, we can describe the notion of *conflict* as a situation in which beings have to fight for some limited resources in order to satisfy their needs. A *cooperation* is a situation in which beings act jointly for a fair division of limited resources. Actually and traditionally, beings are named: *players*, agents, actors, persons, and groups. As their needs may be of non-material nature, they may have also interests, *objectives*, scopes, aims, goals, etc.

V. Ungureanu, *Pareto-Nash-Stackelberg Game and Control Theory*,
Smart Innovation, Systems and Technologies 89,
https://doi.org/10.1007/978-3-319-75151-1_1

There is an entire descriptive science dedicated to conflicts and their resolutions: *conflict theory* and *conflict resolution/reconciliation* (see, e.g., [1, 2]) mostly applied in sociology, social psychology and politics [3–5].

But, what is the role of game theory in such case?

1.1.2 What is Game Theory?

Game theory is characterized by its dominant features as *a theory of mathematical models of decision making in situations and processes of conflict and cooperation between intelligent rational players.*

By intelligent players we mean generally the persons who have exhaustive knowledge. By rational players we assume the persons that maximize their pay-off.

According to Robert Aumann, "**Interactive Decision Theory**" would be a more appropriate name/title/description for a discipline traditionally called "Game Theory" [6]. Sure, an ideal name for a discipline is hard to find. The last one has its shortcomings, e.g. there are some situations [7] that may be called interactive decisions, but they do not fulfil the main characteristics of the situations of conflict and cooperation traditionally considered in classical game theory. Other definitions of game theory may be presented, for example Shoham and Leyton-Brown in [8] defines game theory as a mathematical study of interaction among self-interested decision makers, Osborne in [9] defines game theory as a bag of analytical tools which are created with the aim to support humans to understand phenomena which appear in the process of decision makers interaction, Tadelis in [10] describes game theory as a framework created on the bases of mathematical models of conflict and cooperation phenomena among rational decision makers. Sure, there are other game theory definitions, but all of them have in common the notion of conflict, cooperation, decision makers, interactivity, and mathematical study. So, we propose to limit ourselves to our definition because it includes all highlighted components, reflects the content of this book and lets us get to the notion of decision making.

1.1.3 Decision Making

Decision making may be interpreted as a choice problem that is usually solved by every being at different moments of his life. Intuitively, the simplest decision making problem consists in choosing a decision from a set of admissible decisions. If the beings are rational, they will choose admissible decisions, optimal according to some criteria. Formally, such decision making problems can be formulated as single-objective optimization problems:

$$f(\mathbf{x}) \to \max, \mathbf{x} \in \mathbf{X}, \tag{1.1}$$

where $\mathbf{X}$ is the set of admissible decisions, $f : \mathbf{X} \to \mathbb{R}$ is an objective function that corresponds to a decision making criterion. Evidently, the objective may be to maximize or minimize the value of the objective function. In this context, it must be mention that we do not concern in this book any theoretical aspects both of Zermelo's axiom of choice [11], in particular, and axiomatic set theory [12], in general.

An essential feature of Problem (1.1) consists of its deterministic characteristic — all the data about Problem (1.1) are known and the effect of the decision variable $\mathbf{x}$ on the objective function may be computed with certainty. So, Problem (1.1) is considered and solved in conditions of certainty. Traditional optimization or mathematical programming problems are typical examples of decision making problems in conditions of certainty.

Generally, Problem (1.1) is not a simple problem. There are no general methods for its solving. But for different special classes of Problem (1.1) (e.g., linear programming [13, 14] and convex programming [15–18]), efficient solving algorithms exist.

1.1.4 Multi-Objective Optimization

If in (1.1) a vector criterion and, consequently, a vector objective function is considered, then the decision making problem becomes a multi-objective optimization problem:

$$\mathbf{f}(\mathbf{x}) \to \max, \mathbf{x} \in \mathbf{X}, \tag{1.2}$$

where $\mathbf{X}$ remains the set of admissible decisions, but $\mathbf{f} : \mathbf{X} \to \mathbb{R}^{\mathbf{m}}$ becomes an objective vector-function that corresponds to m criteria of decision making. Problem (1.2) requires a definition of optimality [19] because the traditional notions of maximum and minimum are not sufficient for Problem (1.2) solving. From this point of view, Problem (1.2) is more difficult than Problem (1.1).

The notion of *efficiency* or *Pareto optimality* is the main in studying *multi-objective optimization problems*. Generally, in such problems the solution consists of a set of equivalent efficient points. From the theoretical point of view, such a solution is good enough. But, from the practical point of view only an instance of the efficient points is needed. So the decision makers have to decide which one from the set of efficient points fits better their expectations.

Problem (1.2) may be generalized by considering partial order relations or preferences on the set of admissible decisions, i.e. this means that criteria are not obligatory of numeric types:

$$\mathbf{f}(\mathbf{x}) \to \max_{\mathbf{R}}, \mathbf{x} \in \mathbf{X}, \tag{1.3}$$

where the notation $\max_{\mathbf{R}}$ means maximization according to partial order relation $\mathbf{R} \subseteq \mathbf{X} \times \mathbf{X}$. Evidently, the partial order relation may be not only of binary type, but also of multiple type. Dominant and non-dominated decisions are traditionally considered optimal. But there is a problem with such solutions. Often, there are no dominant decisions according to considered partial order relation. Nevertheless, from practical point of view such an optimal solution must be found. In such cases, on the bases of existent partial order relation another one is constructed on which the decisions can be ordered. Evidently, such a relation is very subjective and depends mainly of decision maker and implied experts. Problem (1.3) is referred to *generalized mathematical programming* [20, 21]. The decision ordering may be achieved by special methods such as: ORESTE, PROMETHEE and ELECTRE [22].

1.1.5 Multi-Agent Decision Problems

In Problems (1.1)–(1.3) decisions are made by a single person. But there are problems in which the values of objective function depends not only of a single person decision $\mathbf{x} \in \mathbf{X}$, but depends of decisions $\mathbf{y} \in \mathbf{Y}$ of a series of persons which have generally different interests:

$$\mathbf{f}(\mathbf{x}, \mathbf{y}) \rightarrow \max, \mathbf{x} \in \mathbf{X}. \tag{1.4}$$

This kind of problems are studied by game theory. It is simply to observe that problem (1.4) may be seen as a parametric programming/optimization problem, where $\mathbf{x}$ is the optimization variable, and $\mathbf{y} \in \mathbf{Y}$ are the parameters. Let us observe, also, that for every particular person an appropriate "parametric" problem must be solved. A new type of "optimization" problems appear (see bellow normal form of the noncooperative game) and a new type of solution concepts are required such as, e.g., Nash equilibrium concept.

A special case of Problem (1.4) is:

$$\mathbf{f}(\mathbf{x}, \xi) \rightarrow \max, \mathbf{x} \in \mathbf{X}, \tag{1.5}$$

where $\xi \in \mathbf{Y}$ is a random variable. Problem (1.5) can be treated as a zero-sum two-player game theory problem in which the first player is the decision maker, and the second is nature.

Evidently, a decision making problem can possess the properties of all Problems (1.1)–(1.5). Such a problem implies at least three players, and one of them is the nature.

Among decision making problems (1.5) we can distinguish problems in conditions of risk and problems in conditions of uncertainty.

1.1.6 Decision Making Problems in Situations of Risk

The random parameter in **decision making problems in situations of risk** transforms the objective function into a random factor. This means that the traditional approach to optimization problem is not justified and we must apply "*statistical*" or "*probability*" criteria:

1. **Mean Criterion**:
$$M[\mathbf{f}(\mathbf{x}, \xi)];$$
2. **Criterion on the Difference Between Average and Dispersion**:
$$M[\mathbf{f}(\mathbf{x}, \xi)] - kD[\mathbf{f}(\mathbf{x}, \xi)],$$
where $k \geq 0$ is a constant that reflects the level of not accepting the risk;
3. **Criterion on the Bounds**: admissible limits/bounds are established for above criteria:
$$\alpha \leq M[\mathbf{f}(\mathbf{x}, \xi)] \leq \beta,$$
$$\alpha \leq M[\mathbf{f}(\mathbf{x}, \xi)] - kD[\mathbf{f}(\mathbf{x}, \xi)] \leq \beta,$$
or for other criteria;
4. **Criterion on the Most Probable Result**: if one value ξ_0 of the random variable ξ has a probability much more greater than the others, the value ξ_0 is setted for ξ, i.e. $\xi = \xi_0$ and Problem (1.5) becomes a deterministic problem of kind (1.1).

Evidently, the application of these criteria and the other of these types needs a rigorous analysis of the opportunity and legality of their application in the problem of decision making in the conditions of risk.

1.1.7 Decision Making Problems Under Uncertainty

Problems of decision making under uncertainty may be treated as problems of choice in situation of uncertainty in two-player games that are not compulsory antagonistic. Applied criteria are subjective and depend on decision making persons in such problems:

1. **Laplace Criterion**: it supposes that values of the variable ξ are achieved with equal probabilities.[1] If the principle of insufficient argumentation is accepted, then the decision making person selects the decision $\mathbf{x}^* \in \mathbf{X}$ that achieves the maximum of the average value:
$$M[\mathbf{f}(\mathbf{x}^*, \xi)] = \max_{\mathbf{x} \in \mathbf{X}} M[\mathbf{f}(\mathbf{x}, \xi)],$$
where ξ has a normal distribution;

[1]The supposition is based on the principle of insufficient argumentation: as the distribution law is unknown, we do not have necessary information to conclude that probabilities are different.

2. **Wald Criterion** (pessimistic, maximin, guaranteed result criterion): decision making person chooses the strategy $\mathbf{x}^* \in \mathbf{X}$ for which:

$$\min_{\xi \in \mathbf{Y}} \mathbf{f}(\mathbf{x}^*, \xi) = \max_{\mathbf{x} \in \mathbf{X}} \min_{\xi \in Y} \mathbf{f}(\mathbf{x}, \xi);$$

3. **Savage Criterion**: if $\mathbf{f}(\mathbf{x}, \xi)$ is the gain function, then the regret function is defined as

$$r(\mathbf{x}, \xi) = \max_{\mathbf{x} \in \mathbf{X}} \mathbf{f}(\mathbf{x}, \xi) - \mathbf{f}(\mathbf{x}, \xi);$$

   if $\mathbf{f}(\mathbf{x}, \xi)$ is the loss function, then the regret function is defined as

$$r(\mathbf{x}, \xi) = \mathbf{f}(\mathbf{x}, \xi) - \max_{\mathbf{x} \in \mathbf{X}} \mathbf{f}(\mathbf{x}, \xi).$$

   The Wald criterion is applied on the regret function, i.e. it is chosen the decision $\mathbf{x}^* \in \mathbf{X}$ for which:

$$\min_{\xi \in \mathbf{Y}} \mathbf{r}(\mathbf{x}^*, \xi) = \max_{\mathbf{x} \in \mathbf{X}} \min_{\xi \in Y} \mathbf{r}(\mathbf{x}, \xi);$$

4. **Hurwics Criterion** (pessimistic-optimistic): it is chosen the strategy $\mathbf{x}^* \in \mathbf{X}$ that solves the optimization problem:

$$\max_{\mathbf{x} \in \mathbf{X}} \left(\alpha \min_{\xi \in \mathbf{Y}} \mathbf{f}(\mathbf{x}, \xi) + (1 - \alpha) \max_{\xi \in \mathbf{Y}} \mathbf{f}(\mathbf{x}, \xi) \right),$$

   where $\alpha \in [0, 1]$. When $\alpha = 0$, the criterion is of extreme pessimism, i.e. it is the maximin criterion. When $\alpha = 1$, the criterion is of extreme optimism, i.e. it is the maximax criterion. Any other value $\alpha \in (0, 1)$ establishes a ration between pessimism and optimism, corresponding to the character of decision making person.

The problems described above are related to normative decision theory which concerns the best decision making. Decision maker is supposed to be fully rational and able to determine the perfect decisions if they exist. In this monograph, the descriptive decision theory, which has the main aim to describe behaviours of decision making persons under known consistent rules, is not considered.

1.1.8 Pascal's Wager

Decision making under uncertainty may be traced historically at least from the 17th century, when Blaise Pascal invoked his famous Wager in Pensées [23] published firstly in 1670. It is nicely described and analysed, e.g., by Steven J. Brams in [24], and Jeff Jordan in [25].

Pascal's Wager searches for a response to the question: is it rational to believe in God? We do not examine the question of proving the existence of God — any of such proofs is somewhat flowed [26]. Let us describe the Pascal's Wager shortly by considering two "*players*": player 1 — a Person, and player 2 — Reality. The Person has two strategies: to believe in God — B, and to not believe in God — N. For the Reality, we highlight two strategies: God exists — E, and God does not exist — N. So, there are four possible outcomes: BE, BN, NE, NN.

We can associate reasonable payoffs with these outcomes for the Person:

$f_1(BE) = \infty$ — if the Person believes and God exists, the award is infinite;
$f_1(BN) = \alpha, -\infty < \alpha \leq 0$ — if the Person believes and God does not exist, the Person has a finite loss associated with the years of faith;
$f_1(NE) = -\infty$ — if the Person does not believe and God exists, the loss is infinite;
$f_1(NE) = \beta, 0 \leq \beta < \infty$ — if the Person does not believe and God does not exist, the Person has a finite gain because of not spending time, money and other things in van.

As the Reality does not distinguish enumerated outcomes, we can associate with all of them a single fixed value of "*payoff*" (Laplace Criterion):

$$f_2(BE) = f_2(BN) = f_2(NE) = f_2(NE) = \gamma, \gamma \in \mathbb{R}.$$

Actually, we have defined above a vector function

$$\mathbf{f} : \{BE, BN, NE, NN\} \to \mathbb{R}^2$$

with values:

$$\mathbf{f}(BE) = (\infty, \gamma), \quad \mathbf{f}(BN) = (\alpha, \gamma),$$
$$\mathbf{f}(NE) = (-\infty, \gamma), \quad \mathbf{f}(NN) = (\beta, \gamma).$$

Let us observe that the payoff functions may be interpreted as 2×2 matrices:

$$f_1 = \begin{bmatrix} \infty & \alpha \\ -\infty & \beta \end{bmatrix}, \quad f_1 \in \mathbb{R}^{2\times 2},$$

$$f_2 = \begin{bmatrix} \gamma & \gamma \\ \gamma & \gamma \end{bmatrix}, \quad f_2 \in \mathbb{R}^{2\times 2},$$

where the matrix lines are associated with the strategies B and N of the first player, and columns are associated with the strategies E and N of the second player. It is possible to represent the vector payoff function in the matrix form, too:

$$\mathbf{f} = \begin{bmatrix} (\infty, \gamma) & (\alpha, \gamma) \\ (-\infty, \gamma) & (\beta, \gamma) \end{bmatrix}, \quad \mathbf{f} \in \mathbb{R}^{2\times 2} \times \mathbb{R}^{2\times 2}.$$

A distinct feature of the decision making problem is highlighted by the payoff function of the second player (Reality): its payoff does not depend on its strategy choice. So, only the payoff of the first player (Person) depends on strategy choices.

What is the best strategy for the Person?

To respond to this question, we can apply the Laplace Criterion and can suppose that the strategies of the second player are realized with equal probabilities (0.5). In such supposition, the average payoff of the first player is:

$$M[f_1(B)] = 0.5 \times \infty + 0.5 \times \alpha = \infty,$$

$$M[f_1(N)] = 0.5 \times (-\infty) + 0.5 \times \beta = -\infty.$$

By comparing the values, we can conclude that the first strategy is the best, i.e. the Person must believe in God.

Sure, this simple model of decision making may be improved [24, 25]. We do not have such a purpose. The example has the aim to highlight the features of the decision making models and problems and the distinctions between decision making and game theory models and problems.

1.1.9 Standard Decision Theory Versus Game Theory

In a standard decision theory, a single **rational** decision maker is involved only. In a game theory model, at least two **rational** players (decision makers) are involved and their payoffs depend both on their own and the opponent strategy choices. More the more, even if at least two players are supposed to have **rational behaviour**, a game theory model may include **irrational** players too, for example the Nature. Usually, in game theory a player is considered rational if he or she is aware of all possible events, may assign probabilities to them and optimizes the value of his payoff function. More the more, it is a difference between individual rationality and group rationality. Generally, group rationality means efficiency/Pareto optimality. A recent original perspective on connections between game theory and classical decision making is presented by Michel Grabisch in [27].

Bertrand Russell prompted in his Nobel Lecture [28] that the presence of payoff functions in game theory models and the enforcement to optimize them are consistent with the human being nature, and human desires, both the essential necessities of life, and the others ones such as acquisitiveness, rivalry, vanity, and love of power.

Sure, Bertrand Russell's characteristics of human beings are mostly related to politics and classical game theory, where rationality and selfishness are prevalent in comparison with sociability and altruism. In this respect, recent game theory studies establish more connections with ethical behaviour of decision makers [29]. Starting with the 1973 article by John Maynard Smith and George Price [30] an evolutionary approach to game theory is developed, based on a concept of evolutionarily stable strategy.

Evolutionary game theory [31] is only one distinct branch of game theory. In evolutionary games players and strategies are or may be associated with genes and payoffs are associated with offsprings. A variety of genes associated with species play evolutionary games. Players with more offsprings wins. **The evolution leads to a Nash equilibrium of a game played by genes**. This conclusion has an important conceptual impact on the evolutionary biology and it changed the way in which the concept of evolution is understood, changing totally the old idea of species maximum fitness. Concerning the human evolution with specific rational players, it must be mentioned that according to Schecter [32] ethical aspects of human behaviour are inserted in game theory models mutually. Generally, evolutionary games form an important branch of game theory with a distinct feature that all players may not be rational. Players with more offsprings are successful in the game that leads to a Nash equilibrium. In such games there is no necessity of rationality and correct beliefs, pre-play communication and self-enforcing agreement, because players/genes are irrational. Trial and errors adjustment is sufficient to achieve Nash equilibrium.

The above conclusion about biological evolution result which leads to a Nash equilibrium is closed to the conclusion made by game theorists concerning social problems: **Nash equilibrium concept is a governing game theory principle, applicable to all social problems**. In contrast to evolutionary games, to play Nash equilibria in social games there are some necessary factors:

- rationality and correct beliefs of the players,
- pre-play communication and self-enforcement agreement,
- dynamic or trial and errors adjustment,

but, sure, Nash equilibrium may emerge because of a variety of other reasons such as, e.g., ethical and moral factors.

Modern game theory studies a variety of models in which players may have hyper-rationality, low-rationality and zero-rationality. Rationality alone is not a sufficient factor to play Nash equilibria. Additionally, players must have correct beliefs (subjective priors) about other players and their incentives, that may be achieved by pre-play communication and reaching a self-enforcing agreement. These conditions mean that in real situations players may behave in ways which differ from that prescribed by game models of classical normative game theory. Namely these arguments and facts served as a bases for creating a descriptive branch of the modern game theory — behavioural game theory, a theory based on bounded rationality, economical and psychological experiments and analysis of experimental results [33–35].

An excellent analysis and survey regarding both the history and theory of self-interest and sympathy concepts in social, politic and economic behaviour were realized by Kjell Hausken in his 1996s article [36]. He highlighted the opposition between concepts of self-interest and sympathy in order to characterize the main principle of human activity and behaviour. Perrow, in his 1986s book [37], points out that sociological, psychological and organization theories tend to focus on self-interest, while political science tends to be based on both self-interest and empathy.

1.1.10 Strategy Notion

In this monograph, the meaning of the term "**strategy**" is considered basically in accordance to the Thomas C. Schelling's "*The strategy of conflict*", published firstly by the author in 1960 [38]. Robert V. Dodge has written a good analysis of the Schelling's game theory view, being also a good survey of the related works [39]. In simplified approach, the strategy meaning may be seen as a plan of actions to achieve the best outcome according to the player's system of values.

The notion of strategy is essential for strategy games. Namely because of this, every game theory textbook and monograph treats it and other related notions especially, see e.g. Carmichael [40], Maschler, Solan, and Zamir [41], Matsumoto and Szidarovszky [42], Mazalov [43], Myerson [44], Shoham and Leyton-Brown [8], Straffin [45], Narahari [46], Nisan, Roughgarden, Tardos, and Vazirani [47], Osborne [9], Petrosyan and Zenkevich [48], Umbhauer [49], Watson [50], and a lot of other works, including modern research papers and monographs.

1.1.11 Brief History of Game Theory and Optimal Control

It is important to mention that the term "*game theory*" looks to be initiated in modern era by **Émile Borel** in his article [51], published in 1921. He is probably the modern inventor of minimax principle as he applied it in several particular games and attempted but failed to prove the minimax theorem for two-player zero-sum games. He proved the minimax theorem only for two-person zero-sum **symmetric matrix** games. It is important to mention that he published also other articles related to the theory of games, e.g. [52, 53].

John von Neumann proved the minimax theorem for the two-matrix zero-sum mixt-strategy games in 1926 and published his results in 1928 [54]. He and **Oskar Morgenstern** are credited to be game theory creators because of their early results and their seminal monograph "*Theory of Games and Economic Behavior*" published in 1944 [55]. An important contribution to game theory earliest development was the book of Luce and Raiffa [56] published in 1957.

A short description of the history of game theory by Anthony Kelly may be found in [57]. Another "*Brief History of Game Theory*" is exposed by Geçkil and Anderson in [58]. A Chronology of Game Theory is maintained in Internet by Paul Walker. A more extensive exposition of the history of game theory is presented by Mary and Robert Dimand in [59], Robert McCain in [60], and Robert Aumann in [6].

Optimal control theory appeared in 1950s and is largely seen as an extension of the calculus of variations, but as it is pointed by different scientists, see e.g. the survey by Bryson [61], optimal control theory has its roots in other domains such as classical control, theory of random processes, linear and nonlinear programming, algorithms and digital computers [61].

Central methods of the optimal control theory are the Hamilton–Jacobi–Bellman equation (a sufficient condition) [62] and Pontryagin's maximum (minimum) principle [63] (a necessary condition). Frank Clarke in [64] points out that the maximum principle was a culmination of a long search for a "tool/method" that may be used not only for research, but for design, too. In this context, the role of Lagrange multipliers is very important and this fact was highlighted earlier in 1939 by McShane [65] for the Lagrange problems in order to extend the calculus of variations to handle control variable inequality constraints.

The integration of game theory and optimal control was inevitable right from the beginning. Rufus Isaacs was the mathematician who sketched the basic ideas of zero-sum dynamic game theory, which included rudimentary precursor ideas of the maximum principle, dynamic programming, and backward analysis, exposed by him at the end of 1954 and the beginning of 1955 in four research memoranda at the RAND Corporation [66]. The memoranda formed the basis of his 1965 outstanding book on Differential Games [67]. A dramatic story of scientific ideas, investigations and scientific areas formations may be felt by reading Breitner historical paper [66].

Rufus Isaacs had outstanding merits in the formation of differential game theory area [67, 68], in stating and solving of some monumental mathematical problems based on such games as, e.g., pursuit-evasion game and Princess and Monster game [67].

Sure, the number of works in joint area of game theory and optimal control is impressively and increasingly large. It is an impossible task to mention only some most important because there are a lot of them. Nevertheless, being subjective we have to mention additionally the Lev Pontryagin's book "*The Maximum Principle*" [69] and Vasil'ev book "*Optimization methods*" [70].

1.2 Game Theory Branches. General and Particular Models. Solution Concepts

Game theory is divided into two large branches: the theory of noncooperative games and the theory of cooperative games. In noncooperative games, overt cooperation between players is forbidden or it is impossible by the essence of games. In cooperative games, cooperation between players exists and they may have mutual or explicit bindings, they may form coalitions with the objective to obtain larger benefits.

1.2.1 Branches. General Models. Solution Concepts

The table presented below has the aim to highlight a general view on the game models considered traditionally by game theorists and practitioners.

It is important to remark that game theorists consider hybrid or integrative models too, with features both of noncooperative and cooperative models, but the last ones may have features of the different game models from the same branch as well.

So, games considered in this monograph are mostly noncooperative games, but they have integrative features both of Nash and Stackelberg games, and the features of multi-objective optimization problems, i.e., in the general model, every player optimize values of more than one objective function simultaneously. The explanation of the name of Pareto–Nash–Stackelberg games comes out from integration of simultaneous — Nash — and sequential — Stackelberg — games with multi-objective — Pareto — optimization problems. It is important to remark that the order of names in the title is important because considered models are treated firstly as sequential — Stackelberg. Second, at each time moment a simultaneous game is played — Nash. Third, each player optimizes simultaneously more than one criterion — Pareto. Finally, in a Pareto–Nash–Stackelberg game, each player is involved simultaneously both in a Stackelberg and Nash game, playing these games from the perspective of multi-objective optimization problems.

To resume the above explanation, we have to mention that a Pareto–Stackelberg–Nash game assumes that the players consider as the fundamental a simultaneous — Nash — game. Second, the strategy in a simultaneous game is selected by a group of players who plays a sequential — Stackelberg — game. Third, the choices are made by solving a multi-objective — Pareto — optimization problem. Evidently, from a general and abstract point of view we can consider six different game theory models. Nevertheless, this monograph deals only with the Pareto–Nash–Stackelberg games.

Why do we not adopt a name like mixed multi-objective simultaneous and sequential games? The response follows from the above explanations and from the fact that such a name involves more general models and solution concepts, which we do not consider in this work. More the more, the adopted name reflects briefly main features of considered models and it makes simultaneously a tribute to the precursors, the scientists who introduced and investigated enumerated games and concepts.

The model of a Pareto–Nash–Stackelberg game and control problem assumes mutually that players have to play a Pareto–Nash–Stackelberg game and simultaneously to control the behaviour of a system that influences values of their payoff functions.

The set of solution concepts in game theory is impressively large. We can recall for example: Nash equilibrium, minimax strategy solution, correlated equilibrium, trembling-hand perfect equilibrium, ε-Nash equilibrium, polymorphic equilibrium, Stackelberg equilibrium, sub-game perfect equilibrium, sequential equilibrium, Bayesian Nash equilibrium, the Shapley value, the core, etc. [8–10, 71]. For every of these solution concepts there is an impressive series of scientific works and a large bibliography. We do not have the purpose to investigate all of them. The monograph is limited mainly to different combinations of Nash equilibrium, Stackelberg equilibrium, minimax strategy solution and Pareto optimality concepts, all of them in the frames of Pareto–Nash–Stackelberg games. More the more, we state and investigate a specific problem to build the entire set of solutions.

Generally, if to compare solutions of game theory problems and that which are studied in physics, it must be mentioned that game theory solutions may be used to predict some real phenomena not exactly, but only as an important force that drives behaviours of players, being also a reference point of the game analysis.

Evidently, the idea to build game theory models by aggregation of various basic entities may conduct to a large diversification of models and areas of game theory application and abstraction, see e.g. mean field games and mean field type control theory in [72], dialogical games in [73], population games in [74], evolutionary game theory in biology [75], developing game theory on the basis of geometry and physical concepts like general relativity and the allied subjects of electromagnetism and gauge theory [76, 77], diplomacy games [78].

The following table presents an integrating overview to the main game theory domains/models and solution concepts.

Game Models

Noncooperative	Cooperative
Player Incentive	
Player acts solely to maximize his own payoff	Players form coalitions and behave jointly to gain mutual benefits
Game Types	
Single-objective or Multi-objective Game	
Static Game	Coalitional Game
Simultaneous Game	Bargaining Game
Nash Game	Matching Game
Dynamic Game	Voting Game
Extensive Game	Transferable Utility Game
Repeated Game	Non-transferable Utility Game
Evolutionary Game	etc.
Markovian Game	
Stackelberg Game	
Auction Game	
Public Goods Game	
Intervention Game	
Supermodular Game	
Security Game	
etc.	
Solution Concepts	
Optimality or Pareto optimality	
Nash Equilibrium	Nash Bargaining Solution
Correlated Equilibrium	Shapley Value
Bayesian Nash Equilibrium	Kalai-Smorodinsky Barg. Sol.
Subgame Perfect Nash Eq.	Egalitarian Bargaining Solution
Evolutionary Stable Strategy	Rubinstein Bargaining Solution
Stackelberg Equilibrium	The nucleolus
Wardrup Equilibrium	The core
Pareto Equilibrium	Imputations
Maximin Solution	Stable sets
etc.	etc.

It must be mentioned once again that the name "*noncooperative*" game means traditionally the absence of communication and not the absolute lack of cooperation. There are examples of games without communication between players which include reciprocal cooperation both mutual and taken for granted, see e.g. Chaps. 3 and 4 in [38]. Regarding the name "*cooperative*" game, it must be mentioned that the traditional meaning is related to a coalition formation and to considering coalitions as the main agents/sides of the game.

1.2.2 Prisoner's Dilemma

The modern game theory comprises a lot of significant examples or particular games that are very important conceptually for different science domains and human activity areas. They capture the essence of different conflict situations and strategic interactions which are frequently met in real life and presented in more complex situations. An outstanding example of this kind is the the Prisoner's Dilemma. There is no a first bibliographic reference published both by its discoverers Merrill Flood and Melvin Dresher, former scientists of the RAND Corporation, and Albert W. Tucker, former consultant of the same corporation, who named the game the prisoner's dilemma based on a story invented by him for a lecture popularizing game theory. William Poundstone [79] characterises the prisoner's dilemma discovery in 1950s by two Rand scientists as the most influential discovery in game theory since its inception. The importance of the Prisoner's Dilemma game is highlighted in the William Poundstone's monograph [79], but the full story of its creation is revealed also by the book of Philip Mirowski [80] describing the RAND Corporation's activity and scientific achievements during the time, and by the book of David McAdams [81], which also presents the story of prisoner's dilemma and highlights its essence and practical application in the series of examples from the worlds of business, medicine, finance, military history, crime, sports, etc., and examines six basic ways to change games like prisoner's dilemma in order to exclude dilemmas: commitment, regulation, cartelization, retaliation, trust, and relationships, under the general book's motto by Zhuge Liang, regent of the Shu kingdom, who lived AD 181–234, and who said that "the wise win before they fight, while the ignorant fight to win".

Let us expose shortly the essence of the prisoner's dilemma [79–82] and let us highlight its main features.

Two criminals have been arrested and imprisoned on charges that carry a detention term of up to two years. The police suspects they also committed a worse crime that carries a term of up to twenty years. Each prisoner is imprisoned in separate cell and does not have a possibility to communicate with other. Both of them are given opportunities either to confess to the worse crime, or to remain silent and to cooperate with other. If only one confesses, he will be freed, and the other will spend twenty years in prison. If neither of them confesses, they will be imprisoned for two years. If both of them confess, they will spend ten years in prison.

A payoff bimatrix may be associated with this two-player game:

$$(A, B) = \begin{bmatrix} -10, -10 & 0, -20 \\ -20, 0 & -2, -2 \end{bmatrix}$$

Conventionally, the players are distinguished by their numbers: the first and the second player. The first player has a payoff matrix A. The second player has the payoff matrix B. They have two strategies: to confess or not. The first player chooses the row, the second — the column, in order to obtain a shorter prison term in correspondence with his payoff matrix, that is to choose a strategy that will guarantee a greater payoff (a shorter term of detention). The first row and column correspond to the confession strategy, the second — to no confession.

It is simply to observe that the players have a dominant rational strategy — to confess/betray. If they confess, i.e. both of them select their first strategy, each of them will spend ten years in prison. The pair $(-10, -10)$ associated/identified with the correspondent outcome (confess, confess) is at the same time a dominant strategy equilibrium, a Nash equilibrium, and a minimax solution in this game.

It is also simply to observe that if both of them behave irrationally and neither of them confesses/cooperates, the irrational result $(-2, -2)$ will be obtained, which is better than the rational one $(-10, -10)$.

If we represent all the results of the game in the Cartesian coordinates, we may observe that the pair $(-2, -2)$ is Pareto optimal and it dominates the pair $(-10, -10)$. Unfortunately, the first pair is unstable because players are susceptible to behave on their own and to betray or to not cooperate with other. So, **the Prisoner's dilemma game has one Nash equilibrium, but it is Pareto dominated**.

Remark. In prisoner's dilemma games, the dilemma for players consists of choosing between two types of behaviour:

1. cooperative/collective/irrational behaviour (group rationality),
2. noncooperative/egoistic/rational behaviour (individual rationality).

Nota bene. *The prisoner's dilemma game has one more interesting feature. It emphasises the value of an altruistic behaviour. If players behave altruistically and choose their strategies both according with the payoff matrix of the partner and according with what is better for the partner, then the pair* $(-2, -2)$*, that corresponds to the outcome (cooperate, cooperate), is simultaneously a dominant strategy equilibrium, a Nash equilibrium, a minimax solution, and more the more, it is an efficient solution, that is it is a Pareto optimal solution. What an extraordinary "proof" of the altruism importance in the prisoner's dilemma game.*

The precedent remark mentions the value of altruism, but it also mentions that the prisoner's dilemma is de facto a trilemma because of the altruistic/self-sacrifice point of view.

The payoff matrix structure of the prisoner's dilemma game may be generalised to describe an entire class of analogical games, e.g.

$$(A, B) = \begin{bmatrix} \gamma, \gamma & \alpha, \delta \\ \delta, \alpha & \beta, \beta \end{bmatrix},$$

where $\alpha > \beta > \gamma > \delta$, or more general

$$(A, B) = \begin{bmatrix} c, \gamma & a, \delta \\ d, \alpha & b, \beta \end{bmatrix},$$

where $a > b > c > d$ and $\alpha > \beta > \gamma > \delta$. More the more, the players may have more than two strategies, for example:

$$(A, B) = \begin{bmatrix} c, \gamma & a, \delta & a, \delta & a, \delta \\ d, \alpha & b, \beta & d, \delta & d, \delta \\ d, \alpha & d, \delta & b, \beta & d, \delta \\ d, \alpha & d, \delta & d, \delta & b, \beta \end{bmatrix}.$$

The multi-dimensional prisoner's dilemma game has new features in comparison with the classical one. First of all, players have dominated incomparable strategies additionally to their dominant strategies. Second, the game has Pareto optimal outcomes, but players must coordinate the choices of their strategies in order to realize a Pareto optimal outcome. So, we deal with a new type of behaviour — coordination, and the multi-dimensional prisoner's dilemma game may be seen as a kind of combination between a classical prisoner's dilemma game and a coordination game, which is exposed bellow.

The number of real life situations that may be modelled as the prisoner's dilemma is extremely large and some examples of this kind may be found in the book by Matin J. Osborne in [9]: working in a joint project, duopoly, the arms race, common property, mating hermaphroditic fish, tariff wars between countries. An interesting Internet variant of the prisoner's dilemma game is a TCP user's game [8] with correct or defective application of a back-off mechanism in the TCP protocol.

A multi-player prisoner's dilemma game may be stated too. The following example is an extension to a three-player case in which every player has an infinite number of strategies.

1.2.3 To Cooperate or Not To Cooperate?

Let us consider a three-player game [82] in which all the players have identical strategy sets:

$$X_1 = X_2 = X_3 = \mathbb{R},$$

and their payoff functions are similar:

$$\begin{aligned} f_1(x) &= 3 - (x_1 - 1)^2 - x_2^2 - x_3^2, \\ f_2(x) &= 3 - x_1^2 - (x_2 - 1)^2 - x_3^2, \\ f_3(x) &= 3 - x_1^2 - x_2^2 - (x_3 - 1)^2. \end{aligned}$$

All players choose their strategy simultaneously and each of them tends to maximise the value of his payoff function on the resulting profile.

All payoff functions have the same global maximal value 3, but every function realises the global value on his own point. For all the players the global value is realised when everyone selects the value of his strategy equal to 1 and the two other players choose their strategies equal to 0. If all of them choose the value of their strategies equal to 1, the final payoff for everyone is equal to 1 on the outcome/profile $x^0 = (1, 1, 1)$. It is simply to observe that the profile x^0 is a dominant strategy equilibrium and a Nash equilibrium.

If the players may cooperate, they may form a common payoff function:

$$F(x) = \lambda_1 f_1(x) + \lambda_2 f_2(x) + \lambda_3 f_3(x),$$

where $\lambda_1 + \lambda_2 + \lambda_3 = 1, \lambda_1 \geq 0, \lambda_2 \geq 0, \lambda_3 \geq 0$. The numbers $\lambda_1, \lambda_2, \lambda_3$, may be interpreted as the player weights or the coefficients used for obtaining the same measurement units. Their values may be established as a result of bargaining and agreement between players.

The function $F(x)$ has a single global solution

$$x^* = (\lambda_1, \lambda_2, \lambda_3).$$

For the profile x^*, the payoff function have the values:

$$\begin{aligned} f_1(x^*) &= 3 - (\lambda_1 - 1)^2 - \lambda_2^2 - \lambda_3^2, \\ f_2(x^*) &= 3 - \lambda_1^2 - (\lambda_2 - 1)^2 - \lambda_3^2, \\ f_3(x^*) &= 3 - \lambda_1^2 - \lambda_2^2 - (\lambda_3 - 1)^2. \end{aligned}$$

It is simply to observe that for a particular profile $x^* = (\frac{1}{3}, \frac{1}{3}, \frac{1}{3})$ the payoffs are

$$f_1(x^*) = f_2(x^*) = f_3(x^*) = 2\frac{1}{3}.$$

In conclusion, everyone may obtain a much more better result through cooperation with other players. Nevertheless, the profile x^* is unstable. The profile x^0 is stable, but it is interesting to observe that if one of the player changes his strategy choosing the value in $\mathbb{R}\backslash[-1, 1]$, he decreases the value of his payoff, but the payoffs of the rest of players decrease even more. So, the player may provoke a minor damage for himself while the damage being weighty for others. Such situations are less studied in game theory, but, evidently, they may be extremely important for different real life situations.

The game may be simply generalised to a general n-player game, the same conclusions remaining available.

1.2.4 Coordination Games

The coordination game has been considered by Thomas Schelling in his outstanding 1960 edition book "*The strategy of conflict*" [38], and republished later repeatedly with Schelling's preface for the second 1980 edition. The game remains of interest because of its distinct important features and various areas of applications. See in this context the 2009 edition book by Osborne [9] and the 1999 edition book by Cooper [83], entirely dedicated to its role in macroeconomics and to various concrete examples. Coordination games on network are considered by Jackson and Zenou in [84].

Conflict situation in coordination game presents a specific contrast to many conflict situations as players' behaviour relies essentially on their confidence and expectations. Its general two-player form has the following payoff bimatrix:

$$(A, B) = \begin{bmatrix} a, \alpha & c, \delta \\ d, \gamma & b, \beta \end{bmatrix},$$

where $a > b > c$, $a > b > d$, and $\alpha > \beta > \gamma$, $\alpha > \beta > \delta$. Sure, these inequalities are mostly formal to reflect a general structure and they admit equality relaxations in some particular models.

The game has two Nash equilibria: (a, α) and (b, β). Both players prefer the equilibrium (a, α) because it dominates (b, β) and it is Pareto optimal.

There is a series of well known generic examples of coordination game, like:

- choosing sides of the road: $\begin{bmatrix} 1, 1 & 0, 0 \\ 0, 0 & 1, 1 \end{bmatrix}$,
- pure coordination: $\begin{bmatrix} 2, 2 & 0, 0 \\ 0, 0 & 1, 1 \end{bmatrix}$,
- battle of the sexes [56]: $\begin{bmatrix} 2, 1 & 0, 0 \\ 0, 0 & 1, 2 \end{bmatrix}$,
- stag hunt: $\begin{bmatrix} 2, 2 & 0, 1 \\ 1, 0 & 1, 1 \end{bmatrix}$.

The first and the second examples are also the examples of so called games of cooperation (team games) in which players have equal payoffs for every profile [8]. The third and fourth examples are also the examples of games that combine cooperation and competition [8].

It must be remarked that a situation which may be formalized as a stag/deer hunt game was discussed in 1755 by the philosopher Jean-Jacques Rousseau [85]. Rousseau's discussion and the above model highlight once again the abstract feature of models that game theory provides.

Martin J. Osborne presents in [9] a "*Security Dilemma*" model — a modification of the stag hunt game, which was considered as an alternative to the Prisoner's dilemma in order to formalize an arms race:

$$\begin{bmatrix} 3, 3 & 0, 2 \\ 2, 0 & 1, 1 \end{bmatrix}.$$

This game doesn't have dominant strategies, but **has two Nash equilibria** $(3, 3)$ and $(1, 1)$, one of them $(3, 3)$ being Pareto optimal and focal, i.e. mutually convenient for both players. Unfortunately, it is not stable as the opponent may choose his strategy provoking for himself an incomparably less damage than for his opponent. So, this is an another model of the dilemma to choose between safety and social cooperation and it may be seen as a kind of combination between the prisoner's dilemma game and the game of coordination.

1.2.5 Anti-Coordination Games

In anti-coordination games players score highest when they choose opposite strategies. Well-known examples of this kind are the Chicken [86] and Hawk–Dove [87] Games.

A canonical version of the Chicken Game was described by Bertrand Russell in his 1959 book "*Common Sense and Nuclear Warfare*" [86] in the context of Cold War and the mutual assured destruction of nuclear warfare.

The Game of Chicken may be simply modelled as a bimatrix game:

$$\begin{bmatrix} -\infty, -\infty & 1, -1 \\ -1, 1 & 0, 0 \end{bmatrix},$$

where the first row and column correspond to strategy "*straight*" and the second — to strategy "*swerve*". **The game has two Pareto optimal Nash equilibria** $(1, -1)$ and $(-1, 1)$. As each player prefers a different equilibrium, an anti-coordination appears — each player want to do the opposite of what the opponent does.

A general form of the Chicken Game may be considered:

$$(A, B) = \begin{bmatrix} d, \delta & a, \gamma \\ c, \alpha & b, \beta \end{bmatrix},$$

where $a > b > c > d$, and $\alpha > \beta > \gamma > \delta$. William Poundstone has provided an exhaustive analysis of the Chicken Game in his book [79].

The Hawk–Dove Game was considered by John Maynard Smith and George Price in 1973 paper "*The logic of animal conflict*" [87]. The traditional form of the payoff matrices are:

$$(A, B) = \begin{bmatrix} \frac{v-c}{2}, \frac{v-c}{2} & v, 0 \\ 0, v & \frac{v}{2}, \frac{v}{2} \end{bmatrix},$$

where $c > v > 0$, v is the value of the contested resource, c is the cost of a fight, the first strategies mean to be a Hawk, and the second strategies means to be a Dove. The game has the same general form as the Chicken Game.

Sure, there are other examples of the anti-coordination games and different areas of their models applications, see e.g. [88, 89].

1.2.6 Antagonistic Games

Antagonistic games or zero-sum games are very well studied. They can be seen as particular cases of the non-zero-sum games. Nevertheless, they have some special own features. We recall them generally for an important characteristic of strategic games — not all strategic games have pure strategy Nash equilibria, but a very large class of them have mixed-strategy Nash equilibria [90]. In this context, let us recall a game of Matching Pennies — a very simple example of antagonistic game.

Two players put a penny on a table simultaneously. If the pennies come up the same side of heads or tails, then one of the players gets both pennies, otherwise its opponent gets the two pennies.

The traditional model of this game is very simple:

$$(A, B) = \begin{bmatrix} 1, -1 & -1, 1 \\ -1, 1 & 1, -1 \end{bmatrix}.$$

It is easy to observe that $A = -B$. The game doesn't have pure strategy Nash equilibria. It has only one mixed-strategy Nash equilibrium.

A general form of the game may be:

$$(A, B) = \begin{bmatrix} a, -a & -a, a \\ -a, a & a, -a \end{bmatrix},$$

where $a > 0$. Sure, it is sufficient to use only one matrix to describe the game:

$$A = \begin{bmatrix} a & -a \\ -a & a \end{bmatrix}.$$

A well known game of Rock–Paper–Scissors [91] is a similar game, but with three strategies. Such games may be generalised in a series of different manners. For example, we can state an infinite strategy game.

Let the first player has a strategy set $X \subseteq \mathbb{R}^m$, the second player has a strategy set $Y \subseteq \mathbb{R}^n$, and their payoff functions are

$$f_1(x, y) : X \times Y \to \mathbb{R}$$

and

$$f_2(x, y) : X \times Y \to \mathbb{R}.$$

The players selects their strategies simultaneously. If both values of the payoff functions have the same sign (positive or negative), the first player wins, otherwise the second player wins.

This game may be non trivial even for some simple payoff functions and strategy sets, e.g.:

$$f_1(x, y) = 1 - \langle x, x\rangle + \langle y, y\rangle,$$

$$f_2(x, y) = 1 + \langle x, x\rangle - \langle y, y\rangle,$$

and X and Y are convex polyhedral sets.

The game of Matching Pennies may be generalized to a multi-objective infinite strategy game if the payoff functions are defined as

$$f_1(x, y) : X \times Y \to \mathbb{R}^k$$

and

$$f_2(x, y) : X \times Y \to \mathbb{R}^k.$$

Players selects their strategies (x, y) simultaneously. If $f_1(x, y) \geqq 0$ and $f_2(x, y) \geqq 0$ or $f_1(x, y) \leqq 0$ and $f_2(x, y) \leqq 0$, the first player wins, otherwise the second player wins.

A suggestion of a less obvious generalization of the precedent games may be found in [92]. The payoff functions and strategy sets are defined as in the precedent case, but the wining criterion is different. For an outcome $(x, y) \in X \times Y$ the first player wins if

$$f_1(x, y) \leqq f_2(x, y),$$

otherwise the second player wins. Geometrically, this Game of Matching Inequalities is connected to problems of set nesting and embedding [92] that may be solved by special algorithms [93, 94] and approaches [95]. Formally, we can define the payoff matrix of the Game of Matching Embeddings in the manner similar with that of defining the payoff matrix A in the Game of Matching Pennies:

$$a_{xy} = a(x, y) = \begin{cases} 1, & \text{if} \quad f_1(x, y) \leqq f_2(x, y), \\ -1, & \text{otherwise.} \end{cases}$$

The term embedding in the name of the game arises from the geometrical interpretation of the system of inequalities in the definition of the payoff functions as a subset of $\mathbb{R}^{n+m}$ and the necessity to satisfy embedding properties of the strategy sets in order to have game solutions.

1.2.7 Game Modelling and Solution Concepts

Analysis of the above game model examples as well as of a lot of other game models exposed in the vast game theory literature and even in the majority of textbooks, see e.g. [10], is enough to suggest us some very important conclusions. The variety of game theory models is impressively large. Some models are very specific and unusual. It must be mentioned that we do not use equation, the traditional modelling tool, to build the above game model examples. More the more, we have used a very known Nash equilibrium concept to solve the above games, but solutions do not satisfy us altogether because they generate dilemmas or even trilemmas. So, having solutions of the game is not enough to deal and solve a real game theory problem. What are the recommendations of game theory specialists? How to treat such situations?

First, based on the Prisoner's Dilemma example, six basic ways to change games with the aim to exclude dilemmas were mentioned: commitment, regulation, cartelization, retaliation, trust, and relationships [96]. There are important works dedicated to these concepts and approaches, e.g. for commitment we can refer the book edited by Randolph M. Nesse [97], and especially the chapters "*Game-Theoretic Interpretations of Commitment*" by Jack Hirshleifer [98] and "*Commitment: Deliberate Versus Involuntary*" by Thomas C. Schelling [96]. Nevertheless, from the practical point of view it is frequently more recommended to change the rules and conditions of the games with the aim to avoid situations that lead to game models generating dilemmas, trilemmas and so on. More the more, there are some theoretical recommendations to achieve good outcomes:

- change/design the game rules,
- use legally binding contracts,
- use long-term relationships.

Mechanism design theory, established by 2007 Nobel Laureates Leonid Hurwicz, Eric Maskin, and Roger Myerson, solves these and other appropriate problems [99–102]. It is called also the reverse game theory because it starts at the game end and goes backward.

1.3 Strategic Form Games

A strategic form game is a simple model considered for noncooperative games. The model includes three basic elements: the set of players, player strategy sets, and player payoff functions. Formally, the strategy form game is denoted

$$\Gamma = \langle N, \{S_p\}_{p\in N}, \{f_p(s)\}_{p\in N}\rangle,$$

where

- $N = \{1, 2, \dots, n\} \subset \mathbb{N}$ is a set of players,
- S_p is a set of strategies of the player $p \in N$,
- $f_p(s)$ is a pth player payoff function for which the Cartesian product $S = \underset{p\in N}{\times} S_p$ is the domain.

Any element $s = (s^1, s^2, \ldots, s^n) \in S$ is called profile of strategies or simply profile. The set $f(S)$ is the image of the domain S. Any element

$$f(s) = (f_1(s), f_2(s), \ldots, f_n(s)) \in f(S)$$

is called profile of payoffs.

The strategic form may be seen as a vector-function that associates to each profile of strategies $s = (s^1, s^2, \ldots, s^n)$ the profile $f(s) = (f_1(s), f_2(s), \ldots, f_n(s))$ of payoff function values. So, we can highlight strategic form as a vector-function:

$$f : S \to f(S).$$

A game is defined when its rules are stated. Such rules include statements about:

- game playing and moves,
- information known by players,
- player criteria and aims,
- payoffs and their calculation.

Based on such rules, various types of strategic games may be considered.

If to suppose that all players maximize their payoffs, then, similar with the multi-objective optimization problem, a solution of the strategic form game means to find such a profile of strategies that solves the problem:

$$\begin{aligned} f_1(s) &\to \max_{s_1 \in S_1}, \\ f_2(s) &\to \max_{s_2 \in S_2}, \\ &\ldots \\ f_n(s) &\to \max_{s_n \in S_n}, \\ s = (s_1, s_2, &\ldots, s_n) \in S. \end{aligned} \tag{1.6}$$

Problem (1.6) has an important feature that every player maximizes his/her payoff function only by means of his/her argument/strategy that influences simultaneously all the payoff function values. Analogically with the need of a new notion of Pareto optimality for multi-optimization optimization problem, Problem (1.6) needs new solution notions, such as, e.g., Nash and Stackelberg equilibria [90, 103]. There are various modifications to this approach, see e.g. [104, 105].

1.4 Simultaneous/Nash Games

Rules of a simultaneous/Nash game presume fulfilment of a series of axioms applied to the strategic form model:

1. **Moves**. All the players choose their strategies simultaneously and confidently;
2. **Information**. All the players have complete information about the all strategy sets and payoff functions;

3. **Rationality**. All the players optimize (maximize or minimize) the value of their functions;
4. **Payoff**. After all strategy selections each player obtains his payoff as the value of his payoff function on the resulting profile.

The first axiom highlights the way in which the game is played (the way in which players do their moves).

The second axiom characterizes players' intelligence associated with their knowledge/information about the game elements and selected strategies.

The third axiom characterizes player rationality — every player formalize his aim and do all to achieve it.

The fourth axiom establishes the rule for payoffs and their computing.

A solution notion that is applied largely for the Nash game is the strategic/Nash equilibrium notion introduced by John Nash in 1951 [90] and different its modifications and variations [90, 106–111]. The notions of multi-objective strategic (Nash) equilibria or "*equilibrium points in games with vector payoffs*" where introduced in 1956 by Lloyd Shapley [112, 113] and David Blackwell [114]. Great potentials of multi-objective games still remain to be explored [115, 116].

1.5 Sequential/Stackelberg Games

Rules of a sequential/Stackelberg game presume fulfilment of the following series of axioms applied on the strategic form model:

1. **Moves**. The players choose their strategies sequentially:

   1. the first player selects his strategy $s_1 \in S_1$ and informs the second player about his choice,
   2. the second player selects his strategy $s_2 \in S_2$ and informs the third player about the choices s_1, s_2, and so on,
   n. at last, the nth player selects his strategy $s^n \in S_n$ being informed about the choices $s_1, \ldots, s_{n-1}$, of the preceding players;

2. **Information**. Each player knows choices of the precedent/leader players and all strategy sets and payoff functions of the followers;
3. **Rationality**. All the players optimize (maximize or minimize) the value of their functions;
4. **Payoff**. After all strategy selections each player obtains his payoff as the value of his payoff function on the resulting profile.

We can observe the first two axioms in the Stackelberg game differ from the first two axioms in the Nash game. Evidently, Stackelberg game solving needs another solution notion — Stackelberg equilibrium concept [103].

Comparing Nash and Stackelberg games we must observe that the Nash game is a single stage game, and the Stackelberg game is a multi-stage game. There are other approaches to multi-level games, e.g. [117], which is appropriate to Kuhn's approach [118, 119].

1.6 Optimal Control Theory

In the both contexts of game theory and optimal control theory, Pareto–Nash–Stackelberg game and control theory may be seen abstractly as an extension for both the initial domains in which a dynamical system led by a multi-agent game and control is studied. Let us recall that a classical optimal control problem in Pontryagin form [63, 69] may be stated as it follows:

$$\begin{array}{lcll} \mathscr{B}_0(\xi(t)) & \longrightarrow & \inf, & \\ \dot{x}(t) & = & \varphi(t, x(t), u(t)), & \\ u(t) & \in & \mathscr{U}, & \text{for all } t \in [t_0, t_1], \\ \mathscr{B}_i\left(\xi(t, t_0, t_1)\right) & \le & 0, & i = 1, \ldots, m_1, \\ \mathscr{B}_i\left(\xi(t, t_0, t_1)\right) & = & 0, & i = m_1, \ldots, m, \end{array}$$

where $\xi(\cdot) = \xi(t, t_0, t_1) = (x(t), u(t), t_0, t_1)$,

$$\mathscr{B}_i\left(\xi(\cdot)\right) = \int_{t_0}^{t_1} f_i\left(t, x(t), u(t)\right) dt + \psi_i\left(t_0, x(t_0), t_1, x(t_1)\right)$$

for $i = 0, 1, \ldots, m$, $t_0, t_1 \in \Delta$, where $\Delta \subset \mathbb{R}$ is a given finite interval in $\mathbb{R}$, and $\mathscr{U} \subset \mathbb{R}^r$ is an arbitrary subset of the vectorial space $\mathbb{R}^r$. In some particular cases, one or both of the extremities t_0 and t_1 may be fixed in the above problem. The system of differential equations, which is called also state equation or first order dynamic constraints, is satisfied in the all continuity points $t \in \Delta$ of the control $u(t)$. Next constraints traditionally may be stated in a simpler form as algebraic path constraints and boundary conditions.

Definition. *Quadruple $\xi(\cdot) = (x(t), u(t), t_0, t_1)$ is called a control process if $x(t) \in \mathbb{PC}^1(\Delta, \mathbb{R}^n)$, $u(t) \in \mathbb{PC}^1(\Delta, \mathbb{R}^r)$, and both the system of differential equations and embedded condition are satisfied.*

Definition. *A control process if called admissible if both the inequality and equality constrains are satisfied.*

Definition. *An admissible control $\hat{\xi}(\cdot) = \left(\hat{x}(t), \hat{u}(t), \hat{t}_0, \hat{t}_1\right)$ is called (local or hard) optimal if there exists such $\delta > 0$ that for every admissible control $\xi(\cdot) = (x(t), u(t), t_0, t_1)$ for which*

$$\| (x(t), t_0, t_1) - \left(\hat{x}(t), \hat{t}_0, \hat{t}_1\right) \|_{\mathbb{PC}^1(\Delta, \mathbb{R}^n) \times \mathbb{R}^2} < \delta$$

the inequality

$$\mathscr{B}_0(\xi(\cdot)) \geq \mathscr{B}_0(\hat{\xi}(\cdot))$$

holds.

A Lagrange functional is defined as:

$$\begin{aligned}&\mathscr{L}(t, x(t), u(t), t_0, x(t_0), t_1, x(t_1), p(t), \lambda) = \\ &= \int_{t_0}^{t_1} \left(\sum_{i=0}^{m} \lambda_i f_i(t, x(t), u(t)) + p(t)\left(\dot{x}(t) - \varphi(t, x(t), u(t))\right) \right) dt + \\ &+ \sum_{i=0}^{m} \lambda_i \psi_i(t_0, x(t_0), t_1, x(t_1)),\end{aligned}$$

where $p(t) \in \mathbb{PC}^1([t_0, t_1], \mathbb{R}^{n*})$, and $\lambda = (\lambda_0, \lambda_1, \ldots, \lambda_m)$. By introducing notation for the integrand and terminal functional:

$$\begin{aligned}L(t, x(t), \dot{x}(t), u(t), p(t), \lambda) &= \sum_{i=0}^{m} \lambda_i f_i(t, x(t), u(t)) + \\ &+ p(t)\left(\dot{x}(t) - \varphi(t, x(t), u(t))\right), \\ l(t_0, x(t_0), t_1, x(t_1), \lambda) &= \sum_{i=0}^{m} \lambda_i \psi_i(t_0, x(t_0), t_1, x(t_1)),\end{aligned}$$

Lagrange functional may be stated as

$$\mathscr{L}(\cdot) = \int_{t_0}^{t_1} L(t, x(t), \dot{x}(t), u(t), p(t), \lambda) dt + l(t_0, x(t_0), t_1, x(t_1), \lambda).$$

We define Hamiltonian functional as

$$\mathscr{H}(t, x(t), u(t), p(t), \lambda) = p(t)\varphi(t, x(t), u(t)) - \sum_{i=0}^{m} \lambda_i f_i(t, x(t), u(t)).$$

For the sake of brevity in the formulation of necessary optimality conditions for the optimal control process $\hat{\xi}(\cdot) = \left(\hat{x}(t), \hat{u}(t), \hat{t}_0, \hat{t}_1\right)$, let us consider notation:

$$\begin{aligned}&\hat{L} = L(t, \hat{x}(t), \dot{\hat{x}}(t), \hat{u}(t), p(t), \lambda), \\ &\hat{L}(t_k) = L(t_k, \hat{x}(t), \dot{\hat{x}}(t), \hat{u}(t), p(t), \lambda),\ k = 0, 1, \\ &\hat{f}_i(t)'_x = \frac{d}{dx} f_i(t, \hat{x}(t), \hat{u}(t)), \\ &\hat{\varphi}(t)'_x = \frac{d}{dx} \varphi(t, \hat{x}(t), \hat{u}(t)), \\ &\hat{\psi}_{i\,t_k}' = \frac{d}{dt_k} \psi_i(\hat{t}_0, \hat{x}(t_0), \hat{t}_1, \hat{x}(t_1)),\ k = 0, 1, \\ &\hat{\psi}_{i\,x(t_k)}' = \frac{d}{dx(t_k)} \psi_i(\hat{t}_0, \hat{x}(t_0), \hat{t}_1, \hat{x}(t_1)),\ k = 0, 1,\end{aligned}$$

$$\begin{aligned}
\hat{l} &= l(\hat{t}_0, \hat{x}(t_0), \hat{t}_1, \hat{x}(t_1), \lambda), \\
\hat{l}'_{t_k} &= \frac{d}{dt_k} l(\hat{t}_0, \hat{x}(t_0), \hat{t}_1, \hat{x}(t_1), \lambda),\ k = 0, 1, \\
\hat{l}'_{x(t_k)} &= \frac{d}{dx(t_k)} l(\hat{t}_0, \hat{x}(t_0), \hat{t}_1, \hat{x}(t_1), \lambda),\ k = 0, 1, \\
\hat{\mathscr{L}}'_{t_k} &= \frac{d}{dt_k} \mathscr{L}(t, \hat{x}(t), \hat{u}(t), \hat{t}_0, \hat{x}(t_0), \hat{t}_1, \hat{x}(t_1)),\ k = 0, 1.
\end{aligned}$$

Based on the above notation, the necessary optimality conditions for the admissible control process $\hat{\xi}(\cdot) = \left(\hat{x}(t), \hat{u}(t), \hat{t}_0, \hat{t}_1\right)$ may be formulated. Let us recall and simplify them with the aim to state the Pontryagin principle theorem in a simple form.

Stationarity with respect to $x(t)$ — Euler–Lagrange equation for the integrand $L(t, x(t), \dot{x}(t), u(t))$:

$$-\frac{d}{dt}\hat{L}'_{\dot{x}} + \hat{L}'_x = 0,$$

which is equivalent to

$$\dot{p}(t) = \sum_{i=0}^{m} \lambda_i\, \hat{f}_i(t)'_x - p(t)\, \hat{\varphi}(t)'_x.$$

Transversality with respect to $x(t)$ **for the terminal** l:

$$\hat{L}'_{\dot{x}}(t_k) = (-1)^k \hat{l}'_{x(t_k)}$$

which is equivalent to

$$p(t_k) = (-1)^k \sum_{i=0}^{m} \lambda_i \hat{\psi}'_{i\,x(t_k)},\ k = 0, 1.$$

Optimality with respect to $u(t)$ **— minimum principle in the Lagrange form**:

$$\hat{L} = \min_{u \in \mathscr{U}} \hat{L}(t, \hat{x}(t), \dot{\hat{x}}(t), u, p(t), \lambda)$$

which is equivalent to

$$\begin{aligned}
&\min_{u \in \mathscr{U}} \sum_{i=0}^{m} \lambda_i f_i(t, \hat{x}(t), u) - p(t)\, \varphi(t, \hat{x}(t), u) = \\
&= \sum_{i=0}^{m} \lambda_i f_i(t, \hat{x}(t), \hat{u}(t)) - p(t)\, \varphi(t, \hat{x}(t), \hat{u}(t)))
\end{aligned}$$

or **maximum principle of Pontryagin** with the Hamiltonian functional:

$$\min_{u\in\mathscr{U}} \mathscr{H}(t, \hat{x}(t), u, p(t), \lambda) = \mathscr{H}(t, \hat{x}(t), \hat{u}(t), p(t), \lambda).$$

Stationarity with respect to t_k which is stated for mobile extremities only:

$$\hat{\mathscr{L}}'_{t_k} = 0, \; k = 0, 1,$$

that is equivalent to

$$(-1)^{k+1}\lambda_i \hat{f}_i(t_k) + \sum_{i=0}^{m} \lambda_i \left(\hat{\psi}'_{i\,t_k} + \hat{\psi}'_{i\,x(t_k)}\dot{\hat{x}}(t_k)\right) = 0, \; k = 0, 1,$$

or to the condition written with the help of Hamiltonian functional:

$$\hat{\mathscr{H}}(t_k) = (-1)^{k+1}\,\hat{l}'_{t_k}, \; k = 0, 1.$$

Complementary slackness:

$$\lambda_i \mathscr{B}_i(\hat{\xi}) = 0, \; i = 1, 2, \ldots, m_1.$$

Dual feasibility:

$$\lambda_i \geq 0, \; i = 0, 1, 2, \ldots, m_1.$$

Theorem (Pontryagin principle) [63, 69]. *Let:*

- *the tuple* $\hat{\xi}(t)$ *be an optimal control process for the optimal control problem,*
- *the functions* $f_i, i = 0, 1, \ldots, m, \varphi$ *and their partial derivatives with respect to* x *be continues on* $\mathscr{V} \times \mathscr{U}$, *where the set* $\mathscr{V}$ *is a neighbourhood of the set* $\{(t, \hat{x}(t)) : t \in [\hat{t}_0, \hat{t}_1]\}$,
- *and the functions* $\psi_i, i = 0, 1, 2, \ldots, m$, *be continuously differentiable in the neighbourhood of the point* $(\hat{t}_0, \hat{x}(t_0), \hat{t}_1, \hat{x}(t_1))$.

Then, there are $\lambda = (\lambda_0, \lambda_1, \ldots, \lambda_m)$ *and* $p(\cdot) \in \mathbb{PC}([\hat{t}_0, \hat{t}_1], \mathbb{R}^n)$ *that are not equal to zero simultaneously such that the following necessary conditions are satisfied:*

- **Stationarity with respect to** x **— Euler–Lagrange equation**:

$$\dot{p}(t) + p(t)\hat{\varphi}(t)'_x = \hat{f}(t)'_x,$$

for all $t \in \mathscr{T}$, *where* $f(t, x, u) = \sum_{i=0}^{m} \lambda_i f_i(t, x, u)$;

- **Transversality with respect to** x:

$$\begin{cases} p(t_0) = \hat{l}'_{x(t_0)}, \\ p(t_1) = \hat{l}'_{x(t_1)}; \end{cases}$$

- **Optimality with respect to** u:

$$f(t, \hat{x}(t), u) - p(t)\varphi(t, \hat{x}(t), u) \geq \hat{f}(t) - p(t)\hat{\varphi}(t),$$

for all $t \in (T), u \in \mathscr{U}$;
- **Stationarity with respect to** t_k, $k = 0, 1$:

$$\begin{aligned} -\hat{f}(\hat{t}_0) - \hat{l}'_{t_0} + \hat{l}'_{x(t_0)}\hat{\varphi}(t_0) = 0, \\ \hat{f}(\hat{t}_1) + \hat{l}'_{t_1} + \hat{l}'_{x(t_1)}\hat{\varphi}(t_1) = 0, \end{aligned}$$

only for mobile extremities.
- **Complementary slackness**:

$$\lambda_i \mathscr{B}_i(\hat{\xi}) = 0, \; i = 1, 2, \ldots, m_1.$$

- **Dual feasibility**:

$$\lambda_i \geq 0, \; i = 0, 1, 2, \ldots, m_1,$$

where $\mathscr{T}$ *is the set of the continuity points of the control* u *in the interval* (t_0, t_1).

A Bellman or Hamilton–Jacobi–Bellman equation offers a sufficient condition for an admissible control to be optimal. It was introduced by Richard Bellman in his 1957 monograph "*Dynamic Programming*" [62]. In order to solve a problem, Dynamic Programming Method divides it into a series of smaller problems and applies to all of them Bellman's Principle of Optimality which states that for any initial state and decision an admissible control is optimal if remains decisions are optimal with regard to previous ones. Bellman's equation is a functional equation because it needs to find a function — an optimal control in the case of optimal control problems. It may be solved by backward induction and from this point of view it is appropriate to concepts of Stackelberg game and solution. A real difficulty in solving Bellman's equations is due to "*the curse of dimensionality*" [62] — a necessity to solve large-dimension problems appearing during the process of solving the functional equation.

1.7 Applications

In [120], Robert Aumann's exposes a view of the first president of the World Game Theory Society on the game theory interdisciplinary character and its various application areas, highlighting the importance of the discipline in the modern scientific era. Let us review further some important contributions to different areas of human modern activities made by game theorists and practitioners. Since its appearance, game theory assumed a preoccupation with the applications especially in economics [55, 56].

Game theory applications in microeconomics are exposed by Mas-Colell and Whinston in [121]. Production games and price dynamics are studied by Sjur Didrik Flåm in [122].

A specialized and profound monograph of Ken Urai presents interesting game theory applications both in economics, social sciences, and in such pure mathematical domains as topology, homology and cohomology theories for general spaces, etc. [123]. An equilibrium model of advertising, production and exchange is studied by K. Hausken in [124]. Applications with really outstanding economical results in the domain of auctions are described by Paul Milgrom in his work "*Putting Auction Theory to Work*" [125]. Michihiro Kandori and Shinya Obayashi present their results concerning studying community unions, a subclass of labor unions that admits individual affiliations, by means ot repeated game theory [126], successfully resolving about 3,000 labor disputes and creating a number of offspring [127].

Game theory applications in computer science are exposed in a collection of works edited by Apt and Grädel [128] and in a monograph by Wynn C. Stirling [129].

Game theory applications in public policy are exposed by McCain in [60].

Game theory applications in social media marketing are presented by Eric Anderson in [130]. A game theory model of an electricity market is analysed in [131]. "*A differential game of advertising for national and store brands*" is studied by Salma Karray and Georges Zaccour in [132]. In [133], Guiomar Martín-Herrän and Sihem Taboubi present a Stackelberg differential game played over an infinite horizon and modelling shelf-space allocation in a marketing channel where two manufacturers compete for a limited shelf-space at a retail store.

An impressive collection of game theory applications in manufacturing and logistics is presented in [134].

A collection of papers edited by Benz, Ebert, Jäger, and Van Rooij presents "*a state-of-the-art survey of current research on language evolution and game theoretic approaches to pragmatics*" [135]. Robin Clark analyses different game theory aspects and models of linguistics in [136].

Andrés Perea published in 2012 a first textbook in the domain of epistemic game theory, that explains its principles, ways of reasoning and their applying [137]. We have to mention also in this domain the book edited by Adam Brandenburger [138] and the Ph.D. dissertation of Andreas Witzel [139].

A book by Wynn C. Stirling provides interesting game theory applications for stakeholders [140].

A series of game theory models applied in biology are analysed by Broom and Rychtář in their book [141] and presented shortly in [142]. Peter Hammerstein, and Olof Leimar apply evolutionary game theory in biology, too [75]. Interesting experiments of a game theoretical approach to microbial coexistence were provided by Monica Abrudan, Li You, Kateřina Staňková, and Frank Thuijsman and presented in [143].

Different applications of evolutionary game theory are presented in the monograph of Jun Tanimoto [31], e.g. modelling the spread of infectious diseases and vaccination behaviour.

Game theory models related to climate change are presented by Christopher Andrey, Olivier Bahn, and Alain Haurie in [144] and models related to negotiations on climate change are presented by Frédéric Babonneau, Alain Haurie, and Marc Vielle in [145]. Alan Haurie present "*a two-timescale stochastic game framework for climate change policy assessment*" in [146].

A series of game theory applications are presented in part 3 of [147]. Simple and various game theory applications are described by Ken Binmore in [148]. A recent description of the state of the art in game theory applications are made by Stephen Schecter and Herbert Gintis [32]. A brief descriptions of different game theory applications may be found in the book of Kim [149].

The book [150], edited by Haunschmied, Veliov, and Wrzaczek, presents a broader view of the state of the art of dynamic games in economics. An impressively large exposition of the state of the art of dynamic games and their applications to economics, ecology, and other areas is realized in the monograph [151] edited by Jørgensen, Quincampoix, and Vincent. The book includes also results related to numerical methods and algorithms. An original survey of dynamic games in economics was realised by Long in [152].

Concerning the optimal control applications we must mention that the area is impressively large and multifaceted including both traditional problems [62, 63, 67, 69, 70, 153] and new and modern ones, e.g., problems from the field of robotics, process engineering, automotive control, power systems, traffic control systems [154], energy storage systems [155], operations strategy and supply chain management [156], real life biomedical problems with an emphasis on cancer treatments [157], thermal engineering [158]. Pursuit-evasion zero-sum dynamic two player games and the princess and monster game may be seen as classical examples of game theory and optimal control integration [67, 159, 160]. Interesting incipient ideas and examples concerning hierarchical control games may be found in the Chap. VII "*Hierarchical Control Systems*" of Moiseev's book [161].

Additionally to mentioned above optimal control topics and problems, optimal control theory has some special subjects that are very important for real applications and which must have a special treatment and investigation. Among them we may refer controllability, observability, algebraic duality of controllability and observability, stability or asymptotic behaviour, bang-bang principle, linearization [162, 163].

1.8 Game Theorists and Nobel Memorial Prize in Economic Sciences

At the moment, there are at least eleven game theorists, laureates of Nobel Memorial Prize in Economic Sciences:

1994 John Charles Harsanyi (May 29, 1920 - August 9, 2000),
John Forbes Nash Jr. (June 13, 1928 - May 23, 2015),
Reinhard Justus Reginald Selten (October 5, 1930 - August 23, 2016),
"for their pioneering analysis of equilibria in the theory of noncooperative games";
2005 Robert John Aumann (June 8, 1930),
Thomas Crombie Schelling (April 14, 1921 - December 13, 2016),
"for having enhanced our understanding of conflict and cooperation through game-theory analysis";
2007 Leonid Hurwicz (August 21, 1917 - June 24, 2008),
Eric Stark Maskin (December 12, 1950),
Roger Bruce Myerson (March 29, 1951),
"for having laid the foundations of mechanism design theory";
2012 Alvin Elliot Roth (December 18, 1951),
Lloyd Stowell Shapley (June 2, 1923 - March 12, 2016),
"for the theory of stable allocations and the practice of market design";
2014 Jean Tirole (August 9, 1953),
"for his analysis of market power and regulation".

Remark. *The prizes of 2007 and 2014 years are not mentioned on the Web-page "Economic Sciences Laureates: Fields" of Nobel Memorial Prize in Economic Sciences site as awards in the fields of game theory. From this point of view only prizes of 1994, 2005, and 2012 years were awarded explicitly for contributions in game theory fields of economic sciences.*

Nobel Prize Awards in Economics to game theorists highlight particularly the importance of game theory in economics only. Sure, the impact of game theory results in other area of research and human activity are not smaller and, in my opinion, game theory has to reach spectacular achievements in the appropriate future.

1.9 Objectives and Outline

The monograph objectives are to survey, synthesize and advance author's results in the field of interactive decision theory, to highlight their importance as a tool for investigation and research of different interactive decision problems of a human activity.

The monograph comprises fifteen chapters. This solitary chapter has as its main objective the task to introduce the reader to the book's content, giving a short historical description of game theory, describing briefly main concepts, notation, problems, and applications, surveying the bibliography of the monograph's main areas of interest.

Numbered Chaps. 2–15 are divided intro three parts. Part I deals with noncooperative games. It consists of seven chapters. Chapter 2 considers normal form simultaneous games — Nash games, defines important concepts of equilibria based on a Nash equilibrium concept, both local and global, and establishes existence conditions for defined equilibria. Equilibrium principles are highlighted for multi-objective strategic games, too.

Chapter 3 deals with polymatrix mixed strategy games. It establishes a general method for computing the entire set of Nash equilibria, firstly considering a special case of bimatrix games.

Next three Chaps. 4–6 apply the proposed method to some very important particular dimension games which have illustrative values, too. Chapter 4 exposes a reduction of the bimatrix 2×3 mixed strategy game to an equivalent game on a triangular prism. For the last game, the Nash equilibrium set is determined by analysing exhaustively all the possible cases.

Chapter 5 reduces a trimatrix $2 \times 2 \times 2$ mixed strategy game to an equivalent game on a unite cube. For the obtained game, the Nash equilibrium set is determined by the same method of graph intersection, considering all the possible cases.

Chapter 6 reduces a bimatrix 2×2 (dyadic) mixed strategy game to an equivalent game on a unite square. For the last game a special Nash equilibrium set function is defined by the means of the Wolfram Language. The Wolfram Language code is used to prove the main result and to define an algorithm and the correspondent program for computing the set of Nash equilibria.

Chapter 7 deals with polymatrix mixed strategy games as sequential (noncooperative) games — Stackelberg games. It introduces main concepts by a general scheme, states and proves main theoretical results. The graphs of best response mappings are largely used as a tool both for defining concepts and for proving the statements.

Chapter 8 considers various strategy games on digraphs and applies minimax principle and solution concept for their solving. Some matrix games on digraphs are considered in details and their properties are revealed. Some special games are revealed, e.g.: maximin directed tree, maximin directed path, maximin traveling salesman problem and maximin cost flow. Polymatrix and dynamic games on digraphs are referred too.

Chapters 9–12 form the second part of the monograph. Part II extends the results of the precedent chapters both to the mixtures of simultaneous and sequential multi-objective games, and to a special class of games with information leaks — corruption games. Chapter 9 defines mains concepts needed to analyse mixed simultaneous and sequential multi-objective games — Pareto–Nash–Stackelberg games, state and prove main theorems.

Chapter 10 emphasises application of the precedent chapter method to computing Pareto–Nash equilibrium sets in finite multi-objective mixed-strategy games. It considers both two-player and multi-player cases. A series of examples are analysed.

Chapter 11 applies the method introduced in Chap. 8 to compute sets of Pareto–Nash equilibria in dyadic two-criterion mixed-strategy games. It is mentioned that the algorithm of the method is realised and published as a Wolfram Language program.

Chapter 12 introduces a new class of the games — games with information leaks or corruption games. A taxonomy of such games is built. A series of solution concepts are introduced. Solution existence conditions are stated and proved.

The remaining three chapters of the monograph — Chapters 13–15, constitutes a final third part. Part III combines game theory models, concepts and methods exposed above with optimal control models, concepts and methods. A new class of interactive multi-agent control problems and models appears, building a foundation for Pareto–

Nash–Stackelberg game and control theory. Chapter 13 exposes new concepts and principles for a case of linear discrete-time Pareto–Nash–Stackelberg control processes. It states and proves theorems based both on a straightforward method and Pontryagin's maximum principle.

Chapter 14 extended the results of the precedent chapter to the linear discrete-time set-valued Pareto–Nash–Stackelberg control processes. The control is defined simultaneously for an entire set.

Chapter 15 deals with a problem of Pareto–Nash–Stackelberg control processes with echoes and retroactive future in the linear discrete-time case.

Each chapter starts with a brief introduction which summarises its content and finalises with conclusions needed for a future work.

The monograph includes also *Bibliography*, *Index* and *List of Symbols*.

References

1. Bartos, O.J., and P. Wehr. 2002. *Using Conflict Theory*. Cambridge: Cambridge University Press, XII+219 pp.
2. Deutch, M., P.T. Coleman, and E.C. Marcus (eds.). 2006. *The Handbook of Conflict Resolution: Theory and Practice*. San Francisco: Jossey-Bass, A Wiley Imprint, XIV+940 pp.
3. Rex, J. 1961. *Key Problems of Sociological Theory*. London: Routledge and Kegan Paul, 145 pp.
4. Turner, J.H. (ed.). 2006. *Handbook of Sociological Theory*. New York: Springer Science+Business Media, XI+745 pp.
5. Dahrendorf, R. 1959. *Class and Class Conflict in Industrial Society*. Stanford: Stanford University Press, XVI+336 pp.
6. Aumann, R.J. 1989. Game theory. In *The New Palgrave: Game Theory*, ed. J. Eatwell, M. Milgate, and P. Newman, 1–53. London: The Macmillan Press Limited.
7. Gaindric, C., V. Ungureanu, and D. Zaporojan. 1993. An interactive decision support system for selection of scientific and technical projects. *Computer Science Journal of Moldova* 1 (2(2)): 105–109.
8. Shoham, Y., and K. Leyton-Brown. 2009. *Multi-agent Systems: Algorithmic, Game-Theoretic, and Logical Foundations*. Cambridge: Cambridge University Press, 532 pp.
9. Osborne, M.J. 2009. *An Introduction to Game Theory*. Oxford: Oxford University Press, International Edition, 685 pp.
10. Tadelis, S. 2013. *Game Theory: An Introduction*. Princeton and Oxford: Princeton University Press, XVI+396 pp.
11. Zermelo, E. 1904. Proof that every set can be well-ordered (Beweis, daßjede Menge wohlgeordnet werden kann). *Mathematische Annalen* 59: 514–516 (in German).
12. Zermelo, E. 1908. Investigations in the foundations of set theory I (Neuer Beweis für die Möglichkeit einer Wohlordnung). *Mathematische Annalen* 65: 107–128 (in German).
13. Dantzig, G.B., and M.N. Thapa. 1997. *Linear Programming 1: Introduction*. New York: Springer, 474 pp.
14. Dantzig, G.B., and M.N. Thapa. 2003. *Linear Programming 2: Theory and Extensions*. New York: Springer, 475 pp.
15. Nemirovsky, A., and D. Yudin. 1983. *Problem Complexity and Method Efficiency in Optimization*. New York: Wiley, 404 pp.
16. Nesterov, Y., and A. Nemirovskii. 1994. *Interior-Point Polynomial Algorithms in Convex Programming*. Philadelphia: SIAM, IX+405 pp.

17. Boyd, S.P., and L. Vandenberghe. 2009. *Convex Optimization*. Cambridge: Cambridge University Press, XIV+716 pp.
18. Bertsekas, D.P., A. Nedic, and A.E. Ozdaglar. 2015. *Convex Analysis and Optimization*. Belmont, Massachusetts: Athena Scientific, XV+534 pp.
19. Pareto, V. 1904. *Manuel d'economie politique*. Paris: Giard, 504 pp. (in French).
20. Yudin, D.B. 1984. Generalized mathematical programming. *Economics and Mathematical Methods* 20 (1): 148–167.
21. Tsoi, E.V., and D.B. Yudin. 1989. Generalized convex programming. *Avtomat. i Telemekh.* 3: 44–55.
22. Figueira, J., S. Greco, and M. Ehrgott (eds.). 2005. *Multiple Criteria Decision Analysis: State of the Art Surveys*. Boston: Springer Science and Business Media, Inc., XXXVI+1045 pp.
23. Pascal, B. 1958. *Pascal's Pensées*. New York: E.P. Dutton & Co., Inc., XXI+297 pp.
24. Brams, S.J. 2011. *Game Theory and the Humanities: Bridging Two Worlds*. Cambridge, Massachusetts: MIT Press, XIV+319 pp.
25. Jordan, J. 2006. *Pascal's Wager: Pragmatic Arguments and Belief in God*. Oxford: Oxford University Press, XII+227 pp.
26. Küng, H. 1980. *Does God Exist? An Answer for Today*. London: SCM Press, XXIV+839 pp.
27. Grabisch, M. 2016. *Set Functions, Games and Capacities in Decision Making*. Switzerland: Springer, XVI+473 pp.
28. Russell, B. 1951. *What Desires Are Politically Important?* 259–270. Nobel Prize Lectures: Literature, 1901–1967, Nobel Lecture, 11 December 1950. Stockholm: P.A. Norstedt.
29. Eastman, W.N. 2015. *Why Business Ethics Matters: The Logic of Moral Emotions in Game Theory*. Hampshire: Palgrave Macmillan, XIX+203 pp.
30. Maynard Smith, J., and G.R. Price. 1973. The logic of animal conflict. *Nature* 246: 15–18.
31. Tanimoto, J. 2015. *Fundamentals of Evolutionary Game Theory and Its Applications*. Tokyo: Springer, XIII+214 pp.
32. Schecter, S., and H. Gintis. 2016. *Game Theory in Action an Introduction to Classical and Evolutionary Models*. Princeton: Princeton University Press, XIV+274 pp.
33. Camerer, C.F. 2003. *Behavioral Game Theory: Experiments in Strategic Interaction*. Princeton: Princeton University Press, 495 pp.
34. Gintis, H. 2009. *The Bounds of Reason: Game Theory and the Unification of the Behavioral Sciences*. Princeton and Oxford: Princeton University Press, XVIII+286 pp.
35. Tremblay, V.J., and C.H. Tremblay. 2012. *New Perspectives on Industrial Organization with Contributions from Behavioral Economics and Game Theory*. New York: Springer, XXVI+811 pp.
36. Hausken, K. 1996. Self-interest and sympathy in economic behaviour. *International Journal of Economics and Business Research* 23 (7): 4–24.
37. Perrow, C. 1986. *Complex Organizations*. New York: Random House, X+307 pp.
38. Schelling, T.C. 1980. *The Strategy of Conflict*. Cambridge, Massachusetts: Harvard University Press, IX+309 pp.
39. Dodge, R.V. 2012. *Sheling's Game Theory: How to Make Decisions*. Oxford: Oxford University Press, XII+292 pp.
40. Carmichael, F. 2005. *A Guide to Game Theory*. Harlow: Pearson Education, XV+286 pp.
41. Maschler, M., E. Solan, and Sh. Zamir. 2013. *Game Theory*. Cambridge: Cambridge University Press, XXVI+979 pp.
42. Matsumoto, A., and F. Szidarovszky. 2016. *Game Theory and Its Applications*. Tokyo: Springer, XIV+268 pp.
43. Mazalov, V. 2014. *Mathematical Game Theory and Its Applications*. Tokyo: Wiley, XIV+414 pp.
44. Myerson, R.B. 1997. *Game Theory: Analysis of Conflict*. Cambridge, Massachusetts: Harvard University Press, XVI+568 pp.
45. Straffin, Ph.D. 1993. *Game Theory and Strategy*. Washington: The Mathematical Association of America, X+244 pp.

46. Narahari, Y. 2014. *Game Theory and Mechanism Design*. New York: IISc Press and World Scientific, XL+492 pp.
47. Nisan, N., T. Roughgarden, E. Tardos, and V.V. Vazirani (eds.). 2007. *Algorithmic Game Theory*. Cambridge: Cambridge University Press, 775 pp.
48. Petrosyan, L.A., and N.A. Zenkevich. 2016. *Game Theory*, 2nd ed. New Jersey: World Scientific, XII+551 pp.
49. Umbhauer, G. 2016. *Game Theory and Exercises*. New York: Routledge, XX+442 pp.
50. Watson, J. 2013. *Strategy: An Introduction to Game Theory*. New York: W. W. Norton & Company, XV+491 pp.
51. Borel, E. 1921. La théorie du jeu les équation intégrales á noyau symétrique gauche. *Compte Rendus Académie des Science* 173: 1304–1308.
52. Borel, E. 1923. Sur les jeu où intervient l'hasard et l'habileté des joueurs. Association Française pour l'Advancement des Sciences, 79–85.
53. Borel, E. 1924. Sur les jeu où intervient l'hasard et l'habileté des joueurs. *Theorie des Probabilités*. Paris: Librairie Scientifique.
54. Von Neumann, J. 1928. Zur Theorie der Gesellschaftsspiele. *Mathematische Annalen* 100: 295–320 (in German).
55. Von Neumann, J., and O. Morgenstern. 1944. *Theory of Games and Economic Behavior*. Princeton: Annals Princeton University Press; 2nd ed., 1947, 674 pp.
56. Luce, D.R., and H. Raiffa. 1957. *Games and Decisions: Introduction and Critical Survey*. New York: Wiley, XX+509 pp.
57. Kelly, A. 2003. *Decision Making Using Game Theory: An Introduction for Managers*. Cambridge: Cambridge University Press, X+204 pp.
58. Geçkil, I.K., and P.L. Anderson. 2010. *Applied Game Theory and Strategic Behaviour*. Boca Raton: CRC Press, XIX+198 pp.
59. Dimand, M.A., and R. Dimand. 2002. *The History of Game Theory, Volume 1: From the Beginning to 1945*. London: Routledge, X+189 pp.
60. McCain, R.A. 2009. *Game Theory and Public Policy*. Cheltenham: Edward Elgar, VI+262 pp.
61. Bryson Jr., A.E. 1996. Optimal control - 1950 to 1985. *IEEE Control Systems* 16 (3): 26–33.
62. Bellman, R. 1957. *Dynamic Programming*. New Jersey: Princeton University Press, 365 pp.
63. Pontryagin, L.S., V.G. Boltyanskii, R.V. Gamkrelidze, and E.F. Mishchenko. 1961. *Mathematical Theory of Optimal Processes*. Moscow: Nauka, 393 pp. (in Russian).
64. Clarke, F.H. 1989. *Methods of Dynamic and Nonsmooth Optimization*. CBMS-NSF Regional Conference Series in Applied Mathematics. Philadelphia: SIAM, VI+91 pp.
65. McShane, E.J. 1939. On multipliers for Lagrange problems. *American Journal of Mathematics* 61 (4): 809–819.
66. Breitner, M.H. 2005. The genesis of differential games in light of Isaacs? Contributions. *Journal of Optimization Theory and Applications* 124 (3): 523–559.
67. Isaacs, R. 1965. *Differential Games: A Mathematical Theory with Applications to Warfare and Pursuit, Control and Optimization*. Berlin: Wiley, XXIII+385 pp.
68. Başar, T., and G.J. Olsder. 1999. *Dynamic Noncooperative Game Theory*. Philadelphia: SIAM: Society for Industrial and Applied Mathematics, 536 pp.
69. Pontryagin, L.S. 1998. *The Maximum Principle*. Moscow: Mathematical Education and Formation Fond, 71 pp. (in Russian).
70. Vasil'ev, F.P. 2002. *Optimization Methods*. Moscow: Factorial Press, 825 pp. (in Russian).
71. Osborne, M.J., and A. Rubinstein. 1994. *A Course in Game Theory*. Cambridge: The MIT Press, 373 pp.
72. Bensoussan, A., J. Frehse, and Ph. Yam. 2013. *Mean Field Games and Mean Field Type Control Theory*. New York: Springer, X+128 pp.
73. Clerbout, N., and Sh. Rahman. 2015. *Linking Game-Theoretical Approaches with Constructive Type Theory: Dialogical Strategies, CTT Demonstrations and the Axiom of Choice*. Heidelberg: Springer, XIX+99 pp.

74. Sandholm, W.H. 2015. Population games and deterministic evolutionary dynamics. In *Handbook of Game Theory with Economic Applications*, vol. 4, ed. H. Peyton Young, and Sh. Zamir, 703–778. Amsterdam: North-Holland.
75. Hammerstein, P., and O. Leimar. 2015. Evolutionary game theory in biology. In *Handbook of Game Theory with Economic Applications*, vol. 4, ed. H. Peyton Young, and Sh. Zamir, 574–617. Amsterdam: North-Holland.
76. Thomas, G.H. 2006. *Geometry, Language and Strategy*, vol. 37. Series on Knots and Everything. New Jersey: World Scientific, 386 pp.
77. Thomas, G.H. 2017. *Geometry, Language and Strategy, Vol. 2: The Dynamics of Decision Processes*. New Jersey: World Scientific, 816 pp.
78. Avenhaus, R., and I.W. Zartman (eds.). 2007. *Diplomacy Games: Formal Models and International Negotiations*. Berlin: Springer, XIX+350 pp.
79. Poundstone, W. 1992. *Prisoner's Dilemma*. New York: Anchor Books, XI+294 pp.
80. Mirowski, P. 2002. *Machine Dreams: Economics Becomes a Cyborg Science*. Cambridge: Cambridge University Press, XIV+655 pp.
81. McAdams, D. 2014. *Game-Changer: Game Theory and the Art of Transforming Strategic Situations*. New York: W. W. Norton & Company, 304 pp.
82. Sagaidac, M., and V. Ungureanu. 2004. *Operational Research*. Chişinău: CEP USM, 296 pp. (in Romanian).
83. Cooper, R.W. 1999. *Coordination Games: Complementarities and Macroeconomics*. Cambridge: Cambridge University Press, XIV+163 pp.
84. Jackson, M.O., and Y. Zenou. 2015. Games on networks. In *Handbook of Game Theory with Economic Applications*, vol. 4, ed. H. Peyton Young, and Sh. Zamir, 95–163. Amsterdam: North-Holland.
85. Rousseau, J.-J. 1987. Discourse on the origin and foundations of inequality among men. Trans. and ed. Donald A. Cress, *The Basic Political Writings of Jean-Jacques Rousseau*, 37–81. Indianapolis: Hackett Publishing Company Inc.
86. Russell, B. 2010. *Common Sense and Nuclear Warfare*, First published 1959. London: Routledge, XXXII+75 pp.
87. Smith, M.J., and G.R. Price. 1973. The logic of animal conflict. *Nature* 246: 15–18.
88. Bramoullé, Y., D. López-Pintadoz, S. Goyalx, and F. Vega-Redondo. 2004. Network formation and anti-coordination games. *International Journal of Game Theory* 33 (1): 1–19.
89. Kun, J., B. Powers, and L. Reyzin. 2013. Anti-coordination games and stable graph colorings. In *Algorithmic Game Theory. SAGT 2013*, vol. 8146, ed. B. Vöcking, 122–133. Lecture Notes in Computer Science. Berlin: Springer.
90. Nash, J. 1951. Noncooperative games. *Annals of Mathematics* 54 (2): 280–295.
91. Fisher, L. 2008. *Rock, Paper, Scissors: Game Theory in Everyday Life*. New York: Basic Books, X+265 pp.
92. Lozovanu, D., and V. Ungureanu. 1989. On the nesting of convex polyhedral sets defined by systems of linear inequalities. *Kibernetika (Kiev)* 135 (1): 104–107.
93. Lozovanu, D., and V. Ungureanu. 1989. An algorithm for testing the embeddability of a k-dimensional unit cube in an orthogonal projection of a polyhedral set defined by a system of linear inequalities. *Mat. Issled., Mathematical Modeling and Optimization* 162 (110): 67–78 (in Russian).
94. Ungureanu, V. 1989. NP-hardness of the problem of testing the embeddability of a k-dimensional unit cube in an orthogonal projection of a convex polyhedral set. *Mat. Issled., Mathematical Modeling and Optimization* 16 (110): 120–128 (in Russian).
95. Gaindric, C., and Ungureanu, V. Systems of linear inequalities with right-hand parametric parts and some optimization problems, *Optimization Methods and Their Applications, Book of Abstracts of the International School and Seminar, Irkutsk, 10–19 September 1989*, 53–54.
96. Schelling, T.C. 2001. Commitment: Deliberate versus involuntary. In *Evolution and the Capacity for Commitment*, ed. R.M. Nesse, 48–56. New York: Russell Sage Press.
97. Nesse, R.M. (ed.). 2001. *Evolution and the Capacity for Commitment*. New York: Russell Sage Press, XVIII+334 pp.

98. Hirshleifer, J. 2001. Game-theoretic interpretations of commitment. In *Evolution and the Capacity for Commitment*, ed. R.M. Nesse, 77–94. New York: Russell Sage Press.
99. Hurwicz, L., and S. Reiter. 2006. *Designing Economic Mechanisms*. Cambridge: Cambridge University Press, X+344 pp.
100. Williams, S.R. 2008. *Communication in Mechanism Design: A Differential Approach*. Cambridge: Cambridge University Press, XV+197 pp.
101. Vohra, R.V. 2011. *Mechanism Design: A Linear Programming Approach*. Cambridge: Cambridge University Press, XV+173 pp.
102. Börgers, T. with a chapter by Daniel, Krähmer, and Roland, Strausz. 2015. *An Introduction to the Theory of Mechanism Design*. Oxford: Oxford University Press, XVI+246 pp.
103. Von Stackelberg, H. *Marktform und Gleichgewicht (Market Structure and Equilibrium)*, Vienna: Springer Verlag, 1934, XIV+134 pp. (in German).
104. Smol'yakov, È.R. 1986. *Equilibrium Models in Which the Participants have Different Interests (Ravnovesnye modeli pri nesovpadayushchikh interesakh uchastnikov)*. Moskva: Nauka, 224 pp. (in Russian).
105. Smol'yakov, È.R. 2000. *Theory of Conflicts and Differential Games (Teoriya antagonizmov i differentsial'nye igry)*. Moskva: Èditorial URSS, 159 pp. (in Russian).
106. Maskin, E. 1999. Nash equilibrium and welfare optimality. *Review of Economic Studies* 66 (1): 23–38.
107. Myerson, R.B. 1999. Nash equilibrium and the history of economic theory. *Journal of Economic Literature* 37 (3): 1067–1082.
108. Govindan, S., and R. Wilson. 2005. Essential equilibria. *Proceedings of the National Academy of Sciences of the United States of America* 102 (43): 15706–15711.
109. Aumann, R.J. 1974. Subjectivity and correlation in randomized strategies. *Journal of Mathematical Economics* 1: 67–96.
110. Hart, S., and A. Mas-Colell. 2013. *Simple Adaptive Strategies: From Regret-Matching to Uncoupled Dynamics*. New Jersey: World Scientific, XXXVIII+296 pp.
111. Apt, K.R., F. Rossi, and K.B. Venable. 2008. Comparing the notions of optimality in CP-nets, strategic games and soft constraints. *Annals of Mathematics and Artificial Intelligence* 52 (1): 25–54.
112. Shapley, L.S. 1956. Equilibrium points in games with vector payoffs, Rand Corporation Research Memorandum RM-1818, I–III, 1–7.
113. Shapley, L.S. 1959. Equilibrium points in games with vector payoffs. *Naval Research Logistics Quarterly* 6: 57–61.
114. Blackwell, D. 1956. An analog of the minimax theorem for vector payoffs. *Pacific Journal of Mathematics* 6: 1–8.
115. Zhao, Y.M. 2017. Three little-known and yet still significant contributions of Lloyd Shapley. *Games and Economic Behavior*, 12 May 2017, 1–26.
116. Mármol, A.M., L. Monroy, M.A. Caraballo, and A. Zapata. 2017. Equilibria with vector-valued utilities and preference information. The analysis of a mixed duopoly. *Theory and Decision*, 27 March 2017, 1–19.
117. Hausken, K., and R. Cressman. 2004. Formalization of multi-level games. *International Game Theory Review* 6 (2): 195–221.
118. Kuhn, H.W. 1950. Extensive games. *Proceedings of the National Academy of Sciences of the United States of America* 36: 570–576.
119. Kuhn, H.W. 1953. Extensive games and the problem of information. In *Contributions to the Theory of Games, Vol. II*, vol. 28, ed. H. Kuhn, and A. Tucker, 217–243. Annals of Mathematics Study. Princeton: Princeton University Press.
120. Aumann, R.J. 2003. Presidential address. *Games and Economic Behavior* 45: 2–14.
121. Mas-Colell, A., M.D. Whinston, and J.R. Green. 1991. *Microeconomics*. Oxford: Oxford University Press, XVII+977 pp.
122. Flåm, S.D. 2005. Production games and price dynamics. In *Dynamic Games: Theory and Applications*, ed. A. Haurie, and G. Zaccour, 79–92. New York: Springer.

123. Urai, K. 2010. *Fixed Points and Economic Equilibria*. New Jersey: World Scientific, XVIII+292 pp.
124. Hausken, K. 2011. An equilibrium model of advertising, production and exchange. *International Journal of Economics and Business Research* 3 (4): 407–442.
125. Milgrom, P. 2004. *Putting Auction Theory to Work*. Cambridge: Cambridge University Press, XXIV+368 pp.
126. Mailath, G.J., and L. Samuelson. 2006. *Repeated Games and Reputations: Long-Run Relationships*. New York: Oxford University Press, XVIII+645 pp.
127. Kandori, M., and S. Obayashi. 2014. Labor union members play an OLG repeated game. *Proceedings of the National Academy of Sciences* 111 (Suppl. 3): 10802–10809.
128. Apt, K.R., and E. Grädel. 2011. *Lectures in Game Theory for Computer Scientists*. Cambridge: Cambridge University Press, XII+295 pp.
129. Stirling, W.C. 2003. *Satisficing Games and Decision Making: With Applications to Engineering and Computer Science*. Cambridge: Cambridge University Press, XVIII+249 pp.
130. Anderson, E. 2010. *Social Media Marketing: Game Theory and the Emergence of Collaboration*. Berlin: Springer, X+188 pp.
131. Bossy, M., N. Maïzi, G.J. Olsder, O. Pourtallier, and E. Tanré. 2005. Electricity prices in a game theory context. In *Dynamic Games: Theory and Applications*, ed. A. Haurie, and G. Zaccour, 135–139. New York: Springer.
132. Karray, S., and G. Zaccour. 2005. A differential game of advertising for national and store brands. In *Dynamic Games: Theory and Applications*, ed. A. Haurie, and G. Zaccour, 213–229. New York: Springer.
133. Martín-Herrän, G., and S. Taboubi. 2005. Incentive strategies for shelf-space allocation in duopolies. In *Dynamic Games: Theory and Applications*, ed. A. Haurie, and G. Zaccour, 231–253. New York: Springer.
134. Benyoucef, L., J.-C. Hennet, and M.K. Tiwari (eds.). 2014. *Applications of Multi-criteria and Game Theory Approaches: Manufacturing and Logistics*. New Jersey: Springer, XVI+408 pp.
135. Benz, A., C. Ebert, G. Jäger, and R. Van Rooij (eds.). 2011. *Language, Games, and Evolution Trends in Current Research on Language and Game Theory*. Berlin: Springer, 188 pp.
136. Clark, R. 2012. *Meaningful Games: Exploring Language with Game Theory*. Cambridge: The MIT Press, XVIII+354 pp.
137. Perea, A. 2012. *Epistemic Game Theory: Reasoning and Choice*. Cambridge: Cambridge University Press, XVIII+561 pp.
138. Brandenburger, A. (ed.). 2014. *The Language of Game Theory: Putting Epistemics into the Mathematics of Games*. New Jersey: World Scientific, XXXIV+263 pp.
139. Witzel, S.A. 2009. Knowledge and games: Theory and implementation, Ph.D. dissertation, Amsterdam: Institute for Logic, Language and Computation, 162 pp.
140. Stirling, W.C. 2012. *Theory of Conditional Games*. Cambridge: Cambridge University Press, XIV+236 pp.
141. Broom, M., and J. Rychtář. 2013. *Game-Theoretical Models in Biology*. Boca Raton: CRC Press, XXVI+488 pp.
142. Broom, M., and J. Rychtář. 2016. Nonlinear and multiplayer evolutionary games. In *Advances in Dynamic and Evolutionary Games: Theory, Applications, and Numerical Methods*, ed. F. Frank Thuijsman, and F. Wagener, 95–115. Annals of the International Society of Dynamic Games. Boston: Birkhäuser.
143. Abrudan, M., L. You, K. Staňková, and F. Thuijsman. 2016. A game theoretical approach to microbial coexistence. In *Advances in Dynamic and Evolutionary Games: Theory, Applications, and Numerical Methods*, vol. 14, ed. F. Frank Thuijsman, and F. Wagener, 267–282. Annals of the International Society of Dynamic Games. Boston: Birkhäuser.
144. Andrey, C., O. Bahn, and A. Haurie. 2016. Computing α-robust equilibria in two integrated assessment models for climate change. In *Advances in Dynamic and Evolutionary Games: Theory, Applications, and Numerical Methods*, vol. 14, ed. F. Frank Thuijsman, and F. Wagener, 283–300. Annals of the International Society of Dynamic Games. Boston: Birkhäuser.

145. Babonneau, F., A. Haurie, and M. Vielle. 2016. A robust noncooperative meta-game for climate negotiation in Europe. In *Advances in Dynamic and Evolutionary Games: Theory, Applications, and Numerical Methods*, vol. 14, ed. F. Frank Thuijsman, and F. Wagener, 301–319. Annals of the International Society of Dynamic Games. Boston: Birkhäuser.
146. Haurie, A. 2005. A two-timescale stochastic game framework for climate change policy assessment. In *Dynamic Games: Theory and Applications*, ed. A. Haurie, and G. Zaccour, 193–211. New York: Springer.
147. Chinchuluun, A., P.M. Pardalos, A. Migdalas, and L. Pitsoulis (eds.). 2008. *Pareto Optimality, Game Theory, and Equilibria*. New York: Springer Science + Business Media, 868 pp.
148. Binmore, K. 2007. *Game Theory: A Very Short Introduction*. Oxford: Oxford University Press, XIV+186 pp.
149. Kim, S. 2014. *Game Theory Applications in Network Design*. Hershey: IGI Global, XXII+500 pp.
150. Haunschmied, J., V.M. Veliov, and S. Wrzaczek (eds.). 2014. *Dynamic Games in Economics*. Berlin: Springer, XII+315 pp.
151. Jørgensen, S., M. Quincampoix, and T.L. Vincent (eds.). 2007. *Advances in Dynamic Game Theory: Numerical Methods, Algorithms, and Applications to Ecology and Economics*. Boston: Birkhäuser, XXII+717 pp.
152. Long, N.V. 2010. *A Survey of Dynamic Games in Economics*. New Jersey: World Scientific, XIV+275 pp.
153. Chikriy, A.A. 1992. *Conflict-Controlled Processes*. Kiev: Naukova Dumka, 384 pp. (in Russian).
154. Böhme, T.J., and B. Frank. 2017. *Hybrid Systems, Optimal Control and Hybrid Vehicles: Theory, Methods and Applications*. Switzerland: Springer, XXXIII+530 pp.
155. Tang, W., and Y.J. Zhang. 2017. *Optimal Charging Control of Electric Vehicles in Smart Grids*. Switzerland: Springer, XI+106 pp.
156. Kim, B. 2017. *Optimal Control Applications for Operations Strategy*. Singapore: Springer, XI+223 pp.
157. Schättler, H., and U. Ledzewicz. 2015. *Optimal Control for Mathematical Models of Cancer Therapies: An Application of Geometric Methods*. New York: Springer, XIX+496 pp.
158. Badescu, V. 2017. *Optimal Control in Thermal Engineering*. Switzerland: Springer, XV+588 pp.
159. Petrosyan, L.A. 2016. Dynamic games with perfect information. In *Advances in Dynamic and Evolutionary Games: Theory, Applications, and Numerical Methods*, ed. F. Thuijsman, and F. Wagener, 1–26. Boston: Birkhäuser.
160. Shinar, J., V.Y. Glizer, and V. Turetsky. 2016. Pursuit-evasion game of kind between hybrid players. In *Advances in Dynamic and Evolutionary Games: Theory, Applications, and Numerical Methods*, ed. F. Thuijsman, and F. Wagener, 187–208. Boston: Birkhäuser.
161. Moiseev, N.N. 1975. *Elements of the Optimal System Theory*. Moscow: Nauka, 526 pp. (in Russian).
162. Terrell, W.J. 1999. Some fundamental control theory I: Controllability, observability, and duality. *The American Mathematical Monthly* 106 (8): 705–719.
163. Terrell, W.J. 1999. Some fundamental control theory II: Feedback linearization of single input nonlinear systems. *The American Mathematical Monthly* 106 (9): 812–828.

Part I
Noncooperative Games

The first part of the monograph is dedicated to non-cooperative games, their solution principles, and methods of finding the full set of solutions.

Chapter 2
Nash Equilibrium Conditions as Extensions of Some Classical Optimisation Theorems

Abstract We apply the notion of Lagrange vector-function to analyse strategic (normal) form games (Ungureanu, Lib Math, XXVII:131–140, 2007, [1]). We have the aim to formulate and prove Nash equilibrium conditions for such games and Pareto-Nash equilibrium conditions for multi-criteria strategic form games. Analytical, theoretical and conceptual foundation for all the results of this chapter stands on domains of normal form games, both simultaneous and sequential, and on domain of optimization theory, both single-criterion and multi-criteria.

2.1 Introduction

Consider a noncooperative (strategic form, normal form) game

$$\Gamma = \langle N, \{X_p\}_{p \in N}, \{f_p(x)\}_{p \in N} \rangle,$$

where

- $N = \{1, 2, \ldots, n\} \subset \mathbb{N}$ is a set of players,
- $X_p = \left\{ x^p \in \mathbb{R}^{n_p} : \begin{array}{c} g_i^p(x^p) \leq 0,\ i = 1, \ldots, m_p \\ h_i^p(x^p) = 0,\ i = 1, \ldots, l_p \\ x^p \in M_p \end{array} \right\}$ is a set of strategies of the player $p \in N$,
- $l_p, m_p, n_p < +\infty$, $p \in N$,
- $g_i^p(x^p), i = 1, \ldots, m_p,\ h_i^p(x^p), i = 1, \ldots, l_p$, are constraint functions with the domain M_p, $p \in N$,
- $f_p(x)$ is a pth player payoff/cost function with the Cartesian product $X = \underset{p \in N}{\times} X_p$ as the domain.

Any element $x = (x^1, x^2, \ldots, x^n) \in X$ is called profile of the game (strategy profile or outcome of the game, global strategy, situation).

Supposition 2.1.1 *Without loss of generality it can be assumed that the players* **minimize** *the values of their payoff functions.*

V. Ungureanu, *Pareto-Nash-Stackelberg Game and Control Theory*,
Smart Innovation, Systems and Technologies 89,
https://doi.org/10.1007/978-3-319-75151-1_2

Definition 2.1 ([2]) The profile $\hat{x} \in X$ of the game Γ is a (global, absolute) Nash equilibrium if

$$f_p(x^p, \hat{x}^{-p}) \geq f_p(\hat{x}^p, \hat{x}^{-p}),$$

for all $x^p \in X_p$ and $p \in N$, where

$$\hat{x}^{-p} = (\hat{x}^1, \hat{x}^2, \dots, \hat{x}^{p-1}, \hat{x}^{p+1}, \dots, \hat{x}^n),$$

$$\hat{x}^{-p} \in X_{-p} = X_1 \times X_2 \times \cdots \times X_{p-1} \times X_{p+1} \times \cdots \times X_n,$$

$$\hat{x}^p || \hat{x}^{-p} = (\hat{x}^p, \hat{x}^{-p}) = (\hat{x}^1, \hat{x}^2, \dots, \hat{x}^{p-1}, \hat{x}^p, \hat{x}^{p+1}, \dots, \hat{x}^n) = \hat{x} \in X.$$

Definition 2.2 The profile $\hat{x} \in X$ of the game Γ is a local (relative) Nash equilibrium if there exists such $\varepsilon > 0$ that

$$f_p(x^p, \hat{x}^{-p}) \geq f_p(\hat{x}^p, \hat{x}^{-p}),$$

for any $x^p \in X_p \cap V_\varepsilon(\hat{x}^p)$, and for all $p \in N$, where

$$V_\varepsilon(\hat{x}^p) = \{x^p \in \mathbb{R}^{n_p} : \|x^p - \hat{x}^p\| \leq \varepsilon\}.$$

Remark 2.1 There are diverse alternative interpretations of a Nash equilibrium (see, e.g., [3]) and a Nash equilibrium set

- *as a fixed point of a best response correspondence,*
- *as a fixed point of a function,*
- *as a solution of a nonlinear complementarity problem,*
- *as a solution of a stationary point problem,*
- *as a minimum of a function on a polytope,*
- *as an element of a semi-algebraic set.*

The Nash equilibrium set may be considered as an **intersection of graphs of best response multivalued mappings**

$$\operatorname{Arg}\min_{\mathbf{x}^p \in X_p} f_p\left(\mathbf{x}^p, \mathbf{x}^{-p}\right) : X_{-p} \multimap X_p, \; p = 1, \dots, n,$$

i.e.

$$NES(\Gamma) = \bigcap_{p \in N} Gr_p,$$

$$Gr_p = \left\{ (\mathbf{x}^p, \mathbf{x}^{-p}) \in X : \begin{array}{c} \mathbf{x}^{-p} \in X_{-p} \\ \mathbf{x}^p \in \operatorname{Arg}\min\limits_{\mathbf{x}^p \in X_p} f_p\left(\mathbf{x}^p, \mathbf{x}^{-p}\right) \end{array} \right\},$$

$p = 1, \dots, n$ [4–7], where Gr_p represents the graph of pth player best response mapping.

In the context of interpreting the Nash equilibrium set as the intersection of the graphs of best response mappings of all players, the chapter represents a part of a series of the author's works [1, 4–9] dedicated to the approaches to Nash equilibrium set computing on the basis of best response mapping graphs. Generally, proofs of the following theorems are similar to the proofs of the optimality conditions for mathematical programming problems (see, e.g., Chap. 7 in [10]).

Let us recall the well known result that every convex continuous compact game Γ has at least one Nash equilibrium, i.e.

$$NES(\Gamma) \neq \emptyset$$

for every convex continuous strategic form game (Nash Theorem [2]). Let us remark additionally that the Nash Theorem constitutes an essential and fundamental result for this chapter.

Let

$$L_p(x, u^p, v^p) = f_p(x) + \sum_{i=1}^{m_p} u_i^p g_i^p(x^p) + \sum_{i=1}^{l_p} v_i^p h_i^p(x^p)$$

be the Lagrange function of the player $p \in N$, where

$$\begin{aligned} u^p &= (u_1^p, u_2^p, \ldots, u_{m_p}^p), \\ v^p &= (v_1^p, v_2^p, \ldots, v_{l_p}^p), \end{aligned}$$

are the Lagrange multipliers, and let

$$L(x, u, v) = \big(L_1(x, u^1, v^1), L_2(x, u^2, v^2), \ldots, L_n(x, u^n, v^n)\big)$$

be the Lagrange vector-function of the game Γ.

Definition 2.3 A point

$$(\hat{x}, \hat{u}, \hat{v}) = (\hat{x}, \hat{u}^1, \hat{v}^1, \ldots, \hat{u}^n, \hat{v}^n) \in X \times \mathbb{R}_{\geq}^{m_1} \times \mathbb{R}^{l_1} \times \cdots \times \mathbb{R}_{\geq}^{m_n} \times \mathbb{R}^{l_n}$$

is a saddle point of the game Γ Lagrange vector function if

$$L_p(\hat{x}, u^p, v^p) \leq L_p(\hat{x}, \hat{u}^p, \hat{v}^p) \leq L_p(x^p, \hat{x}^{-p}, \hat{u}^p, \hat{v}^p) \tag{2.1}$$

for any $x^p \in M_p$, $u^p \in \mathbb{R}_{\geq}^{m_p}$, $v^p \in \mathbb{R}^{l_p}$, $p \in N$.

2.2 The Saddle Point and the Nash Equilibrium General Sufficient Condition

In this section, we establish connections between the saddle point and Nash equilibrium notions in the form of a sufficient condition.

Theorem 2.1 *If* $(\hat{x}, \hat{u}^1, \hat{v}^1, \dots, \hat{u}^n, \hat{v}^n)$ *is a saddle point of the Lagrange vector function* $L(x, u, v)$ *of the game* Γ*, then* $\hat{x}$ *is a Nash equilibrium in* Γ*.*

Proof Assume that $(\hat{x}, \hat{u}^1, \hat{v}^1, \dots, \hat{u}^n, \hat{v}^n)$ is a saddle point of the Lagrange vector-function $L(x, u, v)$. Then, from the left inequality of (2.1), it follows that

$$\begin{aligned} f_p(\hat{x}) + \sum_{i=1}^{m_p} u_i^p g_i^p(\hat{x}^p) & \\ + \sum_{i=1}^{l_p} v_i^p h_i^p(\hat{x}^p) &= L_p(\hat{x}, u^p, v^p) \\ &\le L_p(\hat{x}, \hat{u}^p, \hat{v}^p) \\ &= f_p(\hat{x}) + \sum_{i=1}^{m_p} u_i^p g_i^p(\hat{x}^p) + \sum_{i=1}^{l_p} v_i^p h_i^p(\hat{x}^p). \end{aligned}$$

This inequality is equivalent to

$$f_p(\hat{x}) + \sum_{i=1}^{m_p} (u_i^p - \hat{u}_i^p) g_i^p(\hat{x}^p) + \sum_{i=1}^{l_p} (v_i^p - \hat{v}_i^p) h_i^p(\hat{x}^p) \le 0.$$

The last inequality is true for any $u^p \in \mathbb{R}_{\geq}^{m_p}$, $v^p \in \mathbb{R}^{l_p}$, $p \in N$. Consequently, this means that $\hat{x} \in X$. In addition, it follows that

$$\hat{u}_i^p g_i^p(\hat{x}^p) = 0$$

for any $p \in N$.

By these equalities and the right inequality in (2.1) it follows that

$$\begin{aligned} f_p(\hat{x}^p, \hat{x}^{-p}) &= L_p(\hat{x}, \hat{u}^p, \hat{v}^p) \\ &\le L_p(x^p, \hat{x}^{-p}, \hat{u}^p, \hat{v}^p) \\ &= f_p(x^p, \hat{x}^{-p}) + \sum_{i=1}^{m_p} \hat{u}_i^p g_i^p(x^p) + \sum_{i=1}^{l_p} \hat{v}_i^p h_i^p(x^p) \\ &\le f_p(x^p, \hat{x}^{-p}) \end{aligned}$$

for any $x^p \in X_p$, $p \in N$. This proves that the profile $\hat{x}$ is a Nash equilibrium for the game Γ. □

2.3 Necessary and Sufficient Conditions for Convex Strategic Games

Consider the strategic form convex game Γ, i.e. the game in which for every player $p \in N$ the strategy set is

$$X_p = \left\{x^p \in \mathbb{R}^{n_p} \ : \ g_i^p(x^p) \leq 0, \ i = 1, \ldots, m_p\right\},$$

where the functions $g_i^p(x^p), i = 1, \ldots, m_p$, are convex on $\mathbb{R}^{n_p}$ and the cost function $f_p(\mathbf{x}^p, \mathbf{x}^{-p})$ is convex on X_p for any fixed $x^{-p} \in X_{-p}$.

The Lagrange **regular** vector-function $L(x, u)$ for convex games has the components

$$L_p(x, u^p) = f_p(x) + \sum_{i=1}^{m_p} u_i^p g_i^p(x^p), \ \ p = 1, \ldots, n.$$

Theorem 2.2 *Suppose the strategy set X_p of every player $p \in N$ satisfies Slater regularity conditions.*

The profile $\hat{x} \in X$ is a Nash equilibrium in the game Γ if and only if there exist the Lagrange multipliers

$$\hat{u}^p \geq 0, \ \ p = 1, \ldots, n,$$

such that the point $(\hat{x}, \hat{u}^1, \ldots, \hat{u}^n)$ is a saddle point of Γ's Lagrange vector-function $L(x, u)$.

Proof The *sufficiency's* truth follows as a corollary of Theorem 2.1.

Necessity. By assumption, the profile $\hat{x} \in X$ is a Nash equilibrium in the game Γ. We must prove that there exist such Lagrange multipliers $\hat{u}^1, \ldots, \hat{u}^n$, that the point $(\hat{x}, \hat{u}^1, \ldots, \hat{u}^n)$ forms a saddle point of the Lagrange vector-function.

In view of the Nash equilibrium definition, the hypothesis implies that for the fixed profile $\hat{x}^{-p} \in X_{-p}$ the strategy $\hat{x}^p$ is a global minimum point for the payoff function $f_p(x^p, \hat{x}^{-p})$. Hence we can consider the following two sets

$$Z = \left\{\begin{pmatrix} z_0 \\ z \end{pmatrix} \in \mathbb{R}^{m_p+1} : \begin{pmatrix} z_0 \\ z \end{pmatrix} < \begin{pmatrix} f_p(\hat{x}^p, \hat{x}^{-p}) \\ 0 \end{pmatrix}\right\},$$

$$Y = \left\{\begin{pmatrix} y_0 \\ y \end{pmatrix} \in \mathbb{R}^{m_p+1} : \exists x^p, \begin{pmatrix} f_p(x^p, \hat{x}^{-p}) \\ g^p(x^p) \end{pmatrix} \leq \begin{pmatrix} y_0 \\ y \end{pmatrix}\right\}.$$

The sets Z and Y are convex. Taking into account that the strategy $\hat{x}^p$ is a global minimum point for the function $f_p(x^p, \hat{x}^{-p})$, we obtain that the sets Z and Y don't have common points. Combining this with the separation theorem [11, 12] we get that there exists a separation hyperplane with the normal vector $(c_0, c) \in \mathbb{R}^{m_p+1}$, $(c_0, c) \neq 0$, such that

$$(c_0, c^T)\begin{pmatrix} z_0 \\ z \end{pmatrix} \le (c_0, c^T)\begin{pmatrix} y_0 \\ y \end{pmatrix}$$

for any $(z_0, z) \in Z$, $(y_0, y) \in Y$.

From this inequality and the definition of the set Z it follows that $(c_0, c) \ge 0$.

Since the point $\left(f_p\left(\hat{x}^p, \hat{x}^{-p}\right), 0\right)$ belongs to the boundary of the set Z, it follows that

$$c_0 f_p\left(\hat{x}^p, \hat{x}^{-p}\right) \le (c_0, c^T)\begin{pmatrix} y_0 \\ y \end{pmatrix}$$

for any $(y_0, y) \in Y$. As a consequence

$$c_0 f_p(\hat{x}^p, \hat{x}^{-p}) \le c_0 f_p(x^p, \hat{x}^{-p}) + c^T g^p(x^p) \tag{2.2}$$

for any x^p.

The Slater regularity condition and the inequality $c \ge 0$ implies that $c_0 > 0$. By dividing inequality (2.2) by the number c_0 we obtain

$$\hat{u}^p \ge 0, \tag{2.3}$$

$$f_p(\hat{x}^p, \hat{x}^{-p}) \le f_p(x^p, \hat{x}^{-p}) + (\hat{u}^p)^T g^p(x) \tag{2.4}$$

for any x^p, where $\hat{u}^p = \dfrac{1}{c_0} c$.

After the substitution $x^p = \hat{x}^p$ in (2.4) we obtain $(\hat{u}^p)^T g(x^p) \ge 0$. Since $\hat{u}^p \ge 0$, $g^p(x^p) \le 0$, we have $(\hat{u}^p)^T g^p(\hat{x}^p) \le 0$. The last two relations imply

$$(\hat{u}^p)^T g^p(\hat{x}^p) = 0.$$

It is obvious that $g^p(\hat{x}^p) \le 0$. Therefore

$$f_p(\hat{x}^p, \hat{x}^{-p}) \ge f_p(\hat{x}^p, \hat{x}^{-p}) + u^p g^p(\hat{x}^p) \tag{2.5}$$

for any $u^p \ge 0$.

Combining relations (2.3)–(2.5) we get that the point $(\hat{x}, \hat{u}^1, \ldots, \hat{u}^n)$ is a saddle point for the Lagrange vector-function of the game Γ. □

The following theorem formulates necessary and sufficient conditions for games with differentiable component functions.

Theorem 2.3 *Suppose the functions*

$$\begin{aligned} &f_p(x),\ p = 1, \ldots, n, \\ &g_i^p(x^p),\ i = 1, \ldots, m_p,\ p = 1, \ldots, n, \end{aligned}$$

are differentiable on $\hat{x}$, and every strategy set X_p satisfies the Slater regularity conditions.

The profile $\hat{x} \in X$ is a Nash equilibrium in the game Γ if and only if there are Lagrange multipliers $\hat{u}^p \geq 0,\ p = 1, \dots, n$, such that the following conditions are verified

$$\frac{\partial L_p(\hat{x}, \hat{u}^p)}{\partial x_j^p} = 0,\ j = 1, \dots, n_p,\ p = 1, \dots, n,$$
$$\hat{u}_i^p g_i^p(\hat{x}^p) = 0,\ i = 1, \dots, m_p,\ p = 1, \dots, n.$$

Proof Necessity. By assumption, the profile $\hat{x} \in X$ is a Nash equilibrium. In view of the definition, the strategy $\hat{x}^p$ is a minimum point of the function $f_p(x^p, \hat{x}^{-p})$ for the fixed profile $\hat{x}^{-p} \in X_{-p}$.

Let us associate with the game Γ the equivalent game Γ' with the strategy sets

$$X'_p = \left\{(x^p, s^p) \in \mathbb{R}^{n_p+m_p} \ :\ g_i^p(x^p) + s_i^2 = 0,\ i = 1, \dots, m_p\right\}$$

$p = 1, \dots, n$, and the same payoff functions.

The Lagrange vector-function of the game Γ' has the components

$$\Lambda_p(x, u^p, s^p) = L_p(x, u^p) + \sum_{i=1}^{m_p} s_i^2,\ p = 1, \dots, n.$$

We can apply Lagrange principle to the game Γ'. As a result we obtain the system

$$\frac{\partial \Lambda_p(\hat{x},\ u^p,\ s^p)}{\partial x_j^p} = 0,\ \ j = 1, \dots, n_p,\ \ p = 1, \dots, n,$$

$$\frac{\partial \Lambda_p(\hat{x},\ u^p,\ s^p)}{\partial s_i^p} = 0,\ \ i = 1, \dots, m_p,\ \ p = 1, \dots, n,$$

$$\frac{\partial \Lambda_p(\hat{x},\ u^p,\ s^p)}{\partial u_i^p} = 0,\ \ i = 1, \dots, m_p,\ \ p = 1, \dots, n,$$

verified by some $\hat{u}^p \geq 0,\ p = 1, \dots, n$. The last system is equivalent to the following

$$\frac{\partial L_p(\hat{x},\ u^p)}{\partial x_j^p} = 0,\ \ j = 1, \dots, n_p,\ \ p = 1, \dots, n,$$

$$u_i^p g_i^p(\hat{x}^p) = 0,\ \ i = 1, \dots, m_p,\ \ p = 1, \dots, n,$$

$$g_i^p(\hat{x}^p) \leq 0,\ \ i = 1, \dots, m_p,\ \ p = 1, \dots, n.$$

As a result we have proved that the theorem's conditions are verified.

Sufficiency. Assume that the point $\hat{x}$ and the multipliers

$$\hat{u}^p \geq 0, \ p = 1, \dots, n,$$

verify the theorem's conditions. Since the component $L_p(x^p, x^{-p}, u^p)$ is convex on the strategy x^p for the fixed profile x^{-p} and the multiplier u^p, we can conclude that

$$L_p(x^p, \hat{x}^{-p}, \hat{u}^p) \geq L_p(\hat{x}, \hat{u}^p) + (x^p - \hat{x}^p)^T \frac{\partial L_p(\hat{x}, \hat{u}^p)}{\partial x^p}.$$

It follows that

$$L_p(\hat{x}, \hat{u}^p) \leq L_p(x^p, \hat{x}^{-p}, \hat{u}^p) \tag{2.6}$$

for any x^p.

Since the function $L_p(x, u^p)$ is linear on u^p, we obtain that

$$L_p(\hat{x}, u^p) = L_p(\hat{x}, \hat{u}^p) + (u^p - \hat{u}^p)^T \frac{\partial L_p(\hat{x}, \hat{u}^p)}{\partial u^p}.$$

Under this relation and the strategy set definition we get

$$L_p(\hat{x}, u_p) \leq L_p(\hat{x}, \hat{u}^p), \ u^p \geq 0, \ p = 1, \dots, n. \tag{2.7}$$

Inequalities (2.6)–(2.7) prove that the point $(\hat{x}, \hat{u}^1, \dots, \hat{u}^n)$ is a saddle point for the Lagrange vector function of the game Γ. Consequently, from Theorem 2.2 it follows that $\hat{x}$ is a Nash equilibrium for Γ. □

Theorems 2.2 and 2.3 may be formulated with some modifications for cases when strategy sets include equations and sign constraints, i.e.

$$X_p = \left\{ x^p \in \mathbb{R}^{n_p} \ : \ \begin{array}{r} g_i^p(x^p) \leq 0, \ i = 1, \dots, m_p \\ x_j^p \geq 0, \ j = 1, \dots, n_p \end{array} \right\}$$

where $g_i^p(x^p), i = 1, \dots, m_p$, are convex on $\mathbb{R}^{n_p}$.

Theorem 2.4 *Assume that the functions*

$$f_p(x), \ g_i^p(x^p), \ i = 1, \dots, m_p, \ p = 1, \dots, n,$$

are differentiable on $\hat{x}$, and every strategy set X_p satisfies the Slater regularity conditions.

The outcome $\hat{x} \in X$ is a Nash equilibrium in the game Γ if and only if there exist the Lagrange multipliers $\hat{u}^p \geq 0$, $p = 1, \dots, n$, such that the following conditions are verified

$$\frac{\partial L_p(\hat{x}, \hat{u}^p)}{\partial x_j^p} \geq 0, \ j = 1, \dots, n_p, \ p = 1, \dots, n,$$

$$x_j^p \frac{\partial L_p(\hat{x}, \hat{u}^p)}{\partial x_j^p} = 0, \ j = 1, \dots, n_p, \ p = 1, \dots, n,$$

$$\hat{u}_i^p g_i^p(\hat{x}^p) = 0, \ i = 1, \dots, m_p, \ p = 1, \dots, n.$$

Remark 2.2 Evidently, if the convexity requirement misses in the game Γ statement, then Theorems 2.3 and 2.4 must be formulated only as necessary conditions for a local Nash equilibrium.

Remark 2.3 Generally, from practical and algorithmic points of view, the equilibrium conditions from Theorem 2.4 require to solve

$$2^{n_1+\dots+n_p+m_1+\dots+m_p}$$

systems of nonlinear equalities and inequalities. This is a difficult problem to solve already for games with modest dimensions.

Other equilibrium conditions may be formulated and proved for various statements of the game Γ.

2.4 Equilibrium Principles and Conditions for Multi-criteria Strategic Games

Let us assume that in the multi-criteria game Γ every player $p \in N$ simultaneously minimizes k_p cost functions

$$f_i^p(x), \ i = 1, \dots, k_p, \ p = 1, \dots, n,$$

and the strategy sets are

$$X_p = \left\{ x^p \in \mathbb{R}^{n_p} \ : \ g_i^p(x^p) \leq 0, \ i = 1, \dots, m_p \right\}$$

where $g_i^p(x^p), \ i = 1, \dots, m_p$, are defined on $\mathbb{R}^{n_p}, \ p = 1, \dots, n$.

By optimality notions of multi-criteria optimization problems (see, e.g., the works [4, 13–16]) we may define multi-objective equilibrium notions [17–23].

Definition 2.4 A profile $\hat{x} \in X$ of a multi-criteria game Γ is called a Pareto-Nash (multi-criteria Nash) equilibrium if there are no any $p \in N$ and $x^p \in X_p$ such that

$$f^p(x^p, \hat{x}^{-p}) \leq f^p(\hat{x}^p, \hat{x}^{-p}).$$

Definition 2.5 A profile $\hat{x} \in X$ of a multi-criteria game Γ is called a weak Pareto-Nash (Slater-Nash, weak multi-criteria Nash) equilibrium if there are no any $p \in N$ and $x^p \in X_p$ such that

$$f^p(x^p, \hat{x}^{-p}) < f^p(\hat{x}^p, \hat{x}^{-p}).$$

A set of Pareto-Nash Equilibria (PNES) may be defined as *the intersection of the graphs of Pareto-optimal response multivalued mappings*

$$\operatorname{Arg}\min_{\mathbf{x}^p \in X_p} f^p\left(\mathbf{x}^p, \mathbf{x}^{-p}\right) : X_{-p} \multimap X_p, \ p = 1, \ldots, n,$$

i.e.

$$PNES(\Gamma) = \bigcap_{p \in N} Gr_p,$$

$$Gr_p = \left\{ (\mathbf{x}^p, \mathbf{x}^{-p}) \in X : \begin{array}{c} \mathbf{x}^{-p} \in X_{-p} \\ \mathbf{x}^p \in \operatorname{Arg}\min\limits_{\mathbf{x}^p \in X_p} f^p\left(\mathbf{x}^p, \mathbf{x}^{-p}\right) \end{array} \right\}, \ p \in N.$$

In the same way as above, the set of weak Pareto-Nash Equilibria may be defined as *the intersection of the graphs of weak Pareto-optimal response multivalued mappings.*

Denote by

$$F_p(x, \rho^p) = \sum_{i=1}^{k_p} \rho_i^p f_i^p(x),$$

the synthesis function of pth player, where

$$\rho_i^p \geq 0, \ i = 1, \ldots, k_p, \ \sum_{i=1}^{k_p} \rho_i^p = 1, \ p = 1, \ldots, n.$$

Theorem 2.5 *Let the functions*

$$\begin{aligned} f_i^p(x^p, x^{-p}), \ i = 1, \ldots, k_p, \ \ p = 1, \ldots, n, \\ g_i^p(x^p), \ i = 1, \ldots, m_p, \ p = 1, \ldots, n, \end{aligned}$$

be convex on $\mathbb{R}^{n_p}$ for any fixed $x^{-p} \in X_{-p}$.

If the profile $\hat{x} \in X$ is a Pareto-Nash equilibrium for Γ, then there exist multipliers

$$\hat{\rho}^p \geq 0, \ \sum_{i=1}^{k_p} \rho_i^p = 1, \ p = 1, \ldots, n,$$

such that $\hat{x}^p$ is a solution of the optimization problem

$$F_p\left(x^p, \hat{x}^{-p}, \hat{\rho}^p\right) \to min,\ x^p \in X_p, \tag{2.8}$$

$p = 1, \ldots, n.$

Proof *The proof* is based on Theorem 1 from work [4] (p. 181) and repeats in the most the same kind of reasoning as in above proof of theorems. □

Let us examine the game $\Gamma(\rho)$ based on Γ, but with payoff functions

$$F_p(x, \rho^p),\ p = 1, \ldots, n.$$

Theorem 2.6 *Let the functions*

$$f_i^p(x^p, x^{-p}),\ i = 1, \ldots, k_p,\ \ p = 1, \ldots, n,$$
$$g_i^p(x^p),\ i = 1, \ldots, m_p,\ p = 1, \ldots, n,$$

be convex on $\mathbb{R}^{n_p}$ for any fixed $x^{-p} \in X_{-p}$.

If the outcome $\hat{x} \in X$ is a Pareto-Nash equilibrium for Γ, then there exist

$$\hat{\rho}^p \geq 0,\ \sum_{i=1}^{k_p} \rho_i^p = 1,\ p \in N,$$

such that the profile $\hat{x}$ is a Nash equilibrium for $\Gamma(\rho)$.

Theorem 2.6 is a corollary of Theorem 2.5.

Theorem 2.7 *Let the functions*

$$f_i^p(x^p, x^{-p}),\ i = 1, \ldots, k_p,\ \ p = 1, \ldots, n,$$
$$g_i^p(x^p),\ i = 1, \ldots, m_p,\ p = 1, \ldots, n,$$

be convex on $\mathbb{R}^{n_p}$ for any fixed $x^{-p} \in X_{-p}$, be differentiable on $\hat{x}$, and let us assume that every strategy set X_p satisfies the Slater regularity conditions.

If the outcome $\hat{x} \in X$ is a Pareto-Nash equilibrium for Γ, then there exist

$$\hat{\rho}^p \geq 0,\ \sum_{i=1}^{k_p} \rho_i^p = 1,\ p = 1, \ldots, n,$$

and $\hat{u} \geq 0$ such that the following conditions are verified

$$\frac{\partial L_p(\hat{x}, \hat{\rho}^p, \hat{u}^p)}{\partial x_j^p} = 0, \; j = 1, \ldots, n_p, \; p = 1, \ldots, n,$$

$$\hat{u}_i^p g_i^p(\hat{x}^p) = 0, \; i = 1, \ldots, m_p, \; p = 1, \ldots, n,$$

where $L_p\left(\hat{x}, \hat{\rho}^p, \hat{u}^p\right) = F_p(x, \rho^p) + \sum_{i=1}^{m_p} u_i^p g_i^p(x^p), \; p = 1, \ldots, n.$

Theorem 2.7 follows as a corollary from Theorems 2.5–2.6 and 2.3.

Remark 2.4 Theorems 2.5–2.7 formulate only necessary equilibrium conditions.

Remark 2.5 A statement similar with Remark 2.3 is valid for equilibrium conditions of Theorem 2.7.

Theorem 2.8 *If* $\hat{x}$ *is a solution of (2.8) for some* $\rho^p > 0, \; p \in N$*, then* $\hat{x}$ *is a Pareto-Nash equilibrium in* Γ*.*

If $\hat{x}$ *is the unique solution of (2.8) for some* $\rho^p \geq 0, \; p \in N$*, then* $\hat{x}$ *is a Pareto-Nash equilibrium in* Γ*.*

The proof is based on Theorem 1′ from work [4] (p. 183).

Other Pareto-Nash, weak Pareto-Nash, etc., equilibrium conditions may be formulated and proved for multi-criteria strategic games.

2.5 Conclusions

Theorems 2.1, 2.2–2.4, are extensions on strategic form games of the well known theorems for matrix games.

Theorems 2.3 and 2.4 are Karush–Kuhn–Tucker type theorems for strategic form games.

Theorems 2.5–2.8 are extensions on multi-criteria strategic form games of the well known theorems for multi-criteria optimization problems.

All these results are important both from theoretical and practical points of view. They establish analytic or symbolic methods for equilibrium computing.

We must remark that the results of this chapter may be appreciated as typical solution principles for domains of classical single-objective and multi-objective optimization theory. Methods of equilibria computing are typical too, because they reduce the problems of equilibria finding to classical single-objective and multi-objective optimization problems. For the obtained optimization problems we formulate necessary and/or sufficient conditions which represent systems of equations and inequalities in the majority of cases.

The definitions and results of this chapter must be seen as a theoretical and conceptual foundations for next chapters. Nevertheless, in the next chapters new ideas and approaches are applied to investigate the whole set of solutions.

References

1. Ungureanu, V. 2007. *Nash equilibrium conditions for strategic form games*, 131–140. XXVII: Libertas Mathematica.
2. Nash, J. 1951. Noncooperative games. *Annals of Mathematics* 54 (2): 280–295.
3. McKelvey, R.D., and A. McLenan. 1996. Computation of equilibria in finite games. *Handbook of Computational Economics*, vol. 1, 87–142. Amsterdam: Elsevier.
4. Sagaidac, M., and V. Ungureanu. 2004. Operational research, Chişinău: CEP USM, 296 pp. (in Romanian).
5. Ungureanu, V., and A. Botnari. 2005. Nash equilibrium sets in mixed extension of $2 \times 2 \times 2$ games. *Computer Science Journal of Moldova* 13 (1(37)): 13–28.
6. Ungureanu, V., and A. Botnari. 2005. Nash equilibrium sets in mixed extended 2×3 games. *Computer Science Journal of Moldova* 13 (2 (38)): 136–150.
7. Ungureanu, V. (2006). Nash equilibrium set computing in finite extended games. *Computer Science Journal of Moldova, 14*(3 (42)), 345–365.
8. Ungureanu, V. 2008. Solution principles for simultaneous and sequential games mixture. *ROMAI Journal* 4 (1): 225–242.
9. Ungureanu, V. (2013). Linear discrete-time Pareto-Nash-Stackelberg control problem and principles for its solving. *Computer Science Journal of Moldova, 21*(1 (61)), 65–85.
10. Ungureanu, V. 2001. Mathematical programming, Chişinău: USM, 348 pp. (in Romanian).
11. Rockafellar, T. 1970. *Convex Analysis*, 468. Princeton: Princeton University Press.
12. Soltan, V. (2015). *Lectures on Convex Sets* (p. X+405). USA: George Mason University, World Scientific Publishing Co. Pte. Ltd.
13. Ehrgott, M. 2005. *Multicriteria Optimization*, 328. Berlin: Springer.
14. Ehrgott, M., and X. Gandibleux (eds.). 2002. *Multiple Criteria Optimization: State of the Art Annotated Bibliographic Surveys*, 519. Kluwer Academic Publishers: Dordrecht.
15. Miettinen, K.M. 1999. *Nonlinear Multiobjective Optimization*, 319. Massachusetts: Kluwer Academic Publishers.
16. Podinovskii, V.V., and V.D. Nogin. 1982. *Pareto-Optimal Solutions of the Multi-Criteria Problems*. Moscow: Nauka, 255 pp. (in Russian).
17. Blackwell, D. 1956. An analog of the minimax theorem for vector payoffs. *Pacific Journal of Mathematics* 6: 1–8.
18. Borm, P., S. Tijs, and J. Van der Aarssen. 1988. Pareto equilibria in multiobjective games. *Methods of Operations Research* 60: 302–312.
19. Chinchuluun, A., P.M. Pardalos, A. Migdalas, and L. Pitsoulis (eds.). 2008. *Pareto Optimality, Game Theory, and Equilibria*. New York: Springer Science + Business Media, 868 pp.
20. Ghose, D., and U.R. Prasad. 1989. Solution concepts in two-person multicriteria games. *Journal of Optimization Theory and Applications* 63: 167–189.
21. Shapley, L.S. 1959. Equilibrium points in games with vector payoffs. *Naval Research Logistics Quarterly* 6: 57–61.
22. Wang, S.Y. 1993. Existence of pareto equilibrium. *Journal of Optimization Theory and Applications* 79: 373–384.
23. Wierzbicki, A.P. 1995. Multiple criteria games – theory and applications. *Journal of Systems Engineering and Electronics* 6 (2): 65–81.

Chapter 3
Sets of Nash Equilibria in Polymatrix Mixed-Strategy Games

Abstract The method of intersection of best response mapping graphs is applied to determine a Nash equilibrium set in mixed extensions of noncooperative finite games (polymatrix games). The problem of Nash equilibrium set representation in **bimatrix** mixed strategy game was considered before by Vorob'ev in 1958 (Theory Probab Appl 3:297–309, 1958) [1] and Kuhn in 1961 (Proc Natl Acad Sci USA 47:1657–1662, 1961) [2], but as stressed by different researchers (see e.g. Raghavan (Handbook of game theory with economic applications. Elsevier Science B.V., North-Holland, 2002) [3]) these results have only been of theoretical interest. They where rarely used practically to compute Nash equilibria as well as the results of Mills (J Soc Ind Appl Math 8:397–402, 1960) [4], Mangasarian (J Soc Ind Appl Math 12:778–780, 1964) [5], Winkels (Game theory and related topics. North-Holland, Amsterdam, 1979) [6], Yanovskaya (Lithuanian Mathematical Collection (Litovskii Matematicheskii Sbornik) 8:381–384, 1968) [7], Howson (Manage Sci 18:312–318, 1972) [8], Eaves (SIAM J Appl Math 24:418–423, 1973) [9], Mukhamediev (U.S.S.R. Computational Mathematics and Mathematical Physics 18:60–66, 1978) [10], Savani (Finding Nash Equilibria of Bimatrix Games. London School of Economics, 2006) [11], and Shokrollahi (Palestine J Math 6:301–306, 2017) [12]. The first practical algorithm for Nash equilibrium computing was the algorithm proposed by Lemke and Howson in 1958 (J Soc Ind Appl Math 12:413–423, 1964) [13]. Unfortunately, it doesn't compute Nash equilibrium sets. There are algorithms for **polymatrix** mixed strategy games too, Ungureanu in (Comput Sci J Moldova 42: 345–365, 2006) [14], Audet et al. in (J Optim Theory Appl 129:349–372, 2006) [15]. More the more, the number of publications devoted to the problem of finding the Nash equilibrium set is continuously increasing, see, e.g., 2010's bibliography survey by Avis et al. (Econ Theory 42:9–37, 2010) [16], and the other by Datta (Econ Theory 42:55–96, 2010) [17].

V. Ungureanu, *Pareto-Nash-Stackelberg Game and Control Theory*,
Smart Innovation, Systems and Technologies 89,
https://doi.org/10.1007/978-3-319-75151-1_3

3.1 Introduction

The Nash equilibrium set is determined as the intersection of graphs of best response mappings [18–20]. This idea yields a natural method for Nash equilibrium set computing in mixed extensions of two-player $m \times n$ games and n-player $m_1 \times m_2 \times \cdots \times m_n$ games.

Consider a noncooperative finite strategic game:

$$\Gamma = \langle N, \{S_p\}_{p \in N}, \{a_{\mathbf{s}}^p = a_{s_1 s_2 \dots s_n}^p\}_{p \in N} \rangle,$$

where

- $N = \{1, 2, ..., n\} \subset \mathbb{N}$ is a set of players,
- $S_p = \{1, 2, \dots, m_p\} \subset \mathbb{N}$ is a set of (pure) strategies of the player $p \in N$,
- $\#S_p = m_p < +\infty$, $p \in N$,
- $a_{\mathbf{s}}^p = a_{s_1 s_2 \dots s_n}^p : S \to \mathbb{R}$ is a player's $p \in N$ payoff function,
- $S = \underset{p \in N}{\times} S_p$ is the set of profiles.

Remark 3.1 The payoff function of the player $p \in N$ is associated with the n dimensional matrix $A^p[m_1 \times m_2 \times \cdots \times m_n]$. This means that, e.g., payoff of the pth player is defined by the elements $a_{i_1 i_2 \dots i_{mn}}^p$ of the pth matrix.

Supposition 3.1.1 *Without loss of generality we assume that the players* ***maximize*** *the values of their payoff functions. It is a supposition for convenience purposes only.*

Remark 3.2 The precedent supposition has connections with the simplex method [21–24] traditional exposition and has the aim to facilitate the text legibility. In this chapter, the simplex method [21–24] argues the legitimacy of the method of Nash equilibrium set computing.

A mixed extension of Γ or a mixed-strategy game $\widetilde{\Gamma}$ is

$$\widetilde{\Gamma} = \langle X_p, f_p(\mathbf{x}), p \in N \rangle,$$

where

- $$f_p(\mathbf{x}) = \sum_{s_1=1}^{m_1} \sum_{s_2=1}^{m_2} \cdots \sum_{s_n=1}^{m_n} a_{s_1 s_2 \dots s_n}^p x_{s_1}^1 x_{s_2}^2 \dots x_{s_n}^n = \sum_{s_1=1}^{m_1} \sum_{s_2=1}^{m_2} \cdots \sum_{s_n=1}^{m_n} a_{\mathbf{s}}^p \prod_{p=1}^{n} x_{s_p}^p$$
  is the payoff function of the pth player;
- $\mathbf{x} = (\mathbf{x}^1, \mathbf{x}^2, \dots, \mathbf{x}^n) \in X = \underset{p \in N}{\times} X_p \subset \mathbb{R}^m$ is a global profile;
- $m = m_1 + m_2 + \cdots + m_n$ is the profile space dimension;

- $X_p = \left\{ \mathbf{x}^p = (x_1^p, \dots, x_{m_p}^p) : \begin{array}{l} x_1^p + \dots + x_{m_p}^p = 1, \\ x_1^p \geq 0, \dots, x_{m_p}^p \geq 0 \end{array} \right\}$ is the set of mixed strategies of the player $p \in N$.

Remark 3.3 We emphasize the mixed strategies with bold face to outline their vector nature.

Taking into consideration Definition 2.1 and Supposition 3.1.1, we can adjust the Nash equilibrium definition to current notation.

Definition 3.1 The profile $\hat{\mathbf{x}} \in X$ of the game $\widetilde{\Gamma}$ is a Nash equilibrium if

$$f_p(\mathbf{x}^p, \hat{\mathbf{x}}^{-p}) \leq f_p(\hat{\mathbf{x}}^p, \hat{\mathbf{x}}^{-p})$$

for all $\mathbf{x}^p \in X_p$ and $p \in N$.

It is well known that there are pure strategy games which don't have Nash equilibria, but all mixed strategy noncooperative finite games have Nash equilibria [25], i.e. $NES(\widetilde{\Gamma}) \neq \emptyset$.

From the players points of view not all Nash equilibria are equally attractive. They may be Pareto ranked. So, a Nash equilibrium may dominate or it may be dominated. There are also different other criteria for distinguishing the Nash equilibria such as perfect equilibria, proper equilibria, sequential equilibria, stable sets, etc. [26–28]. Thus the methods that found only a sample of Nash equilibrium don't guarantee that determined Nash equilibrium complies all the players demands and refinement conditions. Evidently, a method that finds all Nash equilibria is useful and required. Sure, there are other theoretical and practical factors that argue for Nash equilibrium set finding too [29].

Even though there are diverse alternative interpretations of a Nash equilibrium set (see Remark 2.1) we will consider it as an intersection of graphs of best response multivalued mappings [14, 18–20, 30] $\operatorname{Arg} \max\limits_{\mathbf{x}^p \in X_p} f_p\left(\mathbf{x}^p, \mathbf{x}^{-p}\right) : X_{-p} \multimap X_p,\ p = 1, \dots, n$:

$$NES(\widetilde{\Gamma}) = \bigcap_{p \in N} Gr_p,$$

$$Gr_p = \left\{ (\mathbf{x}^p, \mathbf{x}^{-p}) \in X : \begin{array}{l} \mathbf{x}^{-p} \in X_{-p}, \\ \mathbf{x}^p \in \operatorname{Arg} \max\limits_{\mathbf{x}^p \in X_p} f_p\left(\mathbf{x}^p, \mathbf{x}^{-p}\right) \end{array} \right\}, \quad p \in N.$$

The most simple solvable sample or individual problems of the Nash equilibrium set finding are problems in mixed extensions of two-person 2×2 games [18, 29–31], 2×3 games [20], and three-person $2 \times 2 \times 2$ games [19]. In this section we consider mixed extensions of bimatrix $m \times n$ games and polymatrix games.

Remark 3.4 For a bimatrix game with maximum payoff sum, Gilboa and Zemel [32] proved that the problem of computing an equilibrium is NP-hard. Consequently, the problem of a Nash equilibrium set computing is NP-hard. And so, from the complexity point of view proposed algorithms are admissible even though they have exponential computational complexity. A relative recent advance in studying the complexity of computing equilibria was done by Christos Papadimitriou [33, 34].

It is evident that the algorithms for a Nash equilibrium set computing in polymatrix mixed extended games contain particularly algorithms that compute Nash equilibrium sets in $m \times n$ bimatrix mixed-strategy games. Nevertheless, bimatrix games has peculiar features that permit to give a more expedient concrete algorithm. Examples has to give the reader the opportunity to grasp easy and prompt the following exposition.

3.2 Nash Equilibrium Sets in Bimatrix Mixed-Strategy Games

A good survey on Nash equilibrium set computing in such games was done by Mathijs Jansen, Peter Jurg, and Dries Vermeulen [35], and T.E.S. Raghavan [3].

Consider a bimatrix $m \times n$ game Γ with matrices

$$A = (a_{ij}),\ \ B = (b_{ij}),\ \ i = 1, \dots, m,\ \ j = 1, \dots, n.$$

Let A^i, $i = 1, \dots, m$, be the rows of the matrix A,
$\mathbf{b}^j$, $j = 1, \dots, n$, be the columns of the matrix B.
Denote by

- $X = \{\mathbf{x} \in \mathbb{R}^m : x_1 + x_2 + \dots + x_m = 1, \mathbf{x} \geq 0\}$ the mixed strategy set of the first player,
- $Y = \{\mathbf{y} \in \mathbb{R}^n : y_1 + y_2 + \dots + y_n = 1, \mathbf{y} \geq 0\}$ the mixed strategy set of the second player,
- $f_1(\mathbf{x}, \mathbf{y}) = \sum\limits_{i=1}^{m}\sum\limits_{j=1}^{n} a_{ij}x_i y_j = (A^1\mathbf{y})x_1 + (A^2\mathbf{y})x_2 + \dots + (A^m\mathbf{y})x_m$ the cost function of the first player,
- $f_2(\mathbf{x}, \mathbf{y}) = \sum\limits_{i=1}^{m}\sum\limits_{j=1}^{n} b_{ij}x_i y_j = (\mathbf{x}^T\mathbf{b}^1)y_1 + (\mathbf{x}^T\mathbf{b}^2)y_2 + \dots + (\mathbf{x}^T\mathbf{b}^n)y_n$ the cost function of the second player.

The game $\widetilde{\Gamma} = \langle X, Y;\ f_1, f_2\rangle$ is a mixed extension of the game Γ.

If the strategy of the second player is fixed, then the first player has to solve a linear programming parametric problem:

$$f_1(\mathbf{x}, \mathbf{y}) = \sum_{i=1}^{m} (A^i\mathbf{y})x_i \to \max, \mathbf{x} \in X, \tag{3.1}$$

Evidently, this problem is a linear programming parametric problem with the parameter-vector $\mathbf{y} \in Y$ (see Remark 3.5).

Analogically, the second player must solve a linear programming parametric problem:

$$f_2(\mathbf{x}, \mathbf{y}) = \sum_{j=1}^{n} (\mathbf{x}^{\mathrm{T}}\mathbf{b}^{\mathbf{j}}) y_j \to \max, \mathbf{y} \in Y, \tag{3.2}$$

with the parameter-vector $\mathbf{x} \in X$.

Let us consider now the notation

$$\mathbf{ex}^T = (1, \dots, 1) \in \mathbb{N}^m,$$
$$\mathbf{ey}^T = (1, \dots, 1) \in \mathbb{N}^n.$$

A linear programming problem realizes its optimal solutions on the vertices of polytopes of feasible solutions. In problems (3.1) and (3.2) the sets X and Y have m and, respectively, n vertices that are the unit vectors $\mathbf{e}^{x_i} \in \mathbb{R}^m, i = 1, \dots, m$, and $\mathbf{e}^{y_j} \in \mathbb{R}^n$, $j = 1, \dots, n$, of the axes. In accordance with the simplex method's optimality criterion [21–23], in parametric problem (3.1) the parameter set Y is partitioned into the such m subsets

$$Y^i = \left\{ \mathbf{y} \in \mathbb{R}^n : \begin{array}{l} (A^k - A^i)\mathbf{y} \le 0, k = 1, \dots, m, \\ \mathbf{ey}^T\mathbf{y} = 1, \\ \mathbf{y} \ge 0 \end{array} \right\}, \quad i = 1, \dots, m,$$

that an optimal solution of (3.1) is some $\mathbf{e}^{x_i}$, i.e. the unit vector of x_i axis.

Let us introduce the notation

$$U = \left\{ i \in \{1, 2, ..., m\} : Y^i \ne \emptyset \right\}.$$

In accordance with the optimality criterion of the simplex method, for all $i \in U$ and for all $I \in \mathscr{P}(U \setminus \{i\})$ all the points of

$$\mathbf{Conv} \left\{ \mathbf{e}^{x_k}, k \in I \cup \{i\} \right\} = \left\{ \mathbf{x} \in \mathbb{R}^m : \begin{array}{l} \mathbf{ex}^T\mathbf{x} = 1, \\ \mathbf{x} \ge \mathbf{0}, \\ x_k = 0, k \notin I \cup \{i\} \end{array} \right\}$$

are optimal for parameters

$$\mathbf{y} \in Y^{iI} = \left\{ \mathbf{y} \in \mathbb{R}^n : \begin{array}{l} (A^k - A^i)\mathbf{y} = 0, k \in I, \\ (A^k - A^i)\mathbf{y} \le 0, k \notin I \cup \{i\}, \\ \mathbf{ey}^T\mathbf{y} = 1, \\ \mathbf{y} \ge 0. \end{array} \right\}$$

Evidently $Y^{i\emptyset} = Y^i$. Hence,

$$Gr_1 = \bigcup_{i \in U,\ I \in \mathscr{P}(U \setminus \{i\})} \mathbf{Conv} \left\{ \mathbf{e}^{x_k}, k \in I \cup \{i\} \right\} \times Y^{iI}.$$

In parametric problem (3.2) the parameter set X is partitioned into the such n subsets

$$X^j = \left\{ \mathbf{x} \in \mathbb{R}^m : \begin{array}{l} (\mathbf{b}^k - \mathbf{b}^j)\mathbf{x} \le 0, k = 1, \dots, n, \\ \mathbf{e}\mathbf{x}^T\mathbf{x} = 1, \\ \mathbf{x} \ge 0, \end{array} \right\}, \quad j = 1, \dots, n,$$

that an optimal solution of (3.2) is some $\mathbf{e}^{y_j}$, i.e. the unit vector of y_j axis.

Let us introduce the notation

$$V = \left\{ j \in \{1, 2, ..., n\} : X^j \ne \emptyset \right\}.$$

By the optimality criterion of the simplex method for all $j \in V$ and for all $J \in \mathscr{P}(V \setminus \{j\})$ all the points of

$$\mathbf{Conv}\left\{\mathbf{e}^{y_k}, k \in J \cup \{j\}\right\} = \left\{ \mathbf{y} \in \mathbb{R}^n : \begin{array}{l} \mathbf{e}\mathbf{y}^T\mathbf{y} = 1, \\ \mathbf{y} \ge \mathbf{0}, \\ y_k = 0, k \notin J \cup \{j\} \end{array} \right\}$$

are optimal for parameters

$$\mathbf{x} \in X^{jJ} = \left\{ \mathbf{x} \in \mathbb{R}^m : \begin{array}{l} (\mathbf{b}^k - \mathbf{b}^j)\mathbf{x} = 0, k \in J, \\ (\mathbf{b}^k - \mathbf{b}^j)\mathbf{x} \le 0, k \notin J \cup \{j\}, \\ \mathbf{e}\mathbf{x}^T\mathbf{x} = 1, \\ \mathbf{x} \ge 0. \end{array} \right\}$$

Evidently $X^{j\emptyset} = X^j$. Hence

$$Gr_2 = \bigcup_{j \in V,\ J \in \mathscr{P}(V \setminus \{j\})} X^{jJ} \times \mathbf{Conv}\left\{\mathbf{e}^{y_k}, k \in J \cup \{j\}\right\}.$$

Finally,

$$NE(\widetilde{\Gamma}) = Gr_1 \bigcap Gr_2 = \bigcup_{\substack{i \in U,\ I \in \mathscr{P}(U \setminus \{i\}) \\ j \in V,\ J \in \mathscr{P}(V \setminus \{j\})}} X_{iI}^{jJ} \times Y_{jJ}^{iI},$$

where $X_{iI}^{jJ} \times Y_{jJ}^{iI}$ is a convex component of the Nash equilibrium set,

$$X_{iI}^{jJ} = \mathbf{Conv}\left\{\mathbf{e}^{x_k}, k \in I \cup \{i\}\right\} \bigcap X^{jJ},$$
$$Y_{jJ}^{iI} = \mathbf{Conv}\left\{\mathbf{e}^{y_k}, k \in J \cup \{j\}\right\} \bigcap Y^{iI},$$

$$X_{iI}^{jJ} = \left\{ \mathbf{x} \in \mathbb{R}^m : \begin{array}{l} (\mathbf{b}^k - \mathbf{b}^j)\mathbf{x} = 0, k \in J, \\ (\mathbf{b}^k - \mathbf{b}^j)\mathbf{x} \le 0, k \notin J \cup \{j\}, \\ \mathbf{e}\mathbf{x}^T\mathbf{x} = 1, \mathbf{x} \ge 0, \\ x_k = 0, k \notin I \cup \{i\} \end{array} \right\}$$

is a set of strategies $\mathbf{x} \in X$ with support from $\{i\} \cup I$ and for which points of $\mathbf{Conv}\,\{\mathbf{e}^{y_k}, k \in J \cup \{j\}\}$ are optimal,

$$Y_{jJ}^{iI} = \left\{ \mathbf{y} \in \mathbb{R}^n : \begin{array}{l} (A^k - A^i)\mathbf{y} = 0, k \in I, \\ (A^k - A^i)\mathbf{y} \le 0, k \notin I \cup \{i\}, \\ \mathbf{e}\mathbf{y}^T\mathbf{y} = 1, \mathbf{y} \ge 0, \\ y_k = 0, k \notin J \cup \{j\} \end{array} \right\}$$

is a set of strategies $\mathbf{y} \in Y$ with support from $\{j\} \cup J$ and for which points of $\mathbf{Conv}\,\{\mathbf{e}^{x_k}, k \in I \cup \{i\}\}$ are optimal.

Based on the above, next Theorem 3.1 has been proved.

Theorem 3.1 *The Nash equilibrium set in $\widetilde{\Gamma}$ is equal to*

$$NES(\widetilde{\Gamma}) = \bigcup_{\substack{i \in U,\ I \in \mathscr{P}(U \setminus \{i\}) \\ j \in V,\ J \in \mathscr{P}(V \setminus \{j\})}} X_{iI}^{jJ} \times Y_{jJ}^{iI}.$$

Theorem 3.2 *If $X_{iI}^{j\emptyset} = \emptyset$, then $X_{iI}^{jJ} = \emptyset$ for all $J \in \mathscr{P}(V)$.*

Proof It is sufficient to specify that $X_{iI}^{jJ} \subseteq X_{iI}^{j\emptyset}$ for $J \ne \emptyset$. □

Theorem 3.3 *If $Y_{jJ}^{i\emptyset} = \emptyset$, then $Y_{jJ}^{iI} = \emptyset$ for all $I \in \mathscr{P}(U)$.*

Proof The theorem is equivalent to precedent one. □

We can now expose an algorithm for Nash equilibrium set finding. For the legibility of exposition, let us consider an alternative notation for a power set

$$2^{UX} = \mathscr{P}(UX),$$
$$2^{VY} = \mathscr{P}(VY).$$

3.2.1 *Algorithm for Nash Equilibrium Sets Computing in Bimatrix Games*

Next algorithm is based on previous theorems and notation.

Algorithm 3.2.1

$$NES = \emptyset;\quad U = \{i \in \{1, 2, ..., m\} : Y^i \ne \emptyset\};\quad UX = U;$$
$$V = \{j \in \{1, 2, ..., n\} : X^j \ne \emptyset\};$$

```
for i ∈ U do
  {
      UX = UX \ {i};
      for I ∈ 2^UX do
        {
            VY = V;
            for j ∈ V do
              {
                  if (X_iI^jØ = Ø) break;
                  VY = VY \ {j};
                  for J ∈ 2^VY do
                      if (Y_jJ^iI ≠ Ø) NES = NES ∪ (X_iI^jJ × Y_jJ^iI);
              }
        }
  }
```

Theorem 3.4 *Algorithm 3.2.1 explores no more than* $(2^m - 1)(2^n - 1)$ *polytopes of the* $X_{iI}^{jJ} \times Y_{jJ}^{iI}$ *type.*

Proof For the proof it is enough to observe that the algorithm executes the interior **if** no more than

$$\begin{aligned} &2^{m-1}(2^{n-1} + 2^{n-2} + \cdots + 2^1 + 2^0) \\ &+2^{m-2}(2^{n-1} + 2^{n-2} + \cdots + 2^1 + 2^0) \\ &\ldots \\ &+2^1(2^{n-1} + 2^{n-2} + \cdots + 2^1 + 2^0) \\ &+2^0(2^{n-1} + 2^{n-2} + \cdots + 2^1 + 2^0) \\ &= (2^m - 1)(2^n - 1) \end{aligned}$$

times.

If all the players' strategies are equivalent, then Nash equilibrium set consists of $(2^m - 1)(2^n - 1)$ polytopes. □

Evidently, for practical reasons Algorithm 3.2.1 may be improved by identifying equivalent, dominant and dominated strategies in pure strategy game [18–20] with the following game simplification as for pure and mixed strategies. It is well known result that *"in a nondegenerate game, both players use the same number of pure strategies in equilibrium, so only supports of equal cardinality need to be examined"* [36]. This property may be used to minimize essentially the number of components $X_{iI}^{jJ} \times Y_{jJ}^{iI}$ examined in an nondegenerate game.

3.2.2 *Examples of Nash Equilibrium Sets Computing in Bimatrix Games*

Let us apply Algorithm 3.2.1 for Nash equilibrium sets computing in some illustrative games. The examples have the meaning to illustrate both algorithm applications and some known theoretical results.

Example 3.1 Matrices of the two-person game [36] are

$$A = \begin{bmatrix} 1 & 0 & 4 \\ 0 & 2 & 3 \end{bmatrix}, B = \begin{bmatrix} 0 & 2 & 3 \\ 6 & 5 & 3 \end{bmatrix}.$$

The exterior cycle is executed for $i = 1$.

As

$$X_{1\emptyset}^{1\emptyset} = \left\{ \mathbf{x} : \begin{array}{l} 2x_1 - x_2 \leq 0, \\ 3x_1 - 3x_2 \leq 0, \\ x_1 + x_2 = 1, \\ x_1 \geq 0, x_2 = 0 \end{array} \right\} = \emptyset$$

then the cycle for $j = 1$ is omitted.

Since

$$X_{1\emptyset}^{2\emptyset} = \left\{ \mathbf{x} : \begin{array}{l} -2x_1 + x_2 \leq 0, \\ x_1 - 2x_2 \leq 0, \\ x_1 + x_2 = 1, \\ x_1 \geq 0, x_2 = 0 \end{array} \right\} = \emptyset,$$

then the cycle for j=2 is omitted.

As

$$X_{1\emptyset}^{3\emptyset} = \left\{ \mathbf{x} : \begin{array}{l} -3x_1 + 3x_2 \leq 0, \\ -x_1 + 2x_2 \leq 0, \\ x_1 + x_2 = 1, \\ x_1 \geq 0, x_2 = 0 \end{array} \right\} = \left\{ \begin{pmatrix} 1 \\ 0 \end{pmatrix} \right\} \neq \emptyset,$$

$$Y_{3\emptyset}^{1\emptyset} = \left\{ \mathbf{y} : \begin{array}{l} -y_1 + 2y_2 - y_3 \leq 0, \\ y_1 + y_2 + y_3 = 1, \\ y_1 = 0, y_2 = 0, y_3 \geq 0 \end{array} \right\} = \left\{ \begin{pmatrix} 0 \\ 0 \\ 1 \end{pmatrix} \right\} \neq \emptyset,$$

the point $\begin{pmatrix} 1 \\ 0 \end{pmatrix} \times \begin{pmatrix} 0 \\ 0 \\ 1 \end{pmatrix}$ is a Nash equilibrium with $f_1 = 4,\ f_2 = 3$.

$$X_{1\{2\}}^{1\emptyset} = \left\{ \mathbf{x} : \begin{array}{l} 2x_1 - x_2 \le 0, \\ 3x_1 - 3x_2 \le 0, \\ x_1 + x_2 = 1, \\ x_1 \ge 0, x_2 \ge 0 \end{array} \right\} \ne \emptyset,$$

$$Y_{1\emptyset}^{1\{2\}} = \left\{ \mathbf{y} : \begin{array}{l} -y_1 + 2y_2 - y_3 = 0, \\ y_1 + y_2 + y_3 = 1, \\ y_1 \ge 0, y_2 = 0, y_3 = 0 \end{array} \right\} = \emptyset,$$

Since

$$X_{1\{2\}}^{1\{2\}} = \left\{ \mathbf{x} : \begin{array}{l} 2x_1 - x_2 = 0, \\ 3x_1 - 3x_2 \le 0, \\ x_1 + x_2 = 1, \\ x_1 \ge 0, x_2 \ge 0 \end{array} \right\} = \begin{pmatrix} 1/3 \\ 2/3 \end{pmatrix} \ne \emptyset,$$

$$Y_{1\{2\}}^{1\{2\}} = \left\{ \mathbf{y} : \begin{array}{l} -y_1 + 2y_2 - y_3 = 0, \\ y_1 + y_2 + y_3 = 1, \\ y_1 \ge 0, y_2 \ge 0, y_3 = 0 \end{array} \right\} = \begin{pmatrix} 2/3 \\ 1/3 \\ 0 \end{pmatrix} \ne \emptyset,$$

the point $\begin{pmatrix} 1/3 \\ 2/3 \end{pmatrix} \times \begin{pmatrix} 2/3 \\ 1/3 \\ 0 \end{pmatrix}$ is Nash equilibrium for which $f_1 = 2/3,\ f_2 = 4.$

$$X_{1\{2\}}^{1\{3\}} = \left\{ \mathbf{x} : \begin{array}{l} 2x_1 - x_2 \le 0, \\ 3x_1 - 3x_2 = 0, \\ x_1 + x_2 = 1, \\ x_1 \ge 0, x_2 \ge 0 \end{array} \right\} = \emptyset,$$

$$X_{1\{2\}}^{1\{2,3\}} = \left\{ \mathbf{x} : \begin{array}{l} 2x_1 - x_2 = 0, \\ 3x_1 - 3x_2 = 0, \\ x_1 + x_2 = 1, \\ x_1 \ge 0, x_2 \ge 0 \end{array} \right\} = \emptyset,$$

$$X_{1\{2\}}^{2\emptyset} = \left\{ \mathbf{x} : \begin{array}{l} -2x_1 + x_2 \le 0, \\ x_1 - 2x_2 \le 0, \\ x_1 + x_2 = 1, \\ x_1 \ge 0, x_2 \ge 0 \end{array} \right\} \ne \emptyset,$$

$$Y_{2\emptyset}^{1\{2\}} = \left\{ \mathbf{y} : \begin{array}{l} -y_1 + 2y_2 - y_3 = 0, \\ y_1 + y_2 + y_3 = 1, \\ y_1 = 0, y_2 \ge 0, y_3 = 0 \end{array} \right\} = \emptyset,$$

As

$$X^{2\{3\}}_{1\{2\}} = \left\{ \mathbf{x} : \begin{array}{l} -2x_1 + x_2 \le 0, \\ x_1 - 2x_2 = 0, \\ x_1 + x_2 = 1, \\ x_1 \ge 0, x_2 \ge 0 \end{array} \right\} = \begin{pmatrix} 2/3 \\ 1/3 \end{pmatrix} \ne \emptyset,$$

$$Y^{1\{2\}}_{2\{3\}} = \left\{ \mathbf{y} : \begin{array}{l} -y_1 + 2y_2 - y_3 = 0, \\ y_1 + y_2 + y_3 = 1, \\ y_1 = 0, y_2 \ge 0, y_3 = 0 \end{array} \right\} = \begin{pmatrix} 0 \\ 1/3 \\ 2/3 \end{pmatrix} \ne \emptyset,$$

the point $\begin{pmatrix} 2/3 \\ 1/3 \end{pmatrix} \times \begin{pmatrix} 0 \\ 1/3 \\ 2/3 \end{pmatrix}$ is a Nash equilibrium for which $f_1 = 8,\ f_2 = 3$.

$$X^{3\emptyset}_{1\{2\}} = \left\{ \mathbf{x} : \begin{array}{l} -2x_1 + x_2 \le 0, \\ x_1 - 2x_2 = 0, \\ x_1 + x_2 = 1, \\ x_1 \ge 0, x_2 \ge 0 \end{array} \right\} \ne \emptyset,$$

$$Y^{1\{2\}}_{3\emptyset} = \left\{ \mathbf{y} : \begin{array}{l} -y_1 + 2y_2 - y_3 = 0, \\ y_1 + y_2 + y_3 = 1, \\ y_1 = 0, y_2 = 0, y_3 \ge 0 \end{array} \right\} = \emptyset.$$

The exterior cycle is executed for $i = 2$.

$$X^{1\emptyset}_{2\emptyset} = \left\{ \mathbf{x} : \begin{array}{l} 2x_1 - x_2 \le 0, \\ 3x_1 - 3x_2 \le 0, \\ x_1 + x_2 = 1, \\ x_1 = 0, x_2 \ge 0 \end{array} \right\} \ne \emptyset,$$

$$Y^{2\emptyset}_{1\emptyset} = \left\{ \mathbf{y} : \begin{array}{l} y_1 - 2y_2 + y_3 \le 0, \\ y_1 + y_2 + y_3 = 1, \\ y_1 \ge 0, y_2 = 0, y_3 = 0 \end{array} \right\} = \emptyset,$$

$$X^{1\{2\}}_{2\emptyset} = \left\{ \mathbf{x} : \begin{array}{l} 2x_1 - x_2 = 0, \\ 3x_1 - 3x_2 \le 0, \\ x_1 + x_2 = 1, \\ x_1 = 0, x_2 \ge 0 \end{array} \right\} = \emptyset,$$

$$X^{1\{3\}}_{2\emptyset} = \left\{ \mathbf{x} : \begin{array}{l} 2x_1 - x_2 \le 0, \\ 3x_1 - 3x_2 = 0, \\ x_1 + x_2 = 1, \\ x_1 = 0, x_2 \ge 0 \end{array} \right\} = \emptyset,$$

$$X_{2\emptyset}^{1\{2,3\}} = \left\{ \mathbf{x} : \begin{array}{c} 2x_1 - x_2 = 0, \\ 3x_1 - 3x_2 = 0, \\ x_1 + x_2 = 1, \\ x_1 = 0, x_2 \geq 0 \end{array} \right\} = \emptyset.$$

Since

$$X_{2\emptyset}^{2\emptyset} = \left\{ \mathbf{x} : \begin{array}{c} -2x_1 + x_2 \leq 0, \\ x_1 - 2x_2 \leq 0, \\ x_1 + x_2 = 1, \\ x_1 = 0, x_2 \geq 0 \end{array} \right\} = \emptyset,$$

the cycle for $j = 2$ is omitted.

$$X_{2\emptyset}^{3\emptyset} = \left\{ \mathbf{x} : \begin{array}{c} -3x_1 + 3x_2 \leq 0, \\ -x_1 + 2x_2 \leq 0, \\ x_1 + x_2 = 1, \\ x_1 = 0, x_2 \geq 0 \end{array} \right\} = \emptyset.$$

Thus, the Nash equilibrium set consists of three elements — one pure and two mixed Nash equilibria. □

Next example illustrates that a simple modification in the first example (by changing one element of the cost matrix of the first player) transforms the Nash equilibrium set so that is consists of one isolated point and one segment, i.e. the the Nash equilibriums set has the power of continuum.

Example 3.2 If in the first example the element a_{23} of the matrix A is modified

$$A = \begin{bmatrix} 1 & 0 & 4 \\ 0 & 2 & \boxed{4} \end{bmatrix}, \quad B = \begin{bmatrix} 0 & 2 & 3 \\ 6 & 5 & 3 \end{bmatrix},$$

then the Nash equilibrium set of the obtained game consist of one distinct point $\begin{pmatrix} 1/3 \\ 2/3 \end{pmatrix} \times \begin{pmatrix} 2/3 \\ 1/3 \\ 0 \end{pmatrix}$ for which $f_1 = 10/9$, $f_2 = 4$ and of one distinct segment $\left[\begin{pmatrix} 2/3 \\ 1/3 \end{pmatrix}, \begin{pmatrix} 1 \\ 0 \end{pmatrix} \right] \times \begin{pmatrix} 0 \\ 0 \\ 1 \end{pmatrix}$ for which $f_1 = 4$, $f_2 = 3$. □

A subsequent modification of the cost matrix of the second player in precedent game transform the Nash equilibrium set into one non-convex set of continuum power.

Example 3.3 If in the second example the first column of the matrix B is equal to the second column

$$A = \begin{bmatrix} 1 & 0 & 4 \\ 0 & 2 & \boxed{4} \end{bmatrix}, \quad B = \begin{bmatrix} \boxed{2} & 2 & 3 \\ \boxed{5} & 5 & 3 \end{bmatrix},$$

then the Nash equilibrium set of the such game consists of four connected segments:

- $\left[\begin{pmatrix}2/3\\1/3\end{pmatrix}, \begin{pmatrix}1\\0\end{pmatrix}\right] \times \begin{pmatrix}0\\0\\1\end{pmatrix} \equiv \begin{pmatrix}1-1/3\lambda\\1/3\lambda\end{pmatrix} \times \begin{pmatrix}0\\0\\1\end{pmatrix}$,
  $\lambda \in [0;1]$, for which $f_1 = 4$, $f_2 = 3$,
- $\begin{pmatrix}2/3\\1/3\end{pmatrix} \times \left[\begin{pmatrix}0\\0\\1\end{pmatrix}, \begin{pmatrix}2/3\\1/3\\0\end{pmatrix}\right] \equiv \begin{pmatrix}2/3\\1/3\end{pmatrix} \times \begin{pmatrix}2/3-2/3\mu\\1/3-1/3\mu\\\mu\end{pmatrix}$,
  $\mu \in [0;1]$, for which $f_1 = 2/3 + 10/3\mu$, $f_2 = 3$,
- $\left[\begin{pmatrix}0\\1\end{pmatrix}, \begin{pmatrix}2/3\\1/3\end{pmatrix}\right] \times \begin{pmatrix}2/3\\1/3\\0\end{pmatrix} \equiv \begin{pmatrix}2/3-2/3\lambda\\1/3+1/3\lambda\end{pmatrix} \times \begin{pmatrix}2/3\\1/3\\0\end{pmatrix}$, $\lambda \in [0;1]$,
  for which $f_1 = 2/3$, $f_2 = 3 + 2\lambda$,
- $\begin{pmatrix}0\\1\end{pmatrix} \times \left[\begin{pmatrix}2/3\\1/3\\0\end{pmatrix}, \begin{pmatrix}0\\1\\0\end{pmatrix}\right] \equiv \begin{pmatrix}0\\1\end{pmatrix} \times \begin{pmatrix}2/3\mu\\1-2/3\mu\\0\end{pmatrix}$,
  $\mu \in [0;1]$, for which $f_1 = 2 - 4/3\mu$, $f_2 = 5$.

□

Next example supplements preceding examples and illustrates that one of the Nash equilibria may strong dominate (may be optimal by Pareto among) all the others Nash equilibria.

Example 3.4 The Nash equilibrium set of the mixed extension of the bimatrix game with matrices

$$A = \begin{bmatrix} 2 & 1 & 6 \\ 3 & 2 & -1 \\ -1 & 2 & 1 \end{bmatrix}, \quad B = \begin{bmatrix} 1 & 0 & 3 \\ -1 & 1 & -2 \\ 2 & -1 & 2 \end{bmatrix},$$

consists of two isolated points and one segment:

- $\begin{pmatrix}1\\0\\0\end{pmatrix} \times \begin{pmatrix}0\\0\\1\end{pmatrix}$, with $f_1 = 6$, $f_2 = 3$.
- $\begin{pmatrix}1/14\\12/14\\2/14\end{pmatrix} \times \begin{pmatrix}1/4\\1/2\\1/4\end{pmatrix}$, with $f_1 = 45/28$, $f_2 = -1/8$.
- $\left[\begin{pmatrix}0\\1\\0\end{pmatrix}, \begin{pmatrix}0\\3/5\\2/5\end{pmatrix}\right] \times \begin{pmatrix}0\\1\\0\end{pmatrix} \equiv \begin{pmatrix}0\\3/5+2/5\lambda\\2/5-2/5\lambda\end{pmatrix} \times \begin{pmatrix}0\\1\\0\end{pmatrix}$, $\lambda \in [0;1]$,
  with $f_1 = 2$, $f_2 = 1/5 + 4/5\lambda$.

Evidently, the Nash equilibrium $\begin{pmatrix}1\\0\\0\end{pmatrix} \times \begin{pmatrix}0\\0\\1\end{pmatrix}$, with $f_1 = 6$, $f_2 = 3$, strong dominates all the others Nash equilibria and it is more preferable than the others Nash equilibria for both the players. □

Finally, we can highlight now the conclusions based on the above examples:

- Example 3.1 confirms the well known property established by Lemke and Howson [13] that the numbers of Nash equilibria in nondegenerate games are odd [3, 13, 29, 36].
- Examples 3.2–3.4 illustrate that in degenerate games the numbers of convex components may be both the even and the odd.
- Examples 3.2–3.4 give us the geometrical samples of Nash equilibrium sets in degenerate games that highlight their possible complicated structure.
- Examples 3.3–3.4 illustrate that in degenerate games the player cannot increase own gain by modifying his Nash strategy, but, by modifying his Nash strategy the player may essentially modify (increase or decrease) the payoff of the opponent.

3.3 Nash Equilibrium Sets in Polymatrix Mixed-Strategy Games

Consider the n-player mixed strategy game $\widetilde{\Gamma} = \langle X_p, f_p(\mathbf{x}), p \in N \rangle$ formulated in the Introduction. The payoff function of the player p is linear if the strategies of the others are fixed, i.e.

$$
\begin{aligned}
f_p(\mathbf{x}) &= \left(\sum_{\mathbf{s}_{-p} \in S_{-p}} a^p_{1\|\mathbf{s}_{-p}} \prod_{\substack{q=1,\ldots,n \\ q \neq p}} x^q_{s_q} \right) x^p_1 \\
&+ \left(\sum_{\mathbf{s}_{-p} \in S_{-p}} a^p_{2\|\mathbf{s}_{-p}} \prod_{\substack{q=1,\ldots,n \\ q \neq p}} x^q_{s_q} \right) x^p_2 \\
&\ldots \\
&+ \left(\sum_{\mathbf{s}_{-p} \in S_{-p}} a^p_{m_p\|\mathbf{s}_{-p}} \prod_{\substack{q=1,\ldots,n \\ q \neq p}} x^q_{s_q} \right) x^p_{m_p}.
\end{aligned}
$$

Thus, the player p has to solve a linear programming parametric problem

$$
f_p\left(\mathbf{x}^p, \mathbf{x}^{-p}\right) \to \max, \quad \mathbf{x}^p \in X_p, \quad p = 1, \ldots, n, \tag{3.3}
$$

with the parameter vector $x^{-p} \in X^{-p}$.

Remark 3.5 Let us remark that linear parametric programming is a distinct domain of linear optimization theory and operational research [37–48]. It is connected with sensitivity analysis and has a lot of real practical applications [48, 49].

Let us begin to prove a theorem for polymatrix mixed-strategy games which generalizes Theorem 3.1 for bimatrix mixed strategy games.

The solution of problem (3.3) is realized on vertices of the polyhedral set X_p. But it has m_p vertices, namely the unit vectors of x_i^p axes

$$\mathbf{e}^{x_i^p} \in \mathbb{R}^{m_p},\ i = 1, \ldots, m_p.$$

By recalling the properties of the simplex method and its optimality criterion [21–23], we can conclude that the parameter set X_{-p} is partitioned into the such m_p subsets

$$X_{-p}\left(i_p\right) = \left\{ \mathbf{x}^{-p} : \begin{array}{c} \sum\limits_{\mathbf{s}_{-p} \in S_{-p}} \left(a^p_{k||\mathbf{s}_{-p}} - a^p_{i_p||\mathbf{s}_{-p}}\right) \prod\limits_{\substack{q=\overline{1,n} \\ q \neq p}} x^q_{s_q} \le 0, \\ k = 1, \ldots, m_p, \\ \sum\limits_{i=1}^{m_q} x_i^q = 1, \quad q = 1, \ldots, n,\ q \neq p, \\ \mathbf{x}^{-p} \ge 0, \\ \mathbf{x}^{-p} \in \mathbb{R}^{m-m_p}, \end{array} \right\},$$

$i_p = 1, \ldots, m_p$, that one of the optimal solution of linear programming problem (3.3) is $\mathbf{e}^{x_i^p}$.

Let us introduce the notation

$$\begin{array}{l} U_p = \{i_p \in \{1, ..., m_p\} : X_{-p}\left(i_p\right) \neq \emptyset\}, \\ \mathbf{ep}^T = (1, \ldots, 1) \in \mathbb{R}^{m_p}. \end{array}$$

In conformity with the optimality criterion of the simplex method for all $i_p \in U_p$ and $I_p \in \mathscr{P}\left(U_p \setminus \{i_p\}\right)$, all the points of

$$\mathbf{Conv}\left\{\mathbf{e}^{x_k^p}, k \in I_p \cup \{i_p\}\right\} = \left\{\mathbf{x} \in \mathbb{R}^m : \begin{array}{l} \mathbf{ep}^T \mathbf{x}^p = 1, \\ \mathbf{x}^p \ge 0, \\ x_k^p = 0, k \notin I_p \cup \{i_p\} \end{array}\right\}$$

are optimal for $\mathbf{x}^{-p} \in X_{-p}\left(i_p I_p\right) \subset \mathbb{R}^{m-m_p}$, where $X_{-p}\left(i_p I_p\right)$ is the set of solutions of the system

$$\begin{cases} \sum\limits_{\mathbf{s}_{-p} \in S_{-p}} \left(a^p_{k||\mathbf{s}_{-p}} - a^p_{i_p||\mathbf{s}_{-p}}\right) \prod\limits_{\substack{q=1,\ldots,n, \\ q \neq p}} x^q_{s_q} = 0, \quad k \in I_p, \\ \sum\limits_{\mathbf{s}_{-p} \in S_{-p}} \left(a^p_{k||\mathbf{s}_{-p}} - a^p_{i_p||\mathbf{s}_{-p}}\right) \prod\limits_{\substack{q=1,\ldots,n \\ q \neq p}} x^q_{s_q} \le 0, \quad k \notin I_p \cup i_p, \\ \mathbf{er}^T \mathbf{x}^r = 1, \quad r = 1, \ldots, n,\ r \neq p, \\ \mathbf{x}^r \ge 0, \quad r = 1, \ldots, n,\ r \neq p. \end{cases}$$

Evidently $X_{-p}\left(i_p\emptyset\right) = X_{-p}\left(i_p\emptyset\right)$. As a result, we have proved the following lemma.

Lemma 3.1 *The graph of best-response mapping of the player* $p \in \{1, 2, \ldots, n\}$, *admits the following representation*

$$Gr_p = \bigcup_{\substack{i_p \in U_p \\ I_p \in \mathscr{P}(U_p \setminus \{i_p\})}} \mathbf{Conv}\left\{\mathbf{e}^{x_k^p}, k \in I_p \cup \{i_p\}\right\} \times X_{-p}\left(i_p I_p\right),$$

where $\mathbf{Conv}\left\{\mathbf{e}^{x_k^p}, k \in I_p \cup \{i_p\}\right\}$ *and* $X_{-p}\left(i_p I_p\right)$ *are defined above.*

Lemma 3.1 give us a general formula for an explicit form representation of the graph of best response mapping of the player $p \in \{1, 2, \ldots, n\}$.

We need to remark that it is important not only for this theorem proof, but may has more general application in the context of mixtures of simultaneous and sequential polymatrix mixed-strategy games.

Consequently, by applying Lemma 3.1, we can obtain the representation of the Nash equilibrium set as the intersection of the player graphs of best response mappings

$$NES(\widetilde{\Gamma}) = \bigcap_{p=1}^{n} Gr_p = \bigcup_{\substack{i_1 \in U_1,\ I_1 \in \mathscr{P}(U_1 \setminus \{i_1\}) \\ \cdots \\ i_n \in U_n,\ I_n \in \mathscr{P}(U_n \setminus \{i_n\})}} X\left(i_1 I_1 \ldots i_n I_n\right)$$

where $X\left(i_1 I_1 \ldots i_n I_n\right) = NE\left(i_1 I_1 \ldots i_n I_n\right)$ is the set of solutions of the system

$$\begin{cases} \sum\limits_{\mathbf{s}_{-r} \in S_{-r}} \left(a^r_{k||\mathbf{s}_{-r}} - a^r_{i_r||\mathbf{s}_{-r}}\right) \prod\limits_{\substack{q=1,\ldots,n \\ q \neq r}} x^q_{s_q} = 0, \ \ k \in I_r, \\ \sum\limits_{\mathbf{s}_{-r} \in S_{-r}} \left(a^r_{k||\mathbf{s}_{-r}} - a^r_{i_r||\mathbf{s}_{-r}}\right) \prod\limits_{\substack{q=1,\ldots,n \\ q \neq r}} x^q_{s_q} \leq 0, \ \ k \notin I_r \cup i_r, \\ r = 1, \ldots, n, \\ \mathbf{e}\mathbf{r}^T \mathbf{x}^r = 1, \ \ \mathbf{x}^r \geq 0, \ r = 1, \ldots, n, \\ x^r_k = 0, \ \ k \notin I_r \cup \{i_r\}, \ r = 1, \ldots, n. \end{cases} \tag{3.4}$$

Expression (3.4) finally leads us to the end of the proof of next Theorem.

Theorem 3.5 *The set of Nash equilibria in a polymatrix mixed-strategy game has the following representation:*

$$NES(\widetilde{\Gamma}) = \bigcup_{\substack{i_1 \in U_1,\ I_1 \in \mathscr{P}(U_1 \setminus \{i_1\}) \\ \cdots \\ i_n \in U_n,\ I_n \in \mathscr{P}(U_n \setminus \{i_n\})}} X\left(i_1 I_1 \ldots i_n I_n\right).$$

Theorem 3.5 is an extension of Theorem 3.1 on the n-player game. The proof has a constructive nature. It permits to develop on its basis both a general method for Nash equilibria set computing and different algorithms based on the method.

Remark 3.6 We have to highlight the nature of the Nash equilibrium set components $X(i_1 I_1 \dots i_n I_n) = NE(i_1 I_1 \dots i_n I_n)$ which are solution sets of system (3.4). System (3.4) represents a system of multilinear equations. Its solving needs a special conceptual and methodological approach both from the perspective of multilinear algebra [50–53] and the perspective of algorithmic, Gröbner bases, tensor, numerical and/or symbolic points of view [54–56].

The Wolfram Programming Language [55], which has a symbolic nature by its origins [57], may be a valuable practical tool for the set of Nash equilibria computing and representation.

Additionally, Stephen Wolfram's concept of modelling the world by means of primitive/atom programs [57] may be another efficient and effective approach to find Nash equilibrium sets in polymatrix mixed-strategy games. Next chapters highlight this assertion for some particular polymatrix mixed-strategy games.

Theorem 3.6 *$NES(\tilde{\Gamma})$ consists of no more than*

$$(2^{m_1} - 1)(2^{m_2} - 1) \dots (2^{m_n} - 1)$$

components of the type $X(i_1 I_1 \dots i_n I_n)$.

Proof Theorem 3.6 may be seen as a corollary of Theorem 3.5, but it may be also directly proved on the basis of expression (3.4) that defines a particular component of the Nash equilibria set. □

In a game for which all the players have equivalent strategies, Nash equilibrium set is partitioned into the maximal number

$$(2^{m_1} - 1)(2^{m_2} - 1) \dots (2^{m_n} - 1)$$

of components.

Generally, if $n \geq 3$, the components $X(i_1 I_1 \dots i_n I_n)$ are non-convex because of multi-linear nature of the systems of equations and inequalities which define them. From the perspective of the system types that define those components, bimatrix games are easier because the components are polyhedral convex sets defined by systems of linear equations and inequalities.

Remark 3.7 Papers [19, 20, 30] have not only the illustrative purposes. They has to exemplify both the difficulty of considered problem, and the structure of Nash equilibria sets in some games which admits solution graphical representation in a traditional Cartesian system of coordinates. This gives the explanation why we limit ourselves with only these illustrative games. Nevertheless, it must be remarked that they has independent as theoretical, as applicative importance (see, for example, the application described in [58]).

3.3.1 Algorithm for Nash Equilibrium Sets Computing in Polymatrix Mixed-Strategy Games

An exponential algorithm for Nash equilibrium set computing in the n-player game simply follows from the expression in Theorem 3.6. The algorithm requires to solve $(2^{m_1} - 1)(2^{m_2} - 1) \dots (2^{m_n} - 1)$ finite systems of multi-linear ($n - 1$-linear) and linear equations and inequalities in m variables. The last problem is itself a difficult one.

The algorithm is based on previous theorems and notation.

Algorithm 3.3.1

$NES = \emptyset$;
for $p \in N$ **do**
  $U_p = \{i \in \{1, 2, ..., m\} : X_{-p}(i_p) \neq \emptyset\}$;

$UX_1 = U_1$;
for $i_1 \in U_1$ **do**
{
  $UX_1 = UX_1 \setminus \{i_1\}$;
  for $I_1 \in 2^{UX_1}$ **do**
  {
    $UX_2 = U_2$;
    for $i_2 \in U_2$ **do**
    {
      $UX_2 = UX_2 \setminus \{i_2\}$;
      for $I_2 \in 2^{UX_2}$ **do**
      {
        ⋮
        $UX_n = U_n$;
        for $i_n \in U_n$ **do**
        {
          $UX_n = UX_n \setminus \{i_n\}$;
          for $I_n \in 2^{UX_n}$ **do**
            $NES = NES \cup X(i_1 I_1 i_2 I_2 \dots i_n I_n)$;
        }
        ⋮
      }
    }
  }
}

Let us illustrate Algorithm 3.3.1 via an example. Remark that in Example 3.5 we use mainly the notation $NE(i_1 I_1 i_2 I_2 \dots i_n I_n)$ instead of $X(i_1 I_1 i_2 I_2 \dots i_n I_n)$ to highlight the components of the Nash equilibrium set.

Example 3.5 Consider a three-player extended $2 \times 2 \times 2$ (dyadic) game [29] with matrices:

$$a_{1**} = \begin{bmatrix} 9 & 0 \\ 0 & 3 \end{bmatrix}, \; a_{2**} = \begin{bmatrix} 0 & 3 \\ 9 & 0 \end{bmatrix},$$

$$b_{*1*} = \begin{bmatrix} 8 & 0 \\ 0 & 4 \end{bmatrix}, \; b_{*2*} = \begin{bmatrix} 0 & 4 \\ 8 & 0 \end{bmatrix},$$

$$c_{**1} = \begin{bmatrix} 12 & 0 \\ 0 & 2 \end{bmatrix}, \; c_{**2} = \begin{bmatrix} 0 & 6 \\ 4 & 0 \end{bmatrix}.$$

$$f_1(\mathbf{x}, \mathbf{y}, \mathbf{z}) = (9y_1z_1 + 3y_2z_2)x_1 + (9y_2z_1 + 3y_1z_2)x_2,$$

$$f_2(\mathbf{x}, \mathbf{y}, \mathbf{z}) = (8x_1z_1 + 4x_2z_2)y_1 + (8x_2z_1 + 4x_1z_2)y_2,$$

$$f_3(\mathbf{x}, \mathbf{y}, \mathbf{z}) = (12x_1y_1 + 2x_2y_2)z_1 + (4x_2y_1 + 6x_1y_2)z_2.$$

Totally, we have to consider $(2^2 - 1)(2^2 - 1)(2^2 - 1) = 27$ components. Further, we will enumerate only non-empty components. Thus, the first component

$$NE(1\emptyset1\emptyset1\emptyset) = (1; 0) \times (1; 0) \times (1; 0)$$

(for which $f_1 = 9$, $f_2 = 8$, $f_3 = 12$) is the solution of the system

$$\begin{cases} 9y_2z_1 + 3y_1z_2 - 9y_1z_1 - 3y_2z_2 = 3(y_2 - y_1)(3z_1 - z_2) \le 0, \\ 8x_2z_1 + 4x_1z_2 - 8x_1z_1 - 4x_2z_2 = 4(x_2 - x_1)(2z_1 - z_2) \le 0, \\ 4x_2y_1 + 6x_1y_2 - 12x_1y_1 - 2x_2y_2 = 2(3x_1 - x_2)(y_2 - 2y_1) \le 0, \\ x_1 + x_2 = 1, x_1 \ge 0, x_2 = 0, \\ y_1 + y_2 = 1, y_1 \ge 0, y_2 = 0, \\ z_1 + z_2 = 1, z_1 \ge 0, z_2 = 0. \end{cases}$$

The second component

$$NE(1\emptyset2\emptyset2\emptyset) = (1; 0) \times (0; 1) \times (0; 1)$$

with $f_1 = 3$, $f_2 = 4$, $f_3 = 6$ is the solution of the system

$$\begin{cases} 3(y_2 - y_1)(3z_1 - z_2) \le 0, \\ -4(x_2 - x_1)(2z_1 - z_2) \le 0, \\ -2(3x_1 - x_2)(y_2 - 2y_1) \le 0, \\ x_1 + x_2 = 1, x_1 \ge 0, x_2 = 0, \\ y_1 + y_2 = 1, y_1 = 0, y_2 \ge 0, \\ z_1 + z_2 = 1, z_1 = 0, z_2 \ge 0. \end{cases}$$

The third component

$$NE(2\emptyset 1\emptyset 2\emptyset) = (0;\ 1) \times (1;\ 0) \times (0;\ 1)$$

with $f_1 = 3,\ f_2 = 4,\ f_3 = 4$ is the solution of the system

$$\begin{cases} -3(y_2 - y_1)(3z_1 - z_2) \le 0, \\ 4(x_2 - x_1)(2z_1 - z_2) \le 0, \\ -2(3x_1 - x_2)(y_2 - 2y_1) \le 0, \\ x_1 + x_2 = 1, x_1 = 0, x_2 \ge 0, \\ y_1 + y_2 = 1, y_1 \ge 0, y_2 = 0, \\ z_1 + z_2 = 1, z_1 = 0, z_2 \ge 0. \end{cases}$$

The fourth component

$$NE(2\emptyset 2\emptyset 1\emptyset) = (0;\ 1) \times (0;\ 1) \times (1;\ 0)$$

with $f_1 = 9,\ f_2 = 8,\ f_3 = 2$ is the solution of the system

$$\begin{cases} -3(y_2 - y_1)(3z_1 - z_2) \le 0, \\ -4(x_2 - x_1)(2z_1 - z_2) \le 0, \\ 2(3x_1 - x_2)(y_2 - 2y_1) \le 0, \\ x_1 + x_2 = 1, x_1 = 0, x_2 \ge 0, \\ y_1 + y_2 = 1, y_1 = 0, y_2 \ge 0, \\ z_1 + z_2 = 1, z_1 \ge 0, z_2 = 0. \end{cases}$$

The fifth component

$$NE(2\emptyset 1\{2\}1\{2\}) = (0;\ 1) \times (1/3;\ 2/3) \times (1/3;\ 2/3)$$

with $f_1 = 8/3$,
$f_2 = 8/3,\ f_3 = 4/3$ is the solution of the system

$$\begin{cases} 3(y_2 - y_1)(3z_1 - z_2) \le 0, \\ 4(x_2 - x_1)(2z_1 - z_2) = 0, \\ 2(3x_1 - x_2)(y_2 - 2y_1) = 0, \\ x_1 + x_2 = 1, x_1 = 0, x_2 \ge 0, \\ y_1 + y_2 = 1, y_1 \ge 0, y_2 \ge 0, \\ z_1 + z_2 = 1, z_1 \ge 0, z_2 \ge 0. \end{cases}$$

The sixth component

$$NE(1\{2\}1\emptyset 1\{2\}) = (1/4; 3/4) \times (1; 0) \times (1/4; 3/4)$$

with $f_1 = 9/4$, $f_2 = 11/4$, $f_3 = 3$ is the solution of the system

$$\begin{cases} 3(y_2 - y_1)(3z_1 - z_2) = 0, \\ -4(x_2 - x_1)(2z_1 - z_2) \le 0, \\ 2(3x_1 - x_2)(y_2 - 2y_1) = 0, \\ x_1 + x_2 = 1, x_1 \ge 0, x_2 \ge 0, \\ y_1 + y_2 = 1, y_1 \ge 0, y_2 = 0, \\ z_1 + z_2 = 1, z_1 \ge 0, z_2 \ge 0. \end{cases}$$

The seventh component

$$NE(1\{2\}1\{2\}1\emptyset) = (1/2; 1/2) \times (1/2; 1/2) \times (1; 0)$$

with $f_1 = 9/2$, $f_2 = 4$, $f_3 = 7/2$ is the solution of the system

$$\begin{cases} 3(y_2 - y_1)(3z_1 - z_2) = 0, \\ -4(x_2 - x_1)(2z_1 - z_2) = 0, \\ 2(3x_1 - x_2)(y_2 - 2y_1) \le 0, \\ x_1 + x_2 = 1, x_1 \ge 0, x_2 \ge 0, \\ y_1 + y_2 = 1, y_1 \ge 0, y_2 \ge 0, \\ z_1 + z_2 = 1, z_1 \ge 0, z_2 = 0. \end{cases}$$

The eighth component

$$NE(1\{2\}1\{2\}1\{2\}) =$$
$$\left\{\begin{pmatrix} 1/2 \\ 1/2 \end{pmatrix} \times \begin{pmatrix} 1/3 \\ 2/3 \end{pmatrix} \times \begin{pmatrix} 1/4 \\ 3/4 \end{pmatrix}, \begin{pmatrix} 1/4 \\ 3/4 \end{pmatrix} \times \begin{pmatrix} 1/2 \\ 1/2 \end{pmatrix} \times \begin{pmatrix} 1/3 \\ 2/3 \end{pmatrix}\right\}$$

for which $f_1 = 9/4$, $f_2 = 5/2$, $f_3 = 8/3$ and $f_1 = 5/2$, $f_2 = 8/3$, $f_3 = 9/4$, has two distinct points that are solutions of the system

$$\begin{cases} 3(y_2 - y_1)(3z_1 - z_2) = 0, \\ -4(x_2 - x_1)(2z_1 - z_2) = 0, \\ 2(3x_1 - x_2)(y_2 - 2y_1) = 0, \\ x_1 + x_2 = 1, x_1 \geq 0, x_2 \geq 0, \\ y_1 + y_2 = 1, y_1 \geq 0, y_2 \geq 0, \\ z_1 + z_2 = 1, z_1 \geq 0, z_2 \geq 0. \end{cases}$$

Remark once again that the last component of the Nash equilibrium set consists of two distinct points. Thus the game has 9 Nash equilibria. Hence, it is non-convex and non-connected. □

3.4 Conclusions

The idea to consider Nash equilibrium set as an intersection of the graphs of best response mappings yields to a simply Nash equilibrium set representation and to a method of Nash equilibrium set computing. Taking into account the computational complexity of the problem, proposed exponential algorithms are pertinent.

Nash equilibrium sets in bimatrix mixed-strategy games may be partitioned into finite number of polytopes, no more than $(2^m - 1)(2^n - 1)$. Generally, numbers of sets $X_{iI}^{jJ} \times Y_{jJ}^{iI}$ examined by the exposed algorithm in concrete games are much more smaller than the theoretical estimation given by formula in Theorem 3.4.

The Nash equilibrium set in a mixed extension of a polymatrix game may be partitioned into finite number of components, no more than

$$(2^{m_1} - 1)(2^{m_2} - 1) \dots (2^{m_n} - 1),$$

but they, generally, are non-convex and non-polytopes. An algorithmic and programming realization of the exposed method is closely related to a problem of solving systems of multi-linear ($n - 1$ – linear and simply linear) equations and inequalities, that represents in itself a serious obstacle to compute efficiently Nash equilibrium sets.

References

1. Vorob'ev, N.N. 1958. Equilibrium points in bimatrix games. *Theory of Probability and its Applications* 3: 297–309.
2. Kuhn, H.W. 1961. An algorithm for equilibrium points in bimatrix games. *Proceedings of the National Academy of Sciences U.S.A* 47: 1657–1662.
3. Raghavan, T.E.S. 2002. Non-zero sum two-person games. In *Handbook of Game Theory with Economic Applications*, ed. R.J. Aumann, and S. Hart, vol. 3, 1687–1721. Elsevier Science B.V.: North-Holland.

4. Mills, H. 1960. Equilibrium points in finite games. *Journal of the Society for Industrial and Applied Mathematics* 8 (2): 397–402.
5. Mangasarian, O.L. 1964. Equilibrium points of bimatrix games. *Journal of the Society for Industrial and Applied Mathematics* 12 (4): 778–780.
6. Winkels, H.-M. 1979. An algorithm to determine all equilibrium points of a bimatrix games. In *Game Theory and Related Topics*, ed. O. Moeschler, and D. Pallschke, 137–148. Amsterdam: North-Holland.
7. Yanovskaya, E.B. 1968. Equilibrium points in polymatrix games. *Lithuanian Mathematical Collection (Litovskii Matematicheskii Sbornik)* 8 (2): 381–384. (in Russian).
8. Howson Jr., J.T. 1972. Equilibria of polymatrix games. *Management Science* 18: 312–318.
9. Eaves, B.C. 1973. Polymatrix games with joint constraints. *SIAM Journal of Applied Mathematics* 24: 418–423.
10. Mukhamediev, B.M. 1978. The solution of bilinear programming problem and finding the equilibrium situations in bimatrix games. *U.S.S.R. Computational Mathematics and Mathematical Physics* 18 (2): 60–66.
11. Savani, R.S.J. 2006. *Finding Nash Equilibria of Bimatrix Games* Ph. D. Thesis, London School of Economics, 115.
12. Shokrollahi, A. 2017. A note on nash equilibrium in bimatrix games with nonnegative matrices. *Palestine Journal of Mathematics* 6 (1): 301–306.
13. Lemke, C.E., and J.T. Howson Jr. 1964. Equilibrium points of bimatrix games. *Journal of the Society for Industrial and Applied Mathematics* 12 (2): 413–423.
14. Ungureanu, V. 2006. Nash equilibrium set computing in finite extended games. *Computer Science Journal of Moldova* 42 (3): 345–365.
15. Audet, C., S. Belhaiza, and P. Hansen. 2006. Enumeration of all the extreme equilibria in game theory: Bimatrix and polymatrix games. *Journal of Optimization Theory and Applications* 129 (3): 349–372.
16. Avis, D., G.D. Rosenberg, R. Savani, and B. Von Stenghel. 2010. Enumeration of Nash equilibria for two player games. *Economic Theory* 42: 9–37.
17. Datta, R.S. 2010. Finding all Nash equilibria of a finite game using polynomial algebra. *Economic Theory* 42: 55–96.
18. Sagaidac, M., and V. Ungureanu. 2004. *Operational research* Chişinău: CEP USM, 296 pp. (in Romanian).
19. Ungureanu, V., and A. Botnari. 2005. Nash equilibrium sets in mixed extension of $2 \times 2 \times 2$ games. *Computer Science Journal of Moldova*13, 1(37): 13–28.
20. Ungureanu, V., and A. Botnari. 2005. Nash equilibrium sets in mixed extended 2×3 games. *Computer Science Journal of Moldova* 13(2)(38): 136–150.
21. Dantzig, G.B., and M.N. Thapa. 1997. *Linear Programming 1: Introduction, 474*. New York: Springer.
22. Dantzig, G.B., and M.N. Thapa. 2003. *Linear Programming 2: Theory and Extensions, 475*. New York: Springer.
23. Ungureanu, V. 2001. *Mathematical Programming*, 348 pp Chişinău: USM. (in Romanian).
24. Rao, S.S. 2009. *Engineering Optimization: Theory and Practice*, XIX+813 pp. New Jersey: John Wiley & Sons, Inc.
25. Nash, J. 1951. Noncooperative games. *Annals of Mathematics* 54 (2): 280–295.
26. Van Damme, E. 1992. *Stability and Perfection of Nash Equilibria*, 2nd ed., 272 pp. Berlin: Springer.
27. Van Damme, E. 2002. Strategic Equilibrium. In *Handbook of Game Theory with Economic Applications*, ed. R.J. Aumann, and S. Hart, 1521–1596. Elsevier Science B.V: North-Holland.
28. Hillas, J., and E. Kohlberg. 2002. Foundations of Strategic Equilibrium. In *Handbook of Game Theory with Economic Applications*, ed. R.J. Aumann, and S. Hart, 1597–1663. Elsevier Science B.V: North-Holland.
29. McKelvey, R.D., and A. McLenan. 1996. Computation of Equilibria in Finite Games. *Handbook of Computational Economics*, vol. 1, 87–142. Elsevier.

30. Ungureanu, V. Nash equilibrium set function in dyadic mixed-strategy games. *Computer Science Journal of Moldova* 25, 1 (73): 3–20.
31. Moulen, H. 1985. *Théorie des jeux pour l'éqonomie et la politique, Paris, 1981 (in Russian: Game Theory, 200.* Moscow: Mir.
32. Gilboa, I., and E. Zemel. 1989. Nash and correlated equilibria: Some complexity considerations. *Games and Economic Behaviour* 1: 80–93.
33. Papadimitriou, C.H. 1994. On the complexity of the parity argument and other inefficient proofs of existence. *Journal of Computer and System Sciences* 48 (3): 498–532.
34. Papadimitriou, C. H. 2015. The complexity of computing equilibria. In *Handbook of Game Theory with Economic Applications*, ed. H. P. Young, and Sh. Zamir, vol. 4, 779–810. Amsterdam: North-Holland.
35. Jansen, M., P. Jurg, and D. Vermeulen. 2002. On the Set of Equilibria of a Bimatrix Game: A Survey. In *Chapters in Game Theory*, ed. P. Borm, and H.J. Peters, 121–142. In Honour of Stef Tijs. Boston: Kluwer Academic Publishers.
36. Von Stengel, B. 2002. Computing equilibria for two-person games. In *Handbook of Game Theory with Economic Applications*, ed. R.J. Aumann, and S. Hart, 1723–1759. Elsevier Science B.V: North-Holland.
37. Gass, S., and T. Saaty. 1955. Parametric objective function (part 2)-generalization. *Operations Research* 3 (4): 395–401.
38. Simons, E. 1962. A note on linear parametric programming. *Management Science* 8 (3): 355–358.
39. Gal, T., and J. Nedoma. 1972. Multiparametric linear programming. *Management Sciences, Theory Series* 18 (7): 406–422.
40. Gal, T. 1979. *Postoptimal Analysis, Parametric Programming, and Related Topics*. New-York: McGraw-Hill.
41. Gal, T. 1995. *Postoptimal Analyses, Parametric Programming, and Related Topics: Degeneracy, Multicriteria Decision Making, Redundancy*. Berlin: Walter de Gruyter.
42. Gal, T., and H.J. Greenberg (eds.). 1997. *Advances in Sensitivity Analysis and Parametric Programming*. Norwell, Massachusetts: Kluwer Academic Publishers.
43. Gal, T. 2016. Parametric programming. *Encyclopedia of Operations Research and Management Science* 23: 1108–1112.
44. Adler, I., and R. Monteiro. 1992. A geometric view of parametric linear programming. *Algorithmica* 8: 161–176.
45. Borelli, F., A. Bemporad, and M. Morari. 2003. Geometric algorithm for multiparametric linear programming. *Journal of Optimization Theory and Applications* 118 (3): 515–540.
46. Monteiro, R.D.C., and S. Mehrotra. 1996. A general parametric analysis approach and its implication to sensitivity analysis in interior point methods. *Mathematical Programming* 72: 65–82.
47. Hladik, M. 2010. Multiparametric linear programming: support set and optimal partition invariancy. *European Journal of Operational Research* 202 (1): 25–31.
48. Taha, H.A. 2007. *Operations Research: An Introduction*, 8th ed., 838 pp. New Jersey, USA: Pearson Education, Inc.
49. Hillier, F.S., and G.J. Lieberman. 2001. *Introduction to Operation Research, 1240*. New-York: McGraw-Hill.
50. Grassmann, H. 2000. *Extension Theory*, 411 pp. American Mathematical Society.
51. Greub, W. 1978. *Multilinear Algebra*, 2nd ed., 315 pp. New-York: Springer.
52. Merriss, R. 1997. *Multilinear Algebra, 340*. Amsterdam: Gordon and Breach Science Publishers.
53. Northcott, D.G. 1984. *Multilinear Algebra, 209*. Cambridge: Cambridge University Press.
54. Ding, W., and Y. Wei. 2016. Solving multi-linear systems with $\mathcal{M}$-tensors. *Journal of Scientific Computing* 8: 1–27.
55. Wolfram, S. 2016. *An elementary Introduction to the Wolfram Language*, XV+324 pp. Champaign, IL: Wolfram Media, Inc.

56. Cox, D.A., J. Little, and D. O'Shea, D. *Ideals, Varieties, and Algorithms: An Introduction to Computational Algebraic Geometry and Commutative Algebra*, XVI+646 pp. Switzerland: Springer International Publishing.
57. Wolfram, S. 2002. *A New Kind of Science, 1197*. Champaign, IL: Wolfram Media Inc.
58. Zhukov, Y.M. 2013. An epidemic model of violence and public support in civil war. *Conflict Management and Peace Science* 30 (1): 24–52.

Chapter 4
Sets of Nash Equilibria in Bimatrix 2 × 3 Mixed-Strategy Games

Abstract The results of the precedent chapter are applied to solve bimatrix 2×3 mixed-strategy games. Such games are important mainly because they admit a simple graphic representation of Nash equilibrium sets. Even though the problem of Nash equilibrium set computing is not so frequently considered in literature, compared to other problems which investigate various aspects concerning the Nash equilibrium concept, some important instances we can refer specially such as (Raghavan, Handbook of Game Theory with Economic Applications. Elsevier Science B.V, North-Holland, pp. 1687–1721, 2002, [1]), (Von Stengel, Handbook of Game Theory with Economic Applications. Elsevier Science B.V, North-Holland, pp. 1723–1759, 2002, [2]), (Avis, Rosenberg, Savani and Von Stenghel, Economic Theory, 42:9–37, 2010, [3]), (Datta, Economic Theory, 42:55–96, 2010, [4]).

4.1 Introduction

Let us apply the method exposed in the precedent chapter to construct Nash equilibrium sets in bimatrix 2×3 mixed strategy games as the intersection of best response graphs [5–7]. The approach may be seen as an illustration exemplifying the method exposed in the precedent chapter and proposed in paper [6], but it summarises the results published in paper [7] too. The results have to illustrate the practical opportunity of the proposed method of Nash equilibrium set computing. Despite its apparent simplicity, we must not forget that many of real world important decision making processes may be modelled by such mathematical models. It is enough to recall a series of well known standard games such as, e.g., prisoner's dilemma [8–10].

So, let us consider, as in precedent chapters, the noncooperative game

$$\Gamma = \langle N, \{X_i\}_{i \in N}, \{f_i(x)\}_{i \in N} \rangle,$$

where

$N = \{1, 2, ..., n\}$ is a set of players,
X_i is a set of strategies of player $i \in N$,

V. Ungureanu, *Pareto-Nash-Stackelberg Game and Control Theory*,
Smart Innovation, Systems and Technologies 89,
https://doi.org/10.1007/978-3-319-75151-1_4

$f_i : X \to R$ is player's $i \in N$ payoff function defined on the Cartesian product $X = \times_{i \in N} X_i$.

We study the Nash equilibrium set as an intersection of best response graphs [5–7], i.e. the intersection of the sets

$$Gr_i = \left\{ (x_i, x_{-i}) \in X : \begin{array}{l} x_{-i} \in X_{-i} \\ x_i \in \operatorname{Arg} \max\limits_{x_i \in X_i} f_i(x_i, x_{-i}) \end{array} \right\}, \quad i \in N.$$

This chapter investigates the problems of a Nash equilibrium set computing in games that permit simple graphic illustrations and that elucidate usefulness of a Nash equilibrium set interpretation as an intersection of best response graphs [5–7].

4.2 Main Results

Consider a two-person matrix game

$$\Gamma = \langle \{1, 2\}; I, J; a_{ij}, b_{ij} \rangle$$

with matrices

$$A = (a_{ij}), \ B = (b_{ij}), \ i \in I, \ j \in J,$$

that define the payoff functions a_{ij}, b_{ij}, of indices i, j, the set of players $N = \{1, 2\}$, and the strategy sets $I = \{1, 2\}$, $J = \{1, 2, 3\}$.

The game

$$\Gamma_m = \langle \{1, 2\}; X, Y; f_1(\mathbf{x}, \mathbf{y}), f_2(\mathbf{x}, \mathbf{y}) \rangle$$

is a mixed extension of Γ, i.e. the game in mixed-strategies, where

$$X = \left\{ \mathbf{x} = (x_1, x_2) \in \mathbb{R}^2 : \begin{array}{l} x_1 + x_2 = 1 \\ x_1 \geq 0, x_2 \geq 0 \end{array} \right\},$$

$$Y = \left\{ \mathbf{y} = (y_1, y_2, y_3) \in \mathbb{R}^3 : \begin{array}{l} y_1 + y_2 + y_3 = 1 \\ y_1 \geq 0, y_2 \geq 0, y_3 \geq 0 \end{array} \right\},$$

$$f_1(\mathbf{x}, \mathbf{y}) = \sum_{i=1}^{2} \sum_{j=1}^{3} a_{ij} x_i y_j,$$

$$f_2(\mathbf{x}, \mathbf{y}) = \sum_{i=1}^{2} \sum_{j=1}^{3} b_{ij} x_i y_j.$$

Let us investigate this game by reducing it to a game on a unit triangular prism. For the last one, we consider a class partition of strategy sets. The Nash equilibrium set is determined for each possible "*subgame*" (see the following propositions). Finally, we unify obtained results by constructing a solving procedure.

4.2.1 Games on a Triangular Prism

By substitutions

$$x_1 = x,\ x_2 = 1 - x,\ x \in [0, 1],$$
$$y_3 = 1 - y_1 - y_2, \quad y_3 \in [0, 1],$$

the game Γ_m is reduced to an equivalent game

$$\Gamma'_m = \langle \{1, 2\}; [0, 1], \Delta; \varphi_1, \varphi_2 \rangle,$$

where

- [0,1] is the set of strategies of the first player,
- $\Delta = \left\{ \mathbf{y} = (y_1, y_2) \in \mathbb{R}^2 : \begin{array}{l} y_1 + y_2 \leq 1 \\ y_1 \geq 0,\ y_2 \geq 0 \end{array} \right\}$ is the set of strategies of the second player,
- $$\begin{aligned} \varphi_1(x, \mathbf{y}) &= (a_{11}y_1 + a_{12}y_2 + a_{13}(1 - y_1 - y_2))x \\ &\quad + (a_{21}y_1 + a_{22}y_2 + a_{23}(1 - y_1 - y_2))(1 - x) \\ &= ((a_{11} - a_{21} + a_{23} - a_{13})y_1 + (a_{12} - a_{22} + a_{23} - a_{13})y_2 + a_{13} - a_{23})x \\ &\quad + (a_{21} - a_{23})y_1 + (a_{22} - a_{23})y_2 + a_{23} \\ &= ((a_{11} - a_{21})y_1 + (a_{12} - a_{22})y_2 + (a_{13} - a_{23})(1 - y_1 - y_2))\,x \\ &\quad + (a_{21} - a_{23})y_1 + (a_{22} - a_{23})y_2 + a_{23} \end{aligned}$$

  is the payoff function of the first player,
- $$\begin{aligned} \varphi_2(x, \mathbf{y}) &= (b_{11}y_1 + b_{12}y_2 + b_{13}(1 - y_1 - y_2))x \\ &\quad + (b_{21}y_1 + b_{22}y_2 + b_{23}(1 - y_1 - y_2))(1 - x) \\ &= ((b_{11} - b_{13} + b_{23} - b_{21})x + b_{21} - b_{23})y_1 \\ &\quad + ((b_{12} - b_{13} + b_{23} - b_{22})x + b_{22} - b_{23})y_2 \\ &\quad + (b_{13} - b_{23})x + b_{23} \\ &= ((b_{11} - b_{13})x + (b_{21} - b_{23})(1 - x))y_1 \\ &\quad + ((b_{12} - b_{13})x + (b_{22} - b_{23})(1 - x))y_2 \\ &\quad + (b_{13} - b_{23})x + b_{23} \end{aligned}$$

  is the payoff function of the second player.

Thus, the game Γ_m is reduced to the game Γ'_m on a triangular prism

$$\Pi = [0, 1] \times \Delta.$$

If the set $NES(\Gamma'_m)$ is known, then it is easy to construct the set $NES(\Gamma_m)$.

Based on strategy properties of each player of the initial pure strategy game Γ, diverse classes of games are considered and for each class the sets $NES(\Gamma'_m)$ are determined.

For convenience, we use notation

$$\Delta_= = \left\{ \mathbf{y} = (y_1, y_2) \in \mathbb{R}^2 : \begin{array}{l} y_1 + y_2 = 1 \\ y_1 \geq 0, y_2 \geq 0 \end{array} \right\}.$$

4.2.2 Both Players Have Either Equivalent Strategies or Dominant Strategies

This section emphasizes cases when both players simultaneously have equivalent or dominant strategies.

Proposition 4.1 *If both players have equivalent strategies, then*

$$NES(\Gamma'_m) = \Pi.$$

Proof From the equivalence of strategies we obtain the following form of the payoff functions

$$\begin{array}{l} \varphi_1(x, \mathbf{y}) = (a_{21} - a_{23})y_1 + (a_{22} - a_{23})y_2 + a_{23}, \\ \varphi_2(x, \mathbf{y}) = (b_{13} - b_{23})x + b_{23}. \end{array}$$

From these expressions the truth of the proposition results. □

Proposition 4.2 *If both players have dominant strategies in Γ, then*

$$NE(\Gamma'_m) = \begin{cases} (0,0,0) & \textit{if strategies (2,3) are dominant,} \\ (0,0,1) & \textit{if strategies (2,2) are dominant,} \\ (0,1,0) & \textit{if strategies (2,1) are dominant,} \\ 0 \times \Delta_= & \textit{if strategies } (2, 1 \sim 2) \textit{ are dominant,} \\ 0 \times [0,1] \times 0 & \textit{if strategies } (2, 1 \sim 3) \textit{ are dominant,} \\ 0 \times 0 \times [0,1] & \textit{if strategies } (2, 2 \sim 3) \textit{ are dominant,} \\ (1,0,0) & \textit{if strategies } (1,3) \textit{ are dominant,} \\ (1,0,1) & \textit{if strategies } (1,2) \textit{ are dominant,} \\ (1,1,0) & \textit{if strategies } (1,1) \textit{ are dominant,} \\ 1 \times \Delta_= & \textit{if strategies } (1, 1 \sim 2) \textit{ are dominant,} \\ 1 \times [0,1] \times 0 & \textit{if strategies } (1, 1 \sim 3) \textit{ are dominant,} \\ 1 \times 0 \times [0,1] & \textit{if strategies } (1, 2 \sim 3) \textit{ are dominant.} \end{cases}$$

Proof It is easy to observe that

$$\operatorname{Arg}\max_{x\in[0,1]} \varphi_1(x, \mathbf{y}) = \begin{cases} 1 & \text{if the 1st strategy is dominant in } \Gamma, \\ 0 & \text{if the 2nd strategy is dominant in } \Gamma, \end{cases}$$

for all $\mathbf{y} \in \triangle$. Hence,

$$Gr_1 = \begin{cases} 1 \times \triangle & \text{if the 1st strategy is dominant,} \\ 0 \times \triangle & \text{if the 2nd strategy is dominant.} \end{cases}$$

For the second player:

$$\operatorname{Arg}\max_{\mathbf{y}\in\triangle} \varphi_2(x, \mathbf{y}) = \begin{cases} (1, 0) & \text{if the 1st strategy dominates in } \Gamma, \\ (0, 1) & \text{if the 2nd strategy dominates in } \Gamma, \\ (0, 0) & \text{if the 3rd strategy dominates in } \Gamma, \\ \Delta_= & \text{if strategies } 1 \sim 2 \text{ dominate 3,} \\ [0, 1] \times 0 & \text{if strategies } 1 \sim 3 \text{ dominate 2,} \\ 0 \times [0, 1] & \text{if strategies } 2 \sim 3 \text{ dominate 1,} \end{cases}$$

for all $\mathbf{x} \in [0, 1]$. Hence,

$$Gr_2 = \begin{cases} [0, 1] \times (1, 0) & \text{if the 1st strategy is dominant,} \\ [0, 1] \times (0, 1) & \text{if the 2nd strategy is dominant,} \\ [0, 1] \times (0, 0) & \text{if the 3rd strategy is dominant.} \\ [0, 1] \times \Delta_= & \text{if strategies } 1 \sim 2 \text{ dominate 3,} \\ [0, 1] \times [0,1] \times 0 & \text{if strategies } 1 \sim 3 \text{ dominate 2,} \\ [0, 1] \times 0 \times [0,1] & \text{if strategies } 2 \sim 3 \text{ dominate 1.} \end{cases}$$

Thus, the Nash equilibrium set contains either only one vertex of a unit prism Π as an intersection of one facet Gr_1 with one edge Gr_2 or only one edge of a unit prism Π as an intersection of one facet Gr_1 with one facet Gr_2. □

4.2.3 One Player Has Dominant Strategy

This section has the aim to investigate the cases when players don't have simultaneously dominant strategies.

Proposition 4.3 *If the first strategy of the first player is dominant, then*

$$NE(\Gamma'_m) = \begin{cases} (1,1,0) & \textit{if } b_{11} > \; max\{b_{12}, b_{13}\}, \\ (1,0,1) & \textit{if } b_{12} > \; max\{b_{11}, b_{13}\}, \\ (1,0,0) & \textit{if } b_{12} > \; max\{b_{11}, b_{13}\}, \\ 1 \times \Delta_{=} & \textit{if } b_{11} = b_{12} > b_{13}, \\ 1 \times [0,1] \times 0 & \textit{if } b_{11} = b_{13} > b_{12}, \\ 1 \times 0 \times [0,1] & \textit{if } b_{12} = b_{13} > b_{11}, \\ 1 \times \Delta & \textit{if } b_{12} = b_{13} = b_{11}, \end{cases}$$

if the second strategy of the first player is dominant, then

$$NE(\Gamma'_m) = \begin{cases} (0,1,0) & \textit{if } b_{11} > \; max\{b_{12}, b_{13}\}, \\ (0,0,1) & \textit{if } b_{12} > \; max\{b_{11}, b_{13}\}, \\ (0,0,0) & \textit{if } b_{12} > \; max\{b_{11}, b_{13}\}, \\ 0 \times \Delta_{=} & \textit{if } b_{11} = b_{12} > b_{13}, \\ 0 \times [0,1] \times 0 & \textit{if } b_{11} = b_{13} > b_{12}, \\ 0 \times 0 \times [0,1] & \textit{if } b_{12} = b_{13} > b_{11}, \\ 0 \times \Delta & \textit{if } b_{12} = b_{13} = b_{11}. \end{cases}$$

Proof If the first player has a dominant strategy, then

$$Gr_1 = \begin{cases} 1 \times \Delta & \text{if the 1st strategy is dominant}, \\ 0 \times \Delta & \text{if the 2nd strategy is dominant} \end{cases}$$

is one triangle facet of the prism.

If the first strategy of the first player is dominant, then

$$\begin{aligned} \varphi_2(1, \mathbf{y}) &= (b_{11} - b_{13})y_1 + (b_{12} - b_{13})y_2 + b_{13} \\ &= b_{11}y_1 + b_{12}y_2 + b_{13}(1 - y_1 - y_2). \end{aligned}$$

From this expression we obtain that

$$\operatorname*{Arg\,max}_{\mathbf{y} \in \Delta} \varphi_2(1, \mathbf{y}) = \begin{cases} (1,0) & \text{if } b_{11} > \max\{b_{12}, b_{13}\}, \\ (0,1) & \text{if } b_{12} > \max\{b_{11}, b_{13}\}, \\ (0,0) & \text{if } b_{12} > \max\{b_{11}, b_{13}\}, \\ \Delta_{=} & \text{if } b_{11} = b_{12} > b_{13}, \\ [0,1] \times 0 & \text{if } b_{11} = b_{13} > b_{12}, \\ 0 \times [0,1] & \text{if } b_{12} = b_{13} > b_{11}, \\ \Delta & \text{if } b_{12} = b_{13} = b_{11} \end{cases}$$

and

$$Gr_2 = 1 \times \operatorname{Arg}\max_{\mathbf{y}\in\triangle} \varphi_2(1, \mathbf{y})$$

is a vertex, edge or triangle facet of the prism Π. Hence, the truth of the first part of the proposition follows.

Analogically, the proposition can be proved when the second strategy is dominant. □

Proposition 4.4 *If the second player has only one dominant strategy, then*

$$NE(\Gamma'_m) = \begin{cases} (0,1,0) & \textit{if } (\cdot,1) \textit{ is dominant and } a_{11} < a_{21}, \\ (1,1,0) & \textit{if } (\cdot,1) \textit{ is dominant and } a_{11} > a_{21}, \\ [0,1]\times 1\times 0 & \textit{if } (\cdot,1) \textit{ is dominant and } a_{11} = a_{21}, \\ (0,0,1) & \textit{if } (\cdot,2) \textit{ is dominant and } a_{12} < a_{22}, \\ (1,0,1) & \textit{if } (\cdot,2) \textit{ is dominant and } a_{12} > a_{22}, \\ [0,1]\times 0\times 1 & \textit{if } (\cdot,2) \textit{ is dominant and } a_{12} = a_{22}, \\ (0,0,0) & \textit{if } (\cdot,3) \textit{ is dominant and } a_{13} < a_{23}, \\ (1,0,0) & \textit{if } (\cdot,3) \textit{ is dominant and } a_{13} > a_{23}, \\ [0,1]\times 0\times 0 & \textit{if } (\cdot,3) \textit{ is dominant and } a_{13} = a_{23}. \end{cases}$$

Proof If the third strategy of the second player is dominant, then

$$\operatorname{Arg}\max_{\mathbf{y}\in\triangle} \varphi_2(x, \mathbf{y}) = \operatorname{Arg}\max_{\mathbf{y}\in\triangle}((b_{11} - b_{13})x + (b_{21} - b_{23})(1 - x))y_1$$

$$+((b_{12} - b_{13})x + (b_{22} - b_{23})(1 - x))y_2 + (b_{13} - b_{23})x + b_{23} = (0, 0),$$

and

$$Gr_2 = [0, 1] \times (0, 0)$$

is an edge of a prism Π.

For the first player we obtain $\varphi_1(x, \mathbf{0}) = a_{13}x + a_{23}(1 - x)$ and

$$Gr_1 = \begin{cases} (1,0,0) & \text{if } a_{13} > a_{13}, \\ (0,0,0) & \text{if } a_{13} < a_{13}, \\ [0,1]\times 0\times 0 & \text{if } a_{13} = a_{13}. \end{cases}$$

Consequently, the Nash equilibrium set represents a vertex or an edge of the prism Π.

Similarly, the remained part of the proposition may be proved in the other two sub-cases. □

Proposition 4.5 *If the second player has two dominant strategies, then*

$$NE(\Gamma'_m) = Gr_1 \cap Gr_2,$$

where

$$Gr_1 = 0 \times Y_{12}^{<} \cup 1 \times Y_{12}^{>} \cup [0, 1] \times Y_{12}^{=}$$

$$Gr_2 = [0, 1] \times \Delta_=$$

if the first and the second strategies are equivalent and they dominate the third strategy;

$$Gr_1 = 0 \times Y_1^{<} \cup 1 \times Y_1^{>} \cup [0, 1] \times Y_1^{=}$$

$$Gr_2 = [0, 1] \times [0, 1] \times 0$$

if the first and the third strategies are equivalent and they dominate the second strategy;

$$Gr_1 = 0 \times Y_2^{<} \cup 1 \times Y_2^{>} \cup [0, 1] \times Y_2^{=}$$

$$Gr_2 = [0, 1] \times 0 \times [0, 1]$$

if the second and the third strategies are equivalent and they dominate the first strategy;

$$Y_{12}^{<} = \{\mathbf{y} \in \Delta_= : (a_{11} - a_{21})y_1 + (a_{12} - a_{22})y_2 < 0\},$$
$$Y_{12}^{>} = \{\mathbf{y} \in \Delta_= : (a_{11} - a_{21})y_1 + (a_{12} - a_{22})y_2 > 0\},$$
$$Y_{12}^{=} = \{\mathbf{y} \in \Delta_= : (a_{11} - a_{21})y_1 + (a_{12} - a_{22})y_2 = 0\},$$

$$Y_1^{<} = \left\{ \mathbf{y} \in \mathbb{R}^2 : \begin{array}{r} (a_{11} - a_{21} + a_{23} - a_{13})y_1 + a_{13} - a_{23} < 0 \\ y_1 \in [0, 1],\ y_2 = 0 \end{array} \right\},$$

$$Y_1^{>} = \left\{ \mathbf{y} \in \mathbb{R}^2 : \begin{array}{r} (a_{11} - a_{21} + a_{23} - a_{13})y_1 + a_{13} - a_{23} > 0 \\ y_1 \in [0, 1],\ y_2 = 0 \end{array} \right\},$$

$$Y_1^{=} = \left\{ \mathbf{y} \in \mathbb{R}^2 : \begin{array}{r} (a_{11} - a_{21} + a_{23} - a_{13})y_1 + a_{13} - a_{23} = 0 \\ y_1 \in [0, 1],\ y_2 = 0 \end{array} \right\},$$

$$Y_2^{<} = \left\{ \mathbf{y} \in \mathbb{R}^2 : \begin{array}{r} (a_{12} - a_{22} + a_{23} - a_{13})y_2 + a_{13} - a_{23} < 0 \\ y_2 \in [0, 1],\ y_1 = 0 \end{array} \right\},$$

$$Y_2^{>} = \left\{ \mathbf{y} \in \mathbb{R}^2 : \begin{array}{r} (a_{12} - a_{22} + a_{23} - a_{13})y_2 + a_{13} - a_{23} > 0 \\ y_2 \in [0, 1],\ y_1 = 0 \end{array} \right\},$$

$$Y_2^{=} = \left\{ \mathbf{y} \in \mathbb{R}^2 : \begin{array}{r} (a_{12} - a_{22} + a_{23} - a_{13})y_2 + a_{13} - a_{23} = 0 \\ y_2 \in [0, 1],\ y_1 = 0 \end{array} \right\}.$$

Proof If the first and the second strategies of the second player are equivalent and they dominate the third strategy, then

$$\operatorname{Arg}\max_{\mathbf{y}\in\triangle} \varphi_2(x,\mathbf{y}) = \operatorname{Arg}\max_{\mathbf{y}\in\triangle}((b_{11}-b_{13})x + (b_{21}-b_{23})(1-x))(y_1+y_2)+$$

$$+(b_{13}-b_{23})x + b_{23} = \Delta_=$$

and

$$Gr_2 = [0,1]\times\Delta_=$$

is a facet of the prism Π.

For the first player we obtain

$$\begin{aligned}\varphi_1(x,\mathbf{y}) = {} & ((a_{11}-a_{21})y_1 + (a_{12}-a_{22})y_2)\,x \\ & + (a_{21}-a_{23})y_1 + (a_{22}-a_{23})y_2 + a_{23};\end{aligned}$$

and

$$Gr_1 = 0\times Y_{\overline{12}}^{<} \cup 1\times Y_{\overline{12}}^{>} \cup [0,1]\times Y_{\overline{12}}^{=}$$

where

$$\begin{aligned}Y_{\overline{12}}^{<} &= \{\mathbf{y}\in\triangle_= : (a_{11}-a_{21})y_1 + (a_{12}-a_{22})y_2 < 0\},\\ Y_{\overline{12}}^{>} &= \{\mathbf{y}\in\triangle_= : (a_{11}-a_{21})y_1 + (a_{12}-a_{22})y_2 > 0\},\\ Y_{\overline{12}}^{=} &= \{\mathbf{y}\in\triangle_= : (a_{11}-a_{21})y_1 + (a_{12}-a_{22})y_2 = 0\}.\end{aligned}$$

Similarly, the remained part of the proposition may be proved in the other two sub-cases. □

Evidently, Propositions 4.3–4.5 elucidate the case when one player has a dominant strategy (strategies) and the other player has equivalent strategies.

4.2.4 One Player Has Equivalent Strategies

In this section we investigate the cases when one of the players has equivalent strategies.

Proposition 4.6 *If the first player has equivalent strategies, then*

$$NE(\Gamma_m') = Gr_2,$$

$$
\begin{aligned}
Gr_2 = {} & X_1 \times (1,0) \cup X_2 \times (0,1) \cup X_3 \times (0,0) \\
& \cup X_{12} \times \Delta_= \cup X_{13} \times [0,1] \times 0 \cup X_{23} \times 0 \times [0,1] \\
& \cup X_{123} \times \Delta,
\end{aligned}
$$

where

$$
X_1 = \left\{ x \in [0,1] : \begin{array}{l} (b_{11} - b_{21})x + b_{21} > (b_{12} - b_{22})x + b_{22} \\ (b_{11} - b_{21})x + b_{21} > (b_{13} - b_{23})x + b_{23} \end{array} \right\},
$$

$$
X_2 = \left\{ x \in [0,1] : \begin{array}{l} (b_{12} - b_{22})x + b_{22} > (b_{11} - b_{21})x + b_{21} \\ (b_{12} - b_{22})x + b_{22} > (b_{13} - b_{23})x + b_{23} \end{array} \right\},
$$

$$
X_3 = \left\{ x \in [0,1] : \begin{array}{l} (b_{13} - b_{23})x + b_{23} > (b_{11} - b_{21})x + b_{21} \\ (b_{13} - b_{23})x + b_{23} > (b_{12} - b_{22})x + b_{22} \end{array} \right\},
$$

$$
X_{12} = \left\{ x \in [0,1] : \begin{array}{l} (b_{11} - b_{21})x + b_{21} = (b_{12} - b_{22})x + b_{22} \\ (b_{11} - b_{21})x + b_{21} > (b_{13} - b_{23})x + b_{23} \end{array} \right\},
$$

$$
X_{13} = \left\{ x \in [0,1] : \begin{array}{l} (b_{11} - b_{21})x + b_{21} > (b_{12} - b_{22})x + b_{22} \\ (b_{11} - b_{21})x + b_{21} = (b_{13} - b_{23})x + b_{23} \end{array} \right\},
$$

$$
X_{23} = \left\{ x \in [0,1] : \begin{array}{l} (b_{12} - b_{22})x + b_{22} > (b_{11} - b_{21})x + b_{21} \\ (b_{12} - b_{22})x + b_{22} = (b_{13} - b_{23})x + b_{23} \end{array} \right\},
$$

$$
X_{123} = \left\{ x \in [0,1] : \begin{array}{l} (b_{11} - b_{21})x + b_{21} = (b_{12} - b_{22})x + b_{22} \\ (b_{11} - b_{21})x + b_{21} = (b_{13} - b_{23})x + b_{23} \end{array} \right\}.
$$

Proof If the strategies of the first player are equivalent, then $Gr_1 = \Pi$.

Assume that $x \in [0,1]$ is fixed. The payoff function of the second player can be represented in the form

$$
\begin{aligned}
\varphi_2(x, \mathbf{y}) = {} & ((b_{11} - b_{21})x + b_{21})y_1 + ((b_{12} - b_{22})x + b_{22})y_2 \\
& + ((b_{13} - b_{23})x + b_{23})(1 - y_1 - y_2).
\end{aligned}
$$

It's evident that for

$x \in X_1$ the minimum of the cost function is realized on $(1,0) \in \Delta$,
$x \in X_2$ the minimum is realized on $(0,1) \in \Delta$,
$x \in X_3$ the minimum is realized on $(0,0) \in \Delta$,

$x \in X_{12}$ the minimum is realized on $\Delta_=$,
$x \in X_{13}$ the minimum is realized on $[0, 1] \times 0 \in \Delta$,
$x \in X_{23}$ the minimum is realized on $0 \times [0, 1] \in \Delta$,
$x \in X_{123}$ the minimum is realized on Δ.

The truth of the proposition follows from the above. □

Proposition 4.7 *If all three strategies of the second player are equivalent, then*

$$NE(\Gamma'_m) = Gr_1 = 1 \times Y_1 \cup 0 \times Y_2 \cup [0, 1] \times Y_{12},$$

where

$$\begin{aligned} Y_1 &= \{\mathbf{y} \in \Delta : \alpha_1 y_1 + \alpha_2 y_2 + \alpha_3 > 0\}, \\ Y_2 &= \{\mathbf{y} \in \Delta : \alpha_1 y_1 + \alpha_2 y_2 + \alpha_3 < 0\}, \\ Y_{12} &= \{\mathbf{y} \in \Delta : \alpha_1 y_1 + \alpha_2 y_2 + \alpha_3 = 0\}, \\ \alpha_1 &= a_{11} - a_{13} + a_{23} - a_{21}, \\ \alpha_2 &= a_{12} - a_{13} + a_{23} - a_{22}, \\ \alpha_3 &= a_{13} - a_{23}. \end{aligned}$$

Proof If all three strategies of the second player are equivalent, then

$$\begin{aligned} b_{11} &= b_{12} = b_{13}, \\ b_{21} &= b_{22} = b_{23}, \end{aligned}$$

and $Gr_2 = \Pi$.

The payoff function of the first player may be represented in the following form

$$\begin{aligned} \varphi_1(x, \mathbf{y}) = {} & ((a_{11} - a_{13})y_1 + (a_{12} - a_{13})y_2 + a_{13})x \\ & + ((a_{21} - a_{23})y_1 + (a_{22} - a_{23})y_2 + a_{23})(1 - x). \end{aligned}$$

It's evident that for

$$\begin{aligned} \mathbf{y} \in Y_1 &= \left\{ \mathbf{y} \in \Delta : \begin{array}{l} a_{11}y_1 + a_{12}y_2 + a_{13}(1 - y_1 - y_2) > \\ > a_{21}y_1 + a_{22}y_2 + a_{23}(1 - y_1 - y_2) \end{array} \right\} \\ &= \left\{ \mathbf{y} \in \Delta : \begin{array}{l} (a_{11} - a_{13} + a_{23} - a_{21})y_1 + \\ +(a_{12} - a_{13} + a_{23} - a_{22})y_2 + a_{13} - a_{23} > 0 \end{array} \right\} \end{aligned}$$

the first strategy of the first player is optimal. For

$$\mathbf{y} \in Y_2 = \left\{ \mathbf{y} \in \triangle : \begin{array}{l} a_{11}y_1 + a_{12}y_2 + a_{13}(1 - y_1 - y_2) < \\ < a_{21}y_1 + a_{22}y_2 + a_{23}(1 - y_1 - y_2) \end{array} \right\}$$

$$= \left\{ \mathbf{y} \in \triangle : \begin{array}{l} (a_{11} - a_{13} + a_{23} - a_{21})y_1 + \\ +(a_{12} - a_{13} + a_{23} - a_{22})y_2 + a_{13} - a_{23} < 0 \end{array} \right\}$$

the second strategy of the first player is optimal, and for

$$\mathbf{y} \in Y_{12} = \left\{ \mathbf{y} \in \triangle : \begin{array}{l} a_{11}y_1 + a_{12}y_2 + a_{13}(1 - y_1 - y_2) = \\ = a_{21}y_1 + a_{22}y_2 + a_{23}(1 - y_1 - y_2) \end{array} \right\}$$

$$= \left\{ \mathbf{y} \in \triangle : \begin{array}{l} (a_{11} - a_{13} + a_{23} - a_{21})y_1 + \\ +(a_{12} - a_{13} + a_{23} - a_{22})y_2 + a_{13} - a_{23} = 0 \end{array} \right\}$$

every strategy $x \in [0, 1]$ of the first player is optimal.

From the last relations, the truth of the proposition follows. □

4.2.5 Players Don't Have Dominant Strategies

Let us elucidate the case when neither the first player, neither the second, doesn't have dominant strategies. In such case it remains to apply the intersection method of best response mapping graphs.

Proposition 4.8 *If both players don't have dominant strategies, then*

$$NE(\Gamma'_m) = Gr_1 \cap Gr_2,$$

where Gr_1, Gr_2, are defined as in introduction, i.e. the graphs of best response mappings:

$$Gr_1 = \left\{ (x, \mathbf{y}) \in \Pi : \begin{array}{l} \mathbf{y} \in \triangle \\ x \in \operatorname{Arg} \max\limits_{x \in [0,1]} \varphi_1(x, \mathbf{y}) \end{array} \right\},$$

$$Gr_2 = \left\{ (x, \mathbf{y}) \in \Pi : \begin{array}{l} \mathbf{x} \in [0, 1] \\ \mathbf{y} \in \operatorname{Arg} \max\limits_{\mathbf{y} \in \Pi} \varphi_2(x, \mathbf{y}) \end{array} \right\}.$$

The proposition truth follows from the above.

4.3 Algorithm for Constructing the Set of Nash Equilibria

Based on obtained results, a simple solving procedure can be constructed.

It's important to observe that in the following algorithm only one item from **1°** to **5°** is executed when a concrete problem is solved.

Remark additionally that the item number combined with letters has the aim to highlight related cases.

Algorithm 4.3.1

- **0°** The game Γ'_m is considered (see Sect. 4.2.1);
- **1°** If both the players have equivalent strategies in Γ, then the Nash equilibrium set in Γ_m is $X \times Y$ (see Proposition 4.1);
- **2°** If both the players have dominant strategies in Γ, then the Nash equilibrium set in Γ_m is constructed in compliance with Proposition 4.2 and substitutions of Sect. 4.2.1;
- **3A°** If only the first player has dominant strategy in Γ, then the Nash equilibrium set in Γ_m is constructed in conformity with Proposition 4.3 and substitutions of Sect. 4.2.1;
- **3B°** If only the second player has only one dominant strategy in Γ, then the Nash equilibrium set in Γ_m is constructed in conformity with Proposition 4.4 and substitutions of Sect. 4.2.1;
- **3C°** If the second player has two dominant strategies that dominate the other strategy in Γ, then the Nash equilibrium set in Γ_m is constructed in conformity with Proposition 4.5 and substitutions of Sect. 4.2.1;
- **4A°** If only the first player has equivalent strategies in Γ, then the Nash equilibrium set in Γ_m is constructed in accordance with Proposition 4.6 and substitutions of Sect. 4.2.1;
- **4B°** If only the second player has equivalent strategies in Γ, then the Nash equilibrium set in Γ_m is constructed in accordance with Proposition 4.7 and substitutions of Sect. 4.2.1;
- **5°** If both the players don't have dominant strategies in Γ, then the Nash equilibrium set in Γ_m is constructed in compliance with Proposition 4.8 and substitutions of Sect. 4.2.1.

4.4 Conclusions

The results presented in this section have the aim to verify practically and simplify the method of the graph intersection in the case of particular 2×3 games. In this context considered games are important because they admit a graphic representation of Nash equilibrium sets in Cartesian coordinates and the general method may be substantially simplified.

Algorithm 4.3.1 is realised in the Wolfram language. The code is published on the Wolfram Demonstrations Project [11]. It may be freely viewed, verified and downloaded from the address [11].

References

1. Raghavan, T.E.S. 2002. Non-zero sum two-person games. In *Handbook of Game Theory with Economic Applications*, ed. R.J. Aumann, and S. Hart, 3, 1687–1721. North-Holland: Elsevier Science B.V.
2. Von Stengel, B. 2002. Computing equilibria for two-person games. In *Handbook of Game Theory with Economic Applications*, ed. R.J. Aumann, and S. Hart, 1723–1759. Elsevier Science B.V: North-Holland.
3. Avis, D., G.D. Rosenberg, R. Savani, and B. Von Stenghel. 2010. Enumeration of Nash equilibria for two player games. *Economic Theory* 42: 9–37.
4. Datta, R.S. 2010. Finding all Nash equilibria of a finite game using polynomial algebra. *Economic Theory* 42: 55–96.
5. Sagaidac, M., and V. Ungureanu, 2004. Operational research, Chişinău: CEP USM, 296 pp. (in Romanian).
6. Ungureanu, V., and A. Botnari. 2005. Nash equilibrium sets in mixed extension of 2 × 2 × 2 games. *Computer Science Journal of Moldova* 13 (1 (37)): 13–28.
7. Ungureanu, V., and A. Botnari. 2005. Nash equilibrium sets in mixed extended 2 × 3 games. *Computer Science Journal of Moldova* 13 (2 (38)): 136–150.
8. Rapoport, A., and A.M. Chammah. 1965. *Prisoner's Dilemma: A Study in Conflict and Cooperation*, USA: University of Michigan Press, XII+258 pp.
9. Poundstone, W. 1992. *Prisoner's Dilemma*, New York: Anchor Books, XI+294 pp.
10. Campbell, R., and L. Sowden (eds.). 1985. *Paradoxes of Rationality and Cooperation: Prisoner's Dilemma and Newcomb's Problem*, Vancouver: The University of British Columbia Press, X+366 pp.
11. Ungureanu, V., and I. Mandric. Set of Nash Equilibria in 2 × 3 Mixed Extended Games, from the Wolfram Demonstrations Project, Published: April 2010. http://demonstrations.wolfram.com/SetOfNashEquilibriaIn2x3MixedExtendedGames/

Chapter 5
Nash Equilibrium Sets in Dyadic Trimatrix Mixed-Strategy Games

Abstract The approach exposed in Chap. 3 is applied to solve three-matrix $2 \times 2 \times 2$ mixed-strategy games. Such games, alike the games considered in the precedent chapter and the result presented in the following chapter, admit a simple graphic representation of Nash equilibrium sets in a Cartesian system of coordinates. The results of this chapter were successfully applied by Y.M. Zhukov to investigate *"An Epidemic Model of Violence and Public Support in Civil War"* in the Department of Government at Harvard University (Zhukov, Conflict Management and Peace Science, 30(1):24–52, 2013, [1]). Evidently, a full graphical representation of the Nash equilibrium sets in the Cartesian system of coordinates is not possible in the case of four and more players. We can only suppose the possibility of such representation for the four-matrix $2 \times 2 \times 2 \times 2$ mixed-strategy games in the Barycentric coordinate system or other coordinate systems. Sure, in the context of results presented in Chap. 6 for dyadic two-matrix games, an important problem for dyadic three-matrix games is that of defining explicitly a Nash equilibrium set function as a piecewise set-valued function. In this chapter, we solve the problem of NES function definition algorithmically as the number of components/pieces of the NES piecewise set-valued function is substantially larger in the case of dyadic three-matrix mixed-strategy games because of more pieces/components of the piecewise set-valued function that defines best response mapping graph of each player.

5.1 Introduction

Let us recall that the problem of Nash equilibrium set computing is enjoying an increased attention of researchers and there are diverse explanations of this fact. Indubitable, one of the main reason is that of obtaining important theoretic results, even though the complexity of this problem [2] involves difficulties. But, there are also practical reasons to solve particular games.

So, special/peculiar cases of the problem have been studied exhaustively, e.g. dyadic games [3, 4], i.e. games in which every player has only two pure strategies. We can refer some instances of real world dyadic games that have been solved analytically [3–5].

V. Ungureanu, *Pareto-Nash-Stackelberg Game and Control Theory*,
Smart Innovation, Systems and Technologies 89,
https://doi.org/10.1007/978-3-319-75151-1_5

Let us consider the noncooperative game

$$\Gamma = \langle N, \{X_i\}_{i \in N}, \{f_i(x)\}_{i \in N} \rangle,$$

where $N = \{1, 2, \ldots, n\}$ is a set of players, X_i is a set of strategies of player $i \in N$ and $f_i : X \rightarrow R$ is a player's $i \in N$ payoff function defined on the Cartesian product $X = \times_{i \in N} X_i$.

Let us recall that the essence of the applied method is highlighted by the following theorem.

Theorem 5.1 *The outcome* $x^* \in X$ *is a Nash equilibrium if and only if* $x^* \in \bigcap_{i \in N} Gr_i$.

The proof follows from the definition of the Nash equilibrium and was especially analysed in Chap. 3.

Corollary 5.1 *The set of Nash equilibria is*

$$NES(\Gamma) = \bigcap_{i \in N} Gr_i.$$

If all the strategy sets $X_i, i \in N$, are finite, then a mixed extension of Γ is

$$\Gamma_m = \langle M_i, f_i^*(\mu), i \in N \rangle,$$

where

$$f_i^*(\mu) = \sum_{x \in X} f_i(x) \mu_1(x_1) \mu_2(x_2) \ldots \mu_n(x_n),$$

$$\mu = (\mu_1, \mu_2, \ldots, \mu_n) \in M = \times_{i \in N} M_i,$$

and M_i is a set of mixed strategies of the player $i \in N$.

Theorem 5.2 *If* X *is a finite set, then the set* $NES(\Gamma_m)$ *is a non-empty compact subset of the set* M*. Moreover, it contains the set of Nash equilibria* $NES(\Gamma)$ *of the pure-strategy game:*

$$NES(\Gamma) \subset NES(\Gamma_m) \neq \emptyset.$$

Proof The theorem is an adaptation of the Nash Theorem [6] to suit conditions of the considered particular game. □

One of the simplest solvable problems of the Nash equilibrium set computing is that in the case of dyadic mixed-strategy two-person 2 × 2 games [2–4, 7, 8] considered in this chapter. The explicit exposition of the NES function as a piecewise set-valued function for that games may be seen as an example or goal to achieve in the future for other types of mixed-strategy games.

Nevertheless, in this chapter we provide a class partition for the all three-person $2 \times 2 \times 2$ mixed-strategy games and the Nash equilibrium set is determined for mixed extensions of each class games. Actually, we construct in an algorithmic manner the NES function for a simplified mixed-strategy game, but we do not give a simple piecewise exposition for it as in Chap. 5.

5.2 Main Results

Consider a dyadic three-matrix game

$$\Gamma = \langle \{1, 2, 3\};\ I, J, K;\ a_{ijk}, b_{ijk}, c_{ijk} \rangle$$

with matrices

$$A = (a_{ijk}),\ B = (b_{ijk}),\ C = (c_{ijk}),\ i \in I,\ j \in J,\ k \in K,$$

that define the payoff functions $a_{ijk}, b_{ijk}, c_{ijk}$, of indices i, j, k, the set of players $N = \{1, 2, 3, \}$ and the set of strategies $I = J = K = \{1, 2\}$.

We need to recall that a dyadic game is a game in which every player has two pure strategies.

A mixed extension of the game Γ is the game

$$\Gamma_m = \langle \{1, 2, 3\};\ X, Y, Z;\ f_1(\mathbf{x}, \mathbf{y}, \mathbf{z}),\ f_2(\mathbf{x}, \mathbf{y}, \mathbf{z}),\ f_3(\mathbf{x}, \mathbf{y}, \mathbf{z}) \rangle,$$

where

$$X = \left\{ \mathbf{x} = (x_1, x_2) \in \mathbb{R}^2 : \begin{array}{c} x_1 + x_2 = 1 \\ x_1 \geq 0, x_2 \geq 0 \end{array} \right\},$$

$$Y = \left\{ \mathbf{y} = (y_1, y_2) \in \mathbb{R}^2 : \begin{array}{c} y_1 + y_2 = 1 \\ y_1 \geq 0, y_2 \geq 0 \end{array} \right\},$$

$$Z = \left\{ \mathbf{z} = (z_1, z_2) \in \mathbb{R}^2 : \begin{array}{c} z_1 + z_2 = 1 \\ z_1 \geq 0, z_2 \geq 0 \end{array} \right\},$$

$$f_1(\mathbf{x}, \mathbf{y}, \mathbf{z}) = \sum_{i=1}^{2} \sum_{j=1}^{2} \sum_{k=1}^{2} a_{ijk} x_i y_j z_k,$$

$$f_2(\mathbf{x}, \mathbf{y}, \mathbf{z}) = \sum_{i=1}^{2} \sum_{j=1}^{2} \sum_{k=1}^{2} b_{ijk} x_i y_j z_k,$$

$$f_3(\mathbf{x}, \mathbf{y}, \mathbf{z}) = \sum_{i=1}^{2} \sum_{j=1}^{2} \sum_{k=1}^{2} c_{ijk} x_i y_j z_k.$$

In the following section, we provide a reduction of this game to an equivalent game on a unit cube.

5.2.1 Games on a Unit Cube

By substitutions

$$\begin{aligned}
&x_1 = x,\ x_2 = 1 - x,\ x \in [0, 1];\\
&y_1 = y,\ y_2 = 1 - y,\ y \in [0, 1];\\
&z_1 = z,\ z_2 = 1 - z,\ z \in [0, 1],
\end{aligned}$$

the game Γ_m is reduced to an equivalent normal form game

$$\Gamma_m' = \langle \{1, 2, 3\}; [0, 1], [0, 1], [0, 1]; \varphi_1(x, y, z), \varphi_2(x, y, z), \varphi_3(x, y, z) \rangle,$$

where

$$\begin{aligned}
&\varphi_1(x, y, z) =\\
&= ((a_{111} - a_{211})yz + (a_{112} - a_{212})y(1 - z) + (a_{121} - a_{221})(1 - y)z\\
&+ (a_{122} - a_{222})(1 - y)(1 - z))x\\
&+ ((a_{211} - a_{221})z + (a_{212} - a_{222})(1 - z))y + (a_{221} - a_{222})z + a_{222};
\end{aligned}$$

$$\begin{aligned}
&\varphi_2(x, y, z) =\\
&= ((b_{111} - b_{121})xz + (b_{112} - b_{122})x(1 - z) + (b_{211} - b_{221})(1 - x)z\\
&+ (b_{212} - b_{222})(1 - x)(1 - z))y\\
&+ ((b_{121} - b_{221})z + (b_{122} - b_{222})(1 - z))x + (b_{221} - b_{222})z + b_{222};
\end{aligned}$$

$$\begin{aligned}
&\varphi_3(x, y, z) =\\
&= ((c_{111} - c_{112})xy + (c_{121} - c_{122})x(1 - y) + (c_{211} - c_{212})(1 - x)y\\
&+ (c_{221} - c_{222})(1 - x)(1 - y))z\\
&+ ((c_{112} - c_{212})y + (c_{122} - c_{222})(1 - y))x + (c_{212} - c_{222})y + c_{222}.
\end{aligned}$$

Thus, the game Γ_m and Γ_m' are equivalent, i.e. if the Nash equilibrium set $NES(\Gamma_m')$ is known, then it is easy to construct the Nash equilibrium set $NES(\Gamma_m)$.

Relying on strategy properties of each player of the initial pure strategies game Γ, diverse classes of games are considered and for any class the Nash equilibrium set $NES(\Gamma_m')$ is determined.

5.2.2 All Players Have Either Equivalent or Dominant Strategies

Proposition 5.1 *If all the players have equivalent strategies, then*

$$NES(\Gamma'_m) = [0, 1]^3.$$

Proof It is sufficient to recall the definition of a Nash equilibrium. □

Remark 5.1 In the case considered in Proposition 5.1, players have the following linear payoff functions

$$\varphi_1(x, y, z) = ((a_{211} - a_{221})z + (a_{212} - a_{222})(1 - z))y + (a_{221} - a_{222})z + a_{222},$$

$$\varphi_2(x, y, z) = ((b_{121} - b_{221})z + (b_{122} - b_{222})(1 - z))x + (b_{221} - b_{222})z + b_{222},$$

$$\varphi_3(x, y, z) = ((c_{112} - c_{212})y + (c_{122} - c_{222})(1 - y))x + (c_{212} - c_{222})y + c_{222}.$$

Every player doesn't influence on his payoff function, but his strategy is essential for payoff values of the rest of the players.

Proposition 5.2 *If all the players have dominant strategies in the game Γ, then the set of Nash equilibria $NES(\Gamma'_m)$ contains only one point:*

$$NES(\Gamma'_m) = \begin{cases} (0, 0, 0) & \text{if strategies (2,2,2) are dominant;} \\ (0, 0, 1) & \text{if strategies (2,2,1) are dominant;} \\ (0, 1, 0) & \text{if strategies (2,1,2) are dominant;} \\ (0, 1, 1) & \text{if strategies (2,1,1) are dominant;} \\ (1, 0, 0) & \text{if strategies (1,2,2) are dominant;} \\ (1, 0, 1) & \text{if strategies (1,2,1) are dominant;} \\ (1, 1, 0) & \text{if strategies (1,1,2) are dominant;} \\ (1, 1, 1) & \text{if strategies (1,1,1) are dominant.} \end{cases}$$

Proof It is easy to observe that graphs coincide with facets of the unite cube.

For the first player

$$\operatorname{Arg} \max_{x \in [0,1]} \varphi_1(x, y, z) = \begin{cases} \{1\} & \text{if the 1st strategy is dominant in } \Gamma, \\ \{0\} & \text{if the 2nd strategy is dominant in } \Gamma, \end{cases}$$

for all $(y, z) \in [0, 1]^2$. Hence,

$$Gr_1 = \begin{cases} 1 \times [0, 1] \times [0, 1] & \text{if the 1st strategy is dominant,} \\ 0 \times [0, 1] \times [0, 1] & \text{if the 2nd strategy is dominant.} \end{cases}$$

For the second player

$$\operatorname{Arg}\max_{y\in[0,1]} \varphi_2(x,y,z) = \begin{cases} \{1\} & \text{if the 1st strategy is dominant in } \Gamma, \\ \{0\} & \text{if the 2nd strategy is dominant in } \Gamma, \end{cases}$$

for all $(x, z) \in [0, 1]^2$. So,

$$Gr_2 = \begin{cases} [0,1] \times 1 \times [0,1] & \text{if the 1st strategy is dominant,} \\ [0,1] \times 0 \times [0,1] & \text{if the 2nd strategy is dominant.} \end{cases}$$

For the third player

$$\operatorname{Arg}\max_{z\in[0,1]} \varphi_3(x,y,z) = \begin{cases} \{1\} & \text{if the 1st strategy is dominant in } \Gamma, \\ \{0\} & \text{if the 2nd strategy is dominant in } \Gamma, \end{cases}$$

for all $(x, y) \in [0, 1]^2$. Hence,

$$Gr_3 = \begin{cases} [0,1] \times [0,1] \times 1 & \text{if the 1st strategy is dominant,} \\ [0,1] \times [0,1] \times 0 & \text{if the 2nd strategy is dominant.} \end{cases}$$

Consequently, the Nash equilibrium set contains only one vertex of the unit cube. □

5.2.3 *Two Players Have Dominant or Equivalent Strategies*

Proposition 5.3 *If the first and the second players have dominant strategies and the third player has incomparable strategies, then*

$$NES(\Gamma'_m) = \begin{cases} (1,1,0) & \text{if } (1,1,\cdot) \text{ are dominant and } c_{111} < c_{112}, \\ (1,1,1) & \text{if } (1,1,\cdot) \text{ are dominant and } c_{111} > c_{112}, \\ 1 \times 1 \times [0,1] & \text{if } (1,1,\cdot) \text{ are dominant and } c_{111} = c_{112}, \\ (0,0,0) & \text{if } (2,2,\cdot) \text{ are dominant and } c_{221} < c_{222}, \\ (0,0,1) & \text{if } (2,2,\cdot) \text{ are dominant and } c_{221} > c_{222}, \\ 0 \times 0 \times [0,1] & \text{if } (2,2,\cdot) \text{ are dominant and } c_{221} = c_{222}, \\ (1,0,0) & \text{if } (1,2,\cdot) \text{ are dominant and } c_{121} < c_{122}, \\ (1,0,1) & \text{if } (1,2,\cdot) \text{ are dominant and } c_{121} > c_{122}, \\ 1 \times 0 \times [0,1] & \text{if } (1,2,\cdot) \text{ are dominant and } c_{121} = c_{122}, \\ (0,1,0) & \text{if } (2,1,\cdot) \text{ are dominant and } c_{211} < c_{212}, \\ (0,1,1) & \text{if } (2,1,\cdot) \text{ are dominant and } c_{211} > c_{212}, \\ 0 \times 1 \times [0,1] & \text{if } (2,1,\cdot) \text{ are dominant and } c_{211} = c_{212}. \end{cases}$$

Proof It is enough to recall the definitions of the involved concepts. □

Similarly, the Nash equilibrium set can be constructed in two other possible cases:

the first and the third players have dominant strategies, and the second player has incomparable strategies;

the second and the third players have dominant strategies, and the first player has incomparable strategies.

So, the Nash equilibrium set is either one vertex or one edge of a unit cube.

Proposition 5.4 *If the first and the second players have dominant strategies and the third one has equivalent strategies, then*

$$NES(\Gamma'_m) = \begin{cases} 1 \times 1 \times [0, 1] \text{ if } (1,1,\cdot) \text{ are dominant,} \\ 0 \times 0 \times [0, 1] \text{ if } (2,2,\cdot) \text{ are dominant,} \\ 1 \times 0 \times [0, 1] \text{ if } (1,2,\cdot) \text{ are dominant,} \\ 0 \times 1 \times [0, 1] \text{ if } (2,1,\cdot) \text{ are dominant.} \end{cases}$$

Proof It is enough to recall the definitions of the concepts which are used in the proposition formulation. □

Similarly the Nash equilibrium set can be constructed in the following cases:

the first and the third players have dominant strategies, and the second player has equivalent strategies,

the second and the third players have dominant strategies, and the first player has equivalent strategies.

Thus, the Nash equilibrium set is an edge of the unit cube.

Proposition 5.5 *If the first and the second players have equivalent strategies, and the third player has dominant strategy, then*

$$NES(\Gamma'_m) = \begin{cases} [0, 1] \times [0, 1] \times 1 \text{ if the 1st strategy is dominant,} \\ [0, 1] \times [0, 1] \times 0 \text{ if the 2nd strategy is dominant.} \end{cases}$$

Proof It is enough to recall the definitions of the concepts which are involved in formulation of the proposition. □

Similarly, the Nash equilibrium set can be constructed in the following cases:

the first and the third players have equivalent strategies, and the second player has dominant strategy,

the second and the third players have equivalent strategies, and the first player has dominant strategy.

In such a way, the Nash equilibrium set is a facet of the unit cube.

5.2.4 Every Player Has Different Types of Strategies: Dominant, Equivalent, or Incomparable

Proposition 5.6 *If the first player has equivalent strategies, the second player has dominant strategy and the third player has incomparable strategies, then*

$$NES(\Gamma'_m) = Gr_3,$$

where in the case when the first strategy of the second player is dominant

$$Gr_3 = [0,1]^3 \cap \begin{cases} \begin{cases} [0; -\frac{\gamma_2}{\gamma_1}) \times 1 \times 0 \\ \cup -\frac{\gamma_2}{\gamma_1} \times 1 \times [0,1] & if\, \gamma_1 > 0, \\ \cup(-\frac{\gamma_2}{\gamma_1}; 1] \times 1 \times 1 \end{cases} \\ \begin{cases} [0; -\frac{\gamma_2}{\gamma_1}) \times 1 \times 1 \\ \cup -\frac{\gamma_2}{\gamma_1} \times 1 \times [0,1] & if\, \gamma_1 < 0, \\ \cup(-\frac{\gamma_2}{\gamma_1}; 1] \times 1 \times 0 \end{cases} \\ [0,1] \times 1 \times 0 & if\, \gamma_1 = 0,\ \gamma_2 < 0, \\ [0,1] \times 1 \times 1 & if\, \gamma_1 = 0,\ \gamma_2 > 0, \\ [0,1] \times 1 \times [0,1] & if\, \gamma_1 = \gamma_2 = 0, \end{cases}$$

and coefficients are calculates by formulas

$$\gamma_1 = c_{111} - c_{112} - c_{211} + c_{212},$$
$$\gamma_2 = c_{211} - c_{212},$$

and in the case when the second strategy of the second player is dominant

$$Gr_3 = [0,1]^3 \cap \begin{cases} \begin{cases} [0; -\frac{\gamma_6}{\gamma_5}) \times 0 \times 0 \\ \cup -\frac{\gamma_6}{\gamma_5} \times 0 \times [0,1] & if\, \gamma_5 > 0, \\ \cup(-\frac{\gamma_6}{\gamma_5}; 1] \times 0 \times 1 \end{cases} \\ \begin{cases} [0; -\frac{\gamma_6}{\gamma_5}) \times 0 \times 1 \\ \cup -\frac{\gamma_6}{\gamma_5} \times 0 \times [0,1] & if\, \gamma_5 < 0, \\ \cup(-\frac{\gamma_6}{\gamma_5}; 1] \times 0 \times 0 \end{cases} \\ [0,1] \times 0 \times 0 & if\, \gamma_5 = 0,\ \gamma_6 < 0, \\ [0,1] \times 0 \times 1 & if\, \gamma_5 = 0,\ \gamma_6 > 0, \\ [0,1] \times 0 \times [0,1] & if\, \gamma_5 = \gamma_6 = 0, \end{cases}$$

and coefficients are calculates by formulas

$$\gamma_5 = c_{121} - c_{122} - c_{221} + c_{222},$$
$$\gamma_6 = c_{221} - c_{222}.$$

Proof If the first strategy of the second player is dominant, then

$$\begin{aligned} \varphi_3(x, y, z) &= (x(c_{111} - c_{112}) + (1-x)(c_{211} - c_{212}))z \\ &\quad + (c_{112} - c_{212})x + c_{212} \\ &= (\gamma_1 x + \gamma_2)z + \gamma_3 x + \gamma_4, \end{aligned}$$

where

$$\gamma_3 = c_{112} - c_{212},$$
$$\gamma_4 = c_{212},$$

From this relation the truth of the first part of the proposition follows.

If the second strategy of the second player is dominant, then

$$\begin{aligned}\varphi_3(x, y, z) &= (x(c_{121} - c_{122}) + (1 - x)(c_{221} - c_{222}))z \\ &\quad + (c_{122} - c_{222})x + c_{222} \\ &= (\gamma_5 x + \gamma_6)z + \gamma_7 x + \gamma_8,\end{aligned}$$

where

$$\gamma_7 = c_{122} - c_{222},$$
$$\gamma_8 = c_{222}.$$

From this last expression, the truth of the second part of the proposition results. □

Similarly, the Nash equilibrium set can be constructed in the following cases:

the first player has equivalent strategies, the third player has dominant strategy, and the second player has incomparable strategies,

the second player has equivalent strategies, the first player has dominant strategy, and the third player has incomparable strategies,

the second player has equivalent strategies, the third player has dominant strategy, and the first player has incomparable strategies,

the third player has equivalent strategies, the first player has dominant strategy, and the second player has incomparable strategies,

the third player has equivalent strategies, the second player has dominant strategy, and the first player has incomparable strategies.

5.2.5 Two Players Have Either Incomparable or Equivalent Strategies

Proposition 5.7 *If the first and the second players have incomparable strategies and the third player has dominant strategy, then*

$$NES(\Gamma'_m) = Gr_1 \cap Gr_2,$$

where in the case when the first strategy of the third player is dominant

$$Gr_1 = [0,1]^3 \cap \begin{cases} \begin{cases} 0 \times [0; -\frac{\alpha_2}{\alpha_1}) \times 1 \\ \cup [0,1] \times -\frac{\alpha_2}{\alpha_1} \times 1 \\ \cup 1 \times (-\frac{\alpha_2}{\alpha_1}; 1] \times 1 \end{cases} & if \alpha_1 > 0, \\ \begin{cases} 1 \times [0; -\frac{\alpha_2}{\alpha_1}) \times 1 \\ \cup [0,1] \times -\frac{\alpha_2}{\alpha_1} \times 1 \\ \cup 0 \times (-\frac{\alpha_2}{\alpha_1}; 1] \times 1 \end{cases} & if \alpha_1 < 0, \\ 0 \times [0,1] \times 1 & if \alpha_1 = 0,\ \alpha_2 < 0, \\ 1 \times [0,1] \times 1 & if \alpha_1 = 0,\ \alpha_2 > 0, \\ [0,1] \times [0,1] \times 1 & if \alpha_1 = \alpha_2 = 0, \end{cases}$$

$$Gr_2 = [0,1]^3 \cap \begin{cases} \begin{cases} [0; -\frac{\beta_2}{\beta_1}) \times 0 \times 1 \\ \cup -\frac{\beta_2}{\beta_1} \times [0,1] \times 1 \\ \cup (-\frac{\beta_2}{\beta_1}; 1] \times 1 \times 1 \end{cases} & if \beta_1 > 0, \\ \begin{cases} [0; -\frac{\beta_2}{\beta_1}) \times 1 \times 1 \\ \cup -\frac{\beta_2}{\beta_1} \times [0,1] \times 1 \\ \cup (-\frac{\beta_2}{\beta_1}; 1] \times 0 \times 1 \end{cases} & if \beta_1 < 0, \\ [0,1] \times 0 \times 1 & if \beta_1 = 0,\ \beta_2 < 0, \\ [0,1] \times 1 \times 1 & if \beta_1 = 0,\ \beta_2 > 0, \\ [0,1] \times [0,1] \times 1 & if \beta_1 = \beta_2 = 0, \end{cases}$$

and coefficients are calculates according to formulae

$$\begin{aligned} \alpha_1 &= a_{111} - a_{211} - a_{121} + a_{221}, \\ \alpha_2 &= a_{121} - a_{221}, \\ \beta_1 &= b_{111} - b_{121} - b_{211} + b_{221}, \\ \beta_2 &= b_{211} - b_{221}, \end{aligned}$$

and in the case when the second strategy of the third player is dominant

$$Gr_1 = [0,1]^3 \cap \begin{cases} \begin{cases} 0 \times [0; -\frac{\alpha_6}{\alpha_5}) \times 0 \\ \cup [0,1] \times -\frac{\alpha_6}{\alpha_5} \times 0 \\ \cup 1 \times (-\frac{\alpha_6}{\alpha_5}; 1] \times 0 \end{cases} & if \alpha_5 > 0, \\ \begin{cases} 1 \times [0; -\frac{\alpha_6}{\alpha_5}) \times 0 \\ \cup [0,1] \times -\frac{\alpha_6}{\alpha_5} \times 0 \\ \cup 0 \times (-\frac{\alpha_6}{\alpha_5}; 1] \times 0 \end{cases} & if \alpha_5 < 0, \\ 0 \times [0,1] \times 1 & if \alpha_5 = 0,\ \alpha_6 < 0, \\ 1 \times [0,1] \times 0 & if \alpha_5 = 0,\ \alpha_6 > 0, \\ [0,1] \times [0,1] \times 0 & if \alpha_5 = \alpha_6 = 0, \end{cases}$$

$$Gr_2 = [0,1]^3 \cap \begin{cases} \begin{cases} [0; -\frac{\beta_6}{\beta_5}) \times 0 \times 0 \\ \cup -\frac{\beta_6}{\beta_5} \times [0,1] \times 0 & if\, \beta_5 > 0, \\ \cup(-\frac{\beta_6}{\beta_5}; 1] \times 1 \times 0 \end{cases} \\ \begin{cases} [0; -\frac{\beta_6}{\beta_5}) \times 1 \times 0 \\ \cup -\frac{\beta_6}{\beta_5} \times [0,1] \times 0 & if\, \beta_5 < 0, \\ \cup(-\frac{\beta_6}{\beta_5}; 1] \times 0 \times 0 \end{cases} \\ [0,1] \times 0 \times 0 & if\, \beta_5 = 0,\ \beta_6 < 0, \\ [0,1] \times 1 \times 0 & if\, \beta_5 = 0,\ \beta_6 > 0, \\ [0,1] \times [0,1] \times 0 & if\, \beta_5 = \beta_6 = 0, \end{cases}$$

and coefficients are calculated according to formulae

$$\begin{aligned} \alpha_5 &= a_{112} - a_{212} - a_{122} + a_{222}, \\ \alpha_6 &= a_{122} - a_{222}, \\ \beta_5 &= b_{112} - b_{122} - b_{212} + b_{222}, \\ \beta_6 &= b_{212} - b_{222}, \end{aligned}$$

Proof If the first strategy of the third player is dominant, then

$$\begin{aligned} \varphi_1(x, y, z) &= (y(a_{111} - a_{211}) + (1 - y)(a_{121} - a_{221}))x \\ &\quad + (a_{211} - a_{221})y + a_{221} \\ &= (\alpha_1 y + \alpha_2)x + \alpha_3 y + \alpha_4, \end{aligned}$$

where α_1, α_2, are defined above,

$$\begin{aligned} \alpha_3 &= a_{211} - a_{221}, \\ \alpha_4 &= a_{221}, \end{aligned}$$

and

$$\begin{aligned} \varphi_2(x, y, z) &= (x(b_{111} - b_{121}) + (1 - x)(b_{211} - b_{221}))y \\ &\quad + (b_{121} - b_{221})x + b_{221} \\ &= (\beta_1 x + \beta_2)y + \beta_3 x + \beta_4, \end{aligned}$$

where β_1, β_2, are defined above and

$$\begin{aligned} \beta_3 &= b_{121} - b_{221}, \\ \beta_4 &= b_{221} \end{aligned}$$

From the above, the truth of the proposition follows.

If the second strategy of the third player is dominant, then

$$\begin{aligned} \varphi_1(x, y, z) &= (y(a_{112} - a_{212}) + (1 - y)(a_{122} - a_{222}))x \\ &\quad + (a_{212} - a_{222})y + a_{222} \\ &= (\alpha_5 y + \alpha_6)x + \alpha_7 y + \alpha_8, \end{aligned}$$

where the expressions for α_5, α_6, are in the proposition formulation,

$$\alpha_7 = a_{212} - a_{222},$$
$$\alpha_8 = a_{222},$$

and

$$\begin{aligned}\varphi_2(x, y, z) &= (x(b_{112} - b_{122}) + (1 - x)(b_{212} - b_{222}))y \\ &\quad + (b_{122} - b_{222})x + b_{222} \\ &= (\beta_5 x + \beta_6)y + \beta_7 x + \beta_8,\end{aligned}$$

where β_5, β_6, are in formulation of the proposition,

$$\beta_7 = b_{122} - b_{222},$$
$$\beta_8 = b_{222}.$$

From these relations, the truth of the second part of the proposition results. □

Similarly, the Nash equilibrium set can be computing in the following cases:
the first and the third players incomparable strategies, the second player has dominant strategy,
the second and the third players have incomparable strategies, the first player has dominant strategy.

Proposition 5.8 *If the first and the second players have equivalent strategies and the third player has incomparable strategies, then*

$$NES(\Gamma'_m) = Gr_3,$$

where

$$Gr_3 = [0, 1]^3 \cap \{X_< \times Y_< \times 0 \cup X_= \times Y_= \times [0, 1] \cup X_> \times Y_> \times 1\},$$

$$X_< \times Y_< = \left\{ (x, y) : \begin{array}{l} x \in [0, 1] \\ y \in [0, 1] \\ \gamma_1 xy + \gamma_2 x + \gamma_3 y + \gamma_4 < 0 \end{array} \right\},$$

$$X_= \times Y_= = \left\{ (x, y) : \begin{array}{l} x \in [0, 1] \\ y \in [0, 1] \\ \gamma_1 xy + \gamma_2 x + \gamma_3 y + \gamma_4 = 0 \end{array} \right\},$$

$$X_> \times Y_> = \left\{ (x, y) : \begin{array}{l} x \in [0, 1], \\ y \in [0, 1], \\ \gamma_1 xy + \gamma_2 x + \gamma_3 y + \gamma_4 > 0 \end{array} \right\},$$

and coefficients are calculated by formulas

$$\begin{aligned}
\gamma_1 &= c_{111} - c_{112} - c_{121} + c_{122} - c_{211} + c_{212} + c_{221} - c_{222},\\
\gamma_2 &= c_{121} - c_{122} - c_{221} + c_{222},\\
\gamma_3 &= c_{211} - c_{212} - c_{221} + c_{222},\\
\gamma_4 &= c_{221} - c_{222}.
\end{aligned}$$

Proof The truth of the proposition results from the following representation of the cost function:

$$\begin{aligned}
\varphi_3(x, y, z) &= (xy(c_{111} - c_{112}) + x(1 - y)(c_{121} - c_{122})\\
&\quad + (1 - x)y(c_{211} - c_{212})\\
&\quad + (1 - x)(1 - y)(c_{221} - c_{222}))z\\
&\quad + (y(c_{112} - c_{212}) + (1 - y)(c_{122} - c_{222}))x\\
&\quad + (c_{212} - c_{222})y + c_{222}\\
&= (\gamma_1 xy + \gamma_2 x + \gamma_3 y + \gamma_4)\, z + \gamma_5 xy + \gamma_6 x + \gamma_7 y + \gamma_8,
\end{aligned}$$

where $\gamma_1, \gamma_2, \gamma_3, \gamma_4$, are in the proposition formulation and

$$\begin{aligned}
\gamma_5 &= c_{112} - c_{212} - c_{122} + c_{222},\\
\gamma_6 &= c_{122} - c_{222},\\
\gamma_7 &= c_{212} - c_{222},\\
\gamma_8 &= c_{222}.
\end{aligned}$$

□

Similarly the Nash equilibrium set can be constructed in the following cases:

the first and the third players have equivalent strategies, the second player has incomparable strategies,

the second and the third players have equivalent strategies, the first player has incomparable strategies.

Proposition 5.9 *If the first and the second players have incomparable strategies and the third player has equivalent strategies, then*

$$NES(\Gamma'_m) = Gr_1 \cap Gr_2,$$

where

$$Gr_1 = [0, 1]^3 \cap \{0 \times Y_< \times Z_< \cup [0, 1] \times Y_= \times Z_= \cup 1 \times Y_> \times Z_>\},$$

$$Gr_2 = [0, 1]^3 \cap \{X_< \times 0 \times Z_< \cup X_= \times [0, 1] \times Z_= \cup X_> \times 1 \times Z_>\},$$

$$Y_< \times Z_< = \left\{ (y, z) : \begin{array}{c} y \in [0, 1] \\ z \in [0, 1] \\ \alpha_1 yz + \alpha_2 y + \alpha_3 z + \alpha_4 < 0 \end{array} \right\},$$

$$Y_= \times Z_= = \left\{ (y, z) : \begin{array}{l} y \in [0, 1] \\ z \in [0, 1] \\ \alpha_1 yz + \alpha_2 y + \alpha_3 z + \alpha_4 = 0 \end{array} \right\},$$

$$Y_> \times Z_> = \left\{ (y, z) : \begin{array}{l} y \in [0, 1] \\ z \in [0, 1] \\ \alpha_1 yz + \alpha_2 y + \alpha_3 z + \alpha_4 > 0 \end{array} \right\},$$

$$X_< \times Z_< = \left\{ (x, z) : \begin{array}{l} x \in [0, 1] \\ z \in [0, 1] \\ \beta_1 xz + \beta_2 x + \beta_3 z + \beta_4 < 0 \end{array} \right\},$$

$$X_= \times Z_= = \left\{ (x, z) : \begin{array}{l} x \in [0, 1] \\ z \in [0, 1] \\ \beta_1 xz + \beta_2 x + \beta_3 z + \beta_4 = 0 \end{array} \right\},$$

$$X_> \times Z_> = \left\{ (x, z) : \begin{array}{l} x \in [0, 1] \\ z \in [0, 1] \\ \beta_1 xz + \beta_2 x + \beta_3 z + \beta_4 > 0 \end{array} \right\},$$

$$\begin{aligned} \alpha_1 &= a_{111} - a_{211} - a_{112} + a_{212} - a_{121} + a_{221} + a_{122} - a_{222}, \\ \alpha_2 &= a_{112} - a_{212} - a_{122} + a_{222}, \\ \alpha_3 &= a_{121} - a_{221} - a_{122} + a_{222}, \\ \alpha_4 &= a_{122} - a_{222}, \end{aligned}$$

$$\begin{aligned} \beta_1 &= b_{111} - b_{121} - b_{112} + b_{122} - b_{211} + b_{221} + b_{212} - b_{222}, \\ \beta_2 &= b_{112} - b_{122} - b_{212} + b_{222}, \\ \beta_3 &= b_{211} - b_{221} - b_{212} + b_{222}, \\ \beta_4 &= b_{212} - b_{222}. \end{aligned}$$

Proof The truth of the proposition results from the following representation of the payoff functions:

$$\begin{aligned} \varphi_1(x, y, z) &= (yz(a_{111} - a_{211}) + y(1 - z)(a_{112} - a_{212}) \\ &\quad + (1 - y)z(a_{121} - a_{221}) \\ &\quad + (1 - y)(1 - z)(a_{122} - a_{222}))x \\ &\quad + (z(a_{211} - a_{221}) + (1 - z)(a_{212} - a_{222}))y \\ &\quad + (a_{221} - a_{222})z + a_{222} \\ &= (\alpha_1 yz + \alpha_2 y + \alpha_3 z + \alpha_4)x + \alpha_5 yz + \alpha_6 y + \alpha_7 z + \alpha_8, \end{aligned}$$

$$\begin{aligned}\varphi_2(x, y, z) &= (xz(b_{111} - b_{121}) + x(1 - z)(b_{112} - b_{122}) \\ &\quad + (1 - x)z(b_{211} - b_{221}) \\ &\quad + (1 - x)(1 - z)(b_{212} - b_{222}))y \\ &\quad + (z(b_{121} - b_{221}) + (1 - z)(b_{122} - b_{222}))x \\ &\quad + (b_{221} - b_{222})z + b_{222} \\ &= (\beta_1 xz + \beta_2 x + \beta_3 z + \beta_4)y + \beta_5 xz + \beta_6 x + \beta_7 z + \beta_8,\end{aligned}$$

where $\alpha_1, \alpha_2, \alpha_3, \alpha_4, \beta_1, \beta_2, \beta_3, \beta_4,$ are defined in the proposition formulation, and

$$\begin{aligned}\alpha_5 &= a_{211} - a_{212} - a_{221} + a_{222}, \\ \alpha_6 &= a_{212} - a_{222}, \\ \alpha_7 &= a_{221} - a_{222}, \\ \alpha_8 &= a_{222}, \\ \beta_5 &= b_{121} - b_{221} - b_{122} + b_{222}, \\ \beta_6 &= b_{122} - b_{222}, \\ \beta_7 &= b_{221} - b_{222}, \\ \beta_8 &= b_{222},\end{aligned}$$

easily can be obtained from the above expressions. □

Similarly the Nash equilibrium set can be constructed in the following cases:

the first and the third players have incomparable strategies, the second player has equivalent strategies,

the second and the third players have incomparable strategies, the first player has equivalent strategies.

5.2.6 All Players Have Incomparable Strategies

Proposition 5.10 *If all players have incomparable strategies, then*

$$NES(\Gamma'_m) = Gr_1 \cap Gr_2 \cap Gr_3,$$

where

$$Gr_1 = [0, 1]^3 \cap \{0 \times Y_< \times Z_< \cup [0, 1] \times Y_= \times Z_= \cup 1 \times Y_> \times Z_>\},$$

$$Gr_2 = [0, 1]^3 \cap \{X_< \times 0 \times Z_< \cup X_= \times [0, 1] \times Z_= \cup X_> \times 1 \times Z_>\},$$

$$Gr_3 = [0, 1]^3 \cap \{X_< \times Y_< \times 0 \cup X_= \times Y_= \times [0, 1] \cup X_> \times Y_> \times 1\},$$

the components of the Gr_1, Gr_2, Gr_3 are defined as precedent propositions.

Proof The truth of proposition results from the representation of the payoff functions used in Propositions 5.8 and 5.9:

$$\begin{aligned}\varphi_1(x,y,z) &= (yz(a_{111}-a_{211})+y(1-z)(a_{112}-a_{212})\\ &+(1-y)z(a_{121}-a_{221})\\ &+(1-y)(1-z)(a_{122}-a_{222}))x\\ &+(z(a_{211}-a_{221})+(1-z)(a_{212}-a_{222}))y\\ &+(a_{221}-a_{222})z+a_{222}\\ &=(\alpha_1 yz+\alpha_2 y+\alpha_3 z+\alpha_4)x+\alpha_5 yz+\alpha_6 y+\alpha_7 z+\alpha_8,\end{aligned}$$

$$\begin{aligned}\varphi_2(x,y,z) &= (xz(b_{111}-b_{121})+x(1-z)(b_{112}-b_{122})\\ &+(1-x)z(b_{211}-b_{221})\\ &+(1-x)(1-z)(b_{212}-b_{222}))y\\ &+(z(b_{121}-b_{221})+(1-z)(b_{122}-b_{222}))x\\ &+(b_{221}-b_{222})z+b_{222}\\ &=(\beta_1 xz+\beta_2 x+\beta_3 z+\beta_4)y+\beta_5 xz+\beta_6 x+\beta_7 z+\beta_8,\end{aligned}$$

$$\begin{aligned}\varphi_3(x,y,z) &= (xy(c_{111}-c_{112})+x(1-y)(c_{121}-c_{122})\\ &+(1-x)y(c_{211}-c_{212})\\ &+(1-x)(1-y)(c_{221}-c_{222}))z\\ &+(y(c_{112}-c_{212})+(1-y)(c_{122}-c_{222}))x\\ &+(c_{212}-c_{222})y+c_{222}\\ &=(\gamma_1 xy+\gamma_2 x+\gamma_3 y+\gamma_4)z+\gamma_5 xy+\gamma_6 x+\gamma_7 y+\gamma_8,\end{aligned}$$

where the expressions for coefficients $\alpha_1, \alpha_2, \alpha_3, \alpha_4, \beta_1, \beta_2, \beta_3, \beta_4$, are the same as in Proposition 5.9, and the expressions for coefficients $\gamma_1, \gamma_2, \gamma_3, \gamma_4, \gamma_5, \gamma_6, \gamma_7, \gamma_8$, are the same as in Proposition 5.8. □

5.3 Algorithm

Based on obtained results, we can construct Algorithm 5.3.1. It is important to observe that only one of the items from **1°** to **5°** is executed by algorithm when a concrete problem is solved.

Remark additionally that as in case of Algorithm 4.3.1, from the precedent chapter, the item numbers combined with letters highlight related cases.

Algorithm 5.3.1

0° The game Γ'_m is considered (see Sect. 5.2.1);

1A° If all the players have equivalent strategies in the game Γ, then the Nash equilibrium set in the game Γ_m is $X \times Y \times Z$ (see Proposition 5.1);

1B° If all the players have dominant strategies in the game Γ, then the Nash equilibrium set in the game Γ_m consists of one point that can be determined in compliance with Proposition 5.2 and substitutions of Sect. 5.2.1;

2A° If two players from three have dominant strategies in the game Γ and the other has incomparable strategies, then the Nash equilibrium set in the game Γ_m consists either of one point or one edge that can be constructed in compliance with Proposition 5.3 and substitutions of Sect. 5.2.1;

2B° If two players from three have dominant strategies in the game Γ and the other has equivalent strategies, then the Nash equilibrium set in the game Γ_m consists of one edge that can be constructed in compliance with Proposition 5.4 and substitutions of Sect. 5.2.1;

2C° If two players from three have equivalent strategies in the game Γ and the other has a dominant strategy, then the Nash equilibrium set in the game Γ_m consists of one facet that can be constructed in compliance with Proposition 5.5 and substitutions of Sect. 5.2.1;

3° If one of the players has a dominant strategy in the game Γ, the other player has equivalent strategies, and the remaining player has incomparable strategies, then the Nash equilibrium set in the game Γ_m coincides with the graph of best responding mapping of the player with incomparable strategies, and can be constructed in conformity with Proposition 5.6 and substitutions of Sect. 5.2.1;

4A° If two players have incomparable strategies, and the other has a dominant strategy in the game Γ, then the Nash equilibrium set in the game Γ_m consists of the intersection of graphs of best response mappings of the players with incomparable strategies and can be constructed in conformity with Proposition 5.7 and substitutions of Sect. 5.2.1;

4B° If two players have equivalent strategies, and the other has incomparable strategies in the game Γ, then the Nash equilibrium set in the game Γ_m coincides with the graph of best responding mapping of the player with incomparable strategies and can be constructed in conformity with Proposition 5.8 and substitutions of Sect. 5.2.1;

4C° If two players have incomparable strategies, and the other has equivalent strategies in the game Γ, then the Nash equilibrium set in the game Γ_m consists of the intersection of graphs of best response mappings of the players with incomparable strategies and can be constructed in conformity with Proposition 5.9 and substitutions of Sect. 5.2.1;

5° If all the players have incomparable strategies in the game Γ, then the Nash equilibrium set in the game Γ_m consists of intersection of all graphs of best response mappings of the players and can be constructed in compliance with Proposition 5.10 and substitutions of Sect. 5.2.1.

5.4 Conclusions

The Nash equilibrium set can be described as the intersection of best response graphs.

The solution of the problem of the set of Nash equilibria construction in the mixed extension of the $2 \times 2 \times 2$ game illustrates that the Nash equilibrium set is not necessarily convex even in convex game. Moreover, the Nash equilibrium set is frequently disconnected. Thus, new conceptual methods *"which derive from the theory of semi-algebraic sets are required for finding all equilibria"* [2].

In this chapter we illustrate that the method constructed in the second chapter works well. It is especially efficient in the case of dyadic game that may be represented graphically in Cartesian coordinates. As in the case of 2×3 games, considered games are important because they admit a graphic representation of Nash equilibrium sets in Cartesian coordinate system.

Algorithm 5.3.1 is realised in the Wolfram language. It is published on the Wolfram Demonstrations Project [9]. The code may be freely viewed, verified and downloaded from the address [9].

References

1. Zhukov, Y.M. 2013. An epidemic model of violence and public support in civil war. *Conflict Management and Peace Science* 30 (1): 24–52.
2. McKelvey, R.D., and A. McLenan. 1996. Computation of equilibria in finite games. *Handbook of Computational Economics*, vol. 1, 87–142. Amsterdam: Elsevier.
3. Vorob'ev, N.N. 1984. *Foundations of Game Theory: Noncooperative Games*. Moscow: Nauka (in Russian); Trans. R.P. Boas, Basel-Boston: Birkhäuser, 1994, 497 pp.
4. Vorob'ev, N.N. 1985. *Game Theory: Lectures for Economists and Systems Scientists*. Moscow: Nauka (in Russian); Trans. and supplemented by S. Kotz, New York: Springer, 1985, 193 pp.
5. Datta, R.S. 2010. Finding all Nash equilibria of a finite game using polynomial algebra. *Economic Theory* 42: 55–96.
6. Nash, J. 1951. Noncooperative games. *Annals of Mathematics* 54 (2): 280–295.
7. Moulen, H. 1981. *Théorie des jeux pour l'éqonomie et la politique*. Paris (in Russian: Game Theory, Moscow: Mir, 1985), 200 pp.
8. Sagaidac, M., and V. Ungureanu. 2004. *Operational Research*. Chişinău: CEP USM, 296 pp. (in Romanian).
9. Ungureanu, V., and I. Mandric. 2010. Set of Nash equilibria in $2 \times 2 \times 2$ mixed extended games, from the Wolfram Demonstrations Project. http://demonstrations.wolfram.com/SetOfNashEquilibriaIn2x2x2MixedExtendedGames/

Chapter 6
Nash Equilibrium Set Function in Dyadic Mixed-Strategy Games

Abstract Dyadic two-person mixed strategy games form the simplest case for which we can determine straightforwardly Nash equilibrium sets. As in the precedent chapters, the set of Nash equilibria in a particular game is determined as an intersection of graphs of optimal reaction mappings of the first and the second players. In contrast to other games, we obtain not only an algorithm, but a set-valued/multi-valued Nash equilibrium set function (Nash function) $NES(A, B)$ that gives directly as its values the Nash equilibrium sets corresponding to the values of payoff matrix instances. The function is piecewise and it has totally 36 distinct pieces, i.e. the domain of the Nash function is divided into 36 distinct subsets for which corresponding Nash sets have a similar pattern. To give an expedient form to such a function definition and to its formula, we use a code written in the Wolfram language (Wolfram, An elementary introduction to the Wolfram language, Wolfram Media, Inc., Champaign, XV+324 pp, 2016, [1]; Hastings et al., Hands-on start to Wolfram mathematica and programming with Wolfram language, Wolfram Media, Inc., Champaign, X+470 pp, 2015, [2]) that constitutes a specific feature of this chapter in comparison with other chapters. To prove the main theoretic result of this chapter, we apply the Wolfram language code too. The Nash function $NES(A, B)$ is a multi-valued/set-valued function that has in the quality of its domain the Cartesian product $\mathbb{R}^{2\times 2} \times \mathbb{R}^{2\times 2}$ of two real spaces of two 2×2 matrices and in the quality of a Nash function image all possible sets of Nash equilibria in dyadic bimatrix mixed-strategy games. These types of games where considered earlier in a series of works, e.g. Vorob'ev (Foundations of game theory: noncooperative games, Nauka, Moscow (in Russian), 1984, [3]; Game theory: lectures for economists and systems scientists, Nauka, Moscow (in Russian), 1985, [4]), Gonzalez-Diaz et al. (An introductory course on mathematical game theory, American Mathematical Society, XIV+324 pp, 2010, [5]), Sagaidac and Ungureanu (Operational research, CEP USM, Chişinău, 296 pp (in Romanian), 2004, [6]), Ungureanu (Set of nash equilibria in 2×2 mixed extended games, from the Wolfram demonstrations project, 2007, [7]), Stahl (A gentle introduction to game theory, American Mathematical Society, XII+176 pp, 1999, [8]), Barron (Game theory: an introduction, 2nd ed, Wiley, Hoboken, XVIII+555 pp, 2013, [9]), Gintis (Game theory evolving: a problem-centered introduction to modeling strategic interaction, 2nd ed, Princeton University Press, Princeton and Oxford, XVIII+390 pp, 2009, [10]). Recently, we found the paper by John Dickhaut and Todd Kaplan that describes

V. Ungureanu, *Pareto-Nash-Stackelberg Game and Control Theory*,
Smart Innovation, Systems and Technologies 89,
https://doi.org/10.1007/978-3-319-75151-1_6

a program written in Wolfram Mathematica for finding Nash equilibria (Dickhaut and Kaplan, Economic and financial modeling with mathematica®, TELOS and Springer, New York, pp 148–166, 1993, [11]). The program is based on the game theory works by Rapoport (Two-person game theory: the essential ideas, University of Michigan Press, 229 pp, 1966, [12]; N-person game theory: concepts and applications, Dover Publications, Mineola, 331 pp, 1970, [13]), Friedman (Game theory with applications to economics, 2nd ed, Oxford University Press, Oxford, XIX+322 pp, 1990, [14]), and Kreps (A course in microeconomic theory, Princeton University Press, Princeton, XVIII+839 pp, 1990, [15]). Some examples from Harsanyi and Selten book (Harsanyi and Selten, General theory of equilibrium selection in games, The MIT Press, Cambridge, XVI+378 pp, 1988, [16]) are selected for tests. Unfortunately, the program and package need to be updated to recent versions of the Wolfram Mathematica and the Wolfram Language.

6.1 Game Statement and Its Simplification

The dyadic two-player mixed strategy game, that we consider in this chapter, is defined by a tuple

$$\Gamma' = \langle N, X, Y, f_1(\mathbf{x}, \mathbf{y}), f_2(\mathbf{x}, \mathbf{y}) \rangle,$$

where

$N = \{1, 2\}$ is a set of players,
$X = \{\mathbf{x} \in \mathbb{R}^2 : x_1 + x_2 = 1, x_1 \geq 0, x_2 \geq 0\}$ is a set of strategies of the first player,
$Y = \{\mathbf{y} \in \mathbb{R}^2 : y_1 + y_2 = 1, y_1 \geq 0, y_2 \geq 0\}$ is a set of strategies of the second player,
$f_i : X \times Y \to R$ is a player's $i \in N$ payoff function,

$$f_1(\mathbf{x}, \mathbf{y}) = \mathbf{x}^T A\mathbf{y}, \;\; A = \begin{bmatrix} a_{11} & a_{12} \\ a_{21} & a_{22} \end{bmatrix}, \; A \in \mathbb{R}^{2\times 2},$$

$$f_2(\mathbf{x}, \mathbf{y}) = \mathbf{x}^{\mathrm{T}} B\mathbf{y}, \;\; B = \begin{bmatrix} b_{11} & b_{12} \\ b_{21} & b_{22} \end{bmatrix}, \; B \in \mathbb{R}^{2\times 2}.$$

We suppose that every player maximizes the value of his payoff function. It is a requirement that is dictated by a need to be consistent with the Wolfram language code.

Remark, the strategy set is a segment for each player, i.e. it is a hypotenuse of the correspondent unit simplex (right triangle) in $\mathbb{R}^2$.

We reduce the game Γ' to a simpler game Γ by substitutions:

$$x_1 = x, \; x_2 = 1 - x, \; 0 \leq x \leq 1,$$
$$y_1 = y, \; y_2 = 1 - y, \; 0 \leq y \leq 1.$$

So, in the game Γ both the players have in the quality of their strategy sets the segment [0, 1] and in the quality of their payoff functions:

$$f_1(x,y) = (\alpha y + \alpha_0)\,x + (a_{21} - a_{22})\,y + a_{22},$$
$$f_2(x,y) = (\beta x + \beta_0)\,y + (b_{21} - b_{22})\,x + b_{22},$$

where

$$\alpha = a_{11} - a_{12} - a_{21} + a_{22},$$
$$\alpha_0 = a_{12} - a_{22},$$
$$\beta = b_{11} - b_{12} - b_{21} + b_{22},$$
$$\beta_0 = b_{21} - b_{22}.$$

Remark 6.1 To compute their optimal strategies, the players can omit the last two members of their payoff functions because the first player chooses a value for the variable x and the second player chooses a value for the variable y. So, they solve the simplified game:

$$\Gamma = \langle [0,1], [0,1];\ \tilde{f}_1(x,y) = (\alpha y - \alpha_0)\,x,\ \tilde{f}_2(x,y) = (\beta x - \beta_0)\,y \rangle.$$

Proposition 6.1 *From the strategy point of view, the games Γ' and Γ are equivalent.*

Proof The truth of the statement follows from the reduction of the game Γ' to the game Γ provided above. □

Further, we will use as the game Γ' as the game Γ to construct the sets of Nash equilibria.

6.2 Optimal Value Functions and Best Response Mappings

The game Γ represents a game on a unit square. The payoff functions are bilinear, i.e. for a fixed value of one variable, the functions become linear functions in relation to the other variable. To choose his best strategy, every player must solve a parametric linear programming problem on a unit segment.

For each player, we can define the optimal value function and the best response mapping:

$\varphi_1(y) = \max\limits_{x\in[0,1]} f_1(x,y)$ — the optimal value function of the first player,

$\gamma_1(y) = \text{Arg}\max\limits_{x\in[0,1]} f_1(x,y) = \text{Arg}\max\limits_{x\in[0,1]} \tilde{f}_1(x,y)$ — the best response mapping of the first player,

$\varphi_2(x) = \max\limits_{y\in[0,1]} f_1(x,y)$ — the optimal value function of the second player,

$\gamma_2(x) = \text{Arg}\max\limits_{y\in[0,1]} f_2(x,y) = \text{Arg}\max\limits_{y\in[0,1]} \tilde{f}_2(x,y)$ — the best response mapping of the second player.

To determine Nash equilibrium sets for particular instances of A and B, we need the graphs of best response mappings:

$$Gr_1(A, B) = \left\{ (x, y) \in [0, 1] \times [0, 1] : \begin{array}{l} y \in [0, 1] \\ x \in \operatorname{Arg\,max}_{x \in [0,1]} \tilde{f}_1(x, y) \end{array} \right\},$$

$$Gr_2(A, B) = \left\{ (x, y) \in [0, 1] \times [0, 1] : \begin{array}{l} x \in [0, 1] \\ y \in \operatorname{Arg\,max}_{y \in [0,1]} \tilde{f}_2(x, y) \end{array} \right\}.$$

As these graphs are functions of the payoff matrices A and B, evidently

$$NES(A, B) = Gr_1(A, B) \bigcap Gr_2(A, B).$$

Remark 6.2 It is important to observe once again that the graphs of best responses $Gr_1(A, B)$ and $Gr_2(A, B)$ are multi-valued (set-valued) functions of the matrices A and B. As the Nash equilibrium set in a particular game is determined as the intersection of these graphs, the Nash equilibrium set function (NES function) is a function of the matrices A and B. Such a function must have eight arguments corresponding to the eight elements of the matrices A and B:

$$A = \begin{bmatrix} a_{11} & a_{12} \\ a_{21} & a_{22} \end{bmatrix}, \quad A \in \mathbb{R}^{2\times 2},$$

$$B = \begin{bmatrix} b_{11} & b_{12} \\ b_{21} & b_{22} \end{bmatrix}, \quad B \in \mathbb{R}^{2\times 2}.$$

Because of the initial game Γ' simplification to the game Γ, we define both the graphs of best response mappings and the NES function as the functions of the four arguments: α, α_0, and β, β_0, the formulas for which are defined above as the functions of the eight matrix elements.

Now, according to the above remark, let us consider the problem of a Nash equilibrium set computing as a problem of defining a Nash equilibrium set function with the payoff matrices A and B (or α, α_0, and β, β_0) in the quality of their arguments, i.e. let us consider analytically the problem of a NES function definition and its computing. In this context, let us consider the graphs as the functions of α, α_0, and β, β_0, correspondingly.

$$\mathrm{g1}(\alpha, \alpha_0) = \left\{ (x, y) \in [0, 1] \times [0, 1] : \begin{array}{l} y \in [0, 1] \\ x \in \operatorname{Arg} \max\limits_{x \in [0,1]} (\alpha y - \alpha_0)\, x \end{array} \right\},$$

$$\mathrm{g2}(\beta, \beta_0) = \left\{ (x, y) \in [0, 1] \times [0, 1] : \begin{array}{l} x \in [0, 1] \\ y \in \operatorname{Arg} \max\limits_{y \in [0,1]} (\beta x - \beta_0)\, y \end{array} \right\}.$$

Optimal solutions of the optimization problems in the expressions for the graph functions $g1(\alpha, \alpha_0)$ and $g2(\beta, \beta_0)$ are attained either on one of the extremities, or on the whole segment [0, 1]. As the results depend of both the values of $g1(\alpha, \alpha_0)$ and $g2(\beta, \beta_0)$, we can establish exactly expressions for these graphs.

The graph of best response mapping of the first player is defined by the function

$$g1(\alpha, \alpha_0) = \begin{cases} [0,1] \times [0,1] & \text{if } \alpha = 0 \& \alpha_0 = 0, \\ [[1,0],[1,1]] & \text{if } (\alpha \geq 0 \& \alpha 0 > 0) \| \\ & (\alpha < 0 \& \alpha + \alpha 0 > 0), \\ [[0,0],[0,1]] & \text{if } (\alpha \leq 0 \& \alpha 0 < 0) \| \\ & (\alpha > 0 \& \alpha + \alpha 0 < 0), \\ [[0,0],[1,0]] \bigcup [[1,0],[1,1]] & \text{if } \alpha > 0 \& \alpha 0 = 0, \\ \left[[0,0],\left[0,-\frac{\alpha 0}{\alpha}\right]\right] \bigcup & \text{if } \alpha > 0 \& \alpha 0 < 0 \& \\ \left[\left[0,-\frac{\alpha 0}{\alpha}\right],\left[1,-\frac{\alpha 0}{\alpha}\right]\right] \bigcup & \alpha + \alpha 0 > 0, \\ \left[\left[1,-\frac{\alpha 0}{\alpha}\right],[1,1]\right] & \\ [[0,0],[0,1]] \bigcup [[0,1],[1,1]] & \text{if } \alpha > 0 \& \alpha + \alpha 0 = 0, \\ [[1,0],[0,0]] \bigcup [[0,0],[0,1]] & \text{if } \alpha < 0 \& \alpha 0 = 0, \\ \left[[1,0],\left[1,-\frac{\alpha 0}{\alpha}\right]\right] \bigcup & \text{if } \alpha < 0 \& \alpha 0 > 0 \& \\ \left[\left[1,-\frac{\alpha 0}{\alpha}\right],\left[0,-\frac{\alpha 0}{\alpha}\right]\right] \bigcup & \alpha + \alpha 0 < 0, \\ \left[\left[0,-\frac{\alpha 0}{\alpha}\right],[0,1]\right] & \\ [[1,0],[1,1]] \bigcup [[1,1],[0,1]] & \text{if } \alpha < 0 \& \alpha + \alpha 0 = 0. \end{cases}$$

Remark 6.3 Above, we denoted the unit square by $[0, 1] \times [0, 1]$ and the segment which connects two points, e.g. [0, 0] and [1, 0], by an expression of the type [[0, 0], [1, 0]].

Remark 6.4 From the expression for the graph function $g1(\alpha, \alpha_0)$, we can conclude that the set of its values or image of the function $g1(\alpha, \alpha_0)$ is formed by a union of the following alternatives:

a unit square,
a unit segment,
a union of two connected segments on the boundary of the unit square (one horizontal and one vertical),
a union of three connected segments (two vertical, on the opposite vertical sides of the unit square, connected by the third interior horizontal segment, from one lateral side to the other).

It is important to observe that the condition specified by the concrete values of α and $\alpha 0$ corresponds to an entire class of matrices $A \in \mathbb{R}^{2\times 2}$. More the more, accordingly to the expressions that define α and α_0, the function $g1(\alpha, \alpha_0)$ is defined on entire space $\mathbb{R}^{2\times 2}$, i.e. it is defined for any numeric dyadic matrix A.

The graph of best response mapping of the second player is defined by the function

$$
g2(\beta, \beta_0) = \begin{cases}
[0,1] \times [0,1] & \text{if } \beta = 0 \& \beta_0 = 0, \\
[[0,1],[1,1]] & \text{if } (\beta \geq 0 \& \beta 0 > 0) \| \\
 & (\beta < 0 \& \beta + \beta 0 > 0), \\
[[0,0],[1,0]] & \text{if } (\beta \leq 0 \& \beta 0 < 0) \| \\
 & (\beta > 0 \& \beta + \beta 0 < 0), \\
[[0,0],[0,1]] \bigcup [[0,1],[1,1]] & \text{if } \beta > 0 \& \beta 0 = 0, \\
\left[[0,0], \left[-\frac{\beta 0}{\beta}, 0\right]\right] \bigcup & \text{if } \beta > 0 \& \beta 0 < 0 \& \\
\left[\left[-\frac{\beta 0}{\beta}, 0\right], \left[-\frac{\beta 0}{\beta}, 1\right]\right] \bigcup & \beta + \beta 0 > 0, \\
\left[\left[-\frac{\beta 0}{\beta}, 1\right], [1,1]\right] & \\
[[0,0],[1,0]] \bigcup [[1,0],[1,1]] & \text{if } \beta > 0 \& \beta + \beta 0 = 0, \\
[[0,1],[0,0]] \bigcup [[0,0],[1,0]] & \text{if } \beta < 0 \& \beta 0 = 0, \\
\left[[0,1], \left[-\frac{\beta 0}{\beta}, 1\right]\right] \bigcup & \text{if } \beta < 0 \& \beta 0 > 0 \& \\
\left[\left[-\frac{\beta 0}{\beta}, 1\right], \left[-\frac{\beta 0}{\beta}, 0\right]\right] \bigcup & \beta + \beta 0 < 0, \\
\left[\left[-\frac{\beta 0}{\beta}, 0\right], [1,0]\right] & \\
[[0,1],[1,1]] \bigcup [[1,1],[1,0]] & \text{if } \beta < 0 \& \beta + \beta 0 = 0.
\end{cases}
$$

For the graph function $g2(\beta, \beta_0)$ analogical conclusions are valid as for the graph function $g1(\alpha, \alpha_0)$.

Remark 6.5 The graph functions $g1(\alpha, \alpha_0)$, $g2(\beta, \beta_0)$, correspond to the graph functions $Gr_1(A, B)$, $Gr_2(A, B)$, respectively. For the NES function $NES(A, B)$ we use a correspondent notation $nes[\alpha, \alpha_0, \beta, \beta_0]$. Additional notation with square brackets is imposed by the syntax of the Wolfram language.

6.3 Nash Equilibria and Nash Equilibrium Set Function

As every graph of best response mappings as a set-valued piecewise function is defined by the means of 9 different possible pieces, for particular instances of the payoff matrices A and B, their intersection abstractly may generate 81 possible instances/cases of Nash equilibrium sets, i.e. the NES function as a piecewise set-valued function may be defined by means of 81 pieces. Some of them may coincide. Next theorem has to highlight distinct cases in the form of the NES function definition based on the matrices A and B. Remark once again that such a function is a set-valued function.

Theorem 6.1 *In a dyadic mixed strategy game, the Nash equilibrium set function* $nes[\alpha, \alpha 0, \beta, \beta 0]$ *may be defined as a piecewise function by means of 36 distinct pieces.*

Proof The proof is constructive. It enumerates distinct pieces of the NES function in the form of the Wolfram language code.

First, the best response mapping graph functions are defined as the Wolfram language functions.

Second, the NES function is defined as the Wolfram language function too.

In the end, it is presented a Wolfram language code to manipulate all the elements together and to highlight results for different initial data. The Wolfram language primitives, such as Rectangle, Line, Point, have clear meanings and are used as a simple language means to expose both the proof, and a Wolfram program for Wolfram Mathematica 11.2.

```
g1[α_, α0_]:=Piecewise[{
{Rectangle[{{0,0}, {1,1}}], α == 0&&α0==0},
{Line[{{1,0},{1,1}}],(α >= 0&&α0 > 0)||(α < 0&&α + α0 > 0)},
{Line[{{0,0},{0,1}}],(α <= 0&&α0 < 0)||(α > 0&&α + α0 < 0)},
{Line[{{0,0},{1,0},{1,1}}], α > 0&&α0 == 0},
{Line[{{0,0}, {0,-α0/α}, {1,-α0/α}, {1,1}}], α > 0&&α0 < 0&&α + α0 > 0},
{Line[{{0,0}, {0,1}, {1,1}}], α > 0&&α + α0 == 0},
{Line[{{1,0}, {0,0}, {0,1}}], α < 0&&α0 == 0},
{Line[{{1,0},{1,-α0/α}, {0, -α0/α}, {0, 1}}], α < 0&&α0 > 0&&α + α0 < 0},
{Line[{{1,0},{1,1},{0,1}}], α < 0&&α + α0 == 0}
}]
```

```
g2[β_, β0_]:= Piecewise[{
{Rectangle[{{0,0},{1,1}}], β == 0&&β0==0},
{Line[{{0,1},{1,1}}], (β >= 0&&β0 > 0)||(β < 0&&β + β0 > 0)},
{Line[{{0,0},{1,0}}], (β <= 0&&β0 < 0)||(β > 0&&β + β0 < 0)},
{Line[{{0,0},{0,1},{1,1}}], β > 0&&β0 == 0},
{Line[{{0,0}, {-β0/β, 0}, {-β0/β,1},{1,1}}], β > 0&&β0 < 0&&β + β0 > 0},
{Line[{{0,0},{1,0},{1,1}}], β > 0&&β + β0 == 0},
{Line[{{0,1},{0,0},{1,0}}], β < 0&&β0 == 0},
{Line[{{0,1},{-β0/β, 1}, {-β0/β, 0}, {1, 0}}], β < 0&&β0 > 0&&β + β0 < 0},
{Line[{{0,1},{1,1},{1,0}}], β < 0&&β + β0 == 0}
}]
```

```
nes[α_, α0_, β_, β0_]:=Piecewise[{

(*1*) { {Point[{{0,0},{1,0},{1,1},{0,1}}],Rectangle[{0,0},{1,1}]},
   (α == 0&&α0 == 0)&&(β == 0&&β0==0) },

(*2*) { {Point[{{0,1},{1,1}}],Line[{{0,1},{1,1}}]},
   (α == 0&&α0 == 0)&&((β ≥ 0&&β0 > 0)||
(β < 0&&β + β0 > 0)))||((α > 0&&α + α0 == 0)&&((β ≥ 0&&β0 > 0)||
(β < 0&&β + β0 > 0)))||
((α > 0&&α + α0 == 0)&&(β < 0&&β + β0 == 0))||
((α < 0&&α + α0 == 0)&&(β > 0&&β0 == 0))||
```

$((\alpha < 0\&\&\alpha + \alpha 0 == 0)\&\&((\beta \geq 0\&\&\beta 0 > 0)\|$
$(\beta < 0\&\&\beta + \beta 0 > 0)))$ },

(*3*) { {Point[{{0,0},{1,0}}], Line[{{0,0},{1,0}}]},
$((\alpha == 0\&\&\alpha 0 == 0)\&\&((\beta \leq 0\&\&\beta 0 < 0)\|$
$(\beta > 0\&\&\beta + \beta 0 < 0)))\|$
$((\alpha > 0\&\&\alpha 0 == 0)\&\&((\beta \leq 0\&\&\beta 0 < 0)\|(\beta > 0\&\&\beta + \beta 0 < 0)))\|$
$((\alpha > 0\&\&\alpha 0 == 0)\&\&(\beta < 0\&\&\beta 0 == 0))\|$
$((\alpha < 0\&\&\alpha 0 == 0)\&\&(\beta > 0\&\&\beta + \beta 0 == 0))\|$
$((\alpha < 0\&\&\alpha 0 == 0)\&\&((\beta \leq 0\&\&\beta 0 < 0)\|(\beta > 0\&\&\beta + \beta 0 < 0)))$ },

(*4*) { {Point[{{0,0},{0,1},{1,1}}], Line[{{0,0},{0,1},{1,1}}]},
$((\alpha == 0\&\&\alpha 0 == 0)\&\&(\beta > 0\&\&\beta 0 == 0))\|$
$((\alpha > 0\&\&\alpha + \alpha 0 == 0)\&\&(\beta == 0\&\&\beta 0 == 0))\|$
$((\alpha > 0\&\&\alpha + \alpha 0 == 0)\&\&(\beta > 0\&\&\beta 0 == 0))$ },

(*5*) { {Point[g2[β, β0][[1]]], g2[β, β0]},
$((\alpha == 0\&\&\alpha 0 == 0)\&\&(\beta > 0\&\&\beta 0 < 0\&\&\beta + \beta 0 > 0))\|$
$((\alpha == 0\&\&\alpha 0 == 0)\&\&(\beta < 0\&\&\beta 0 > 0\&\&\beta + \beta 0 < 0))$ },

(*6*) { {Point[{{0,0},{1,0},{1,1}}], Line[{{0,0},{1,0},{1,1}}]},
$((\alpha == 0\&\&\alpha 0 == 0)\&\&(\beta > 0\&\&\beta + \beta 0 == 0))\|$
$((\alpha > 0\&\&\alpha 0 == 0)\&\&(\beta == 0\&\&\beta 0 == 0))\|$
$((\alpha > 0\&\&\alpha 0 == 0)\&\&(\beta > 0\&\&\beta + \beta 0 == 0))$ },

(*7*) { {Point[{{0,1},{0,0},{1,0}}], Line[{{0,1},{0,0},{1,0}}]},
$((\alpha == 0\&\&\alpha 0 == 0)\&\&(\beta < 0\&\&\beta 0 == 0))\|$
$((\alpha < 0\&\&\alpha 0 == 0)\&\&(\beta == 0\&\&\beta 0 == 0))\|$
$((\alpha < 0\&\&\alpha 0 == 0)\&\&(\beta < 0\&\&\beta 0 == 0))$ },

(*8*) { {Point[{{0,1},{1,1},{1,0}}], Line[{{0,1},{1,1},{1,0}}]},
$((\alpha == 0\&\&\alpha 0 == 0)\&\&(\beta < 0\&\&\beta + \beta 0 == 0))\|$
$((\alpha < 0\&\&\alpha + \alpha 0 == 0)\&\&(\beta == 0\&\&\beta 0 == 0))\|$
$((\alpha < 0\&\&\alpha + \alpha 0 == 0)\&\&(\beta < 0\&\&\beta + \beta 0 == 0))$ },

(*9*) { {Point[{{1,0},{1,1}}], Line[{{1,0},{1,1}}]},
$(((\alpha \geq 0\&\&\alpha 0 > 0)\|$
$(\alpha < 0\&\&\alpha + \alpha 0 > 0))\&\&(\beta == 0\&\&\beta 0 == 0))\|$
$(((\alpha \geq 0\&\&\alpha 0 > 0)\|(\alpha < 0\&\&\alpha + \alpha 0 > 0))\&\&(\beta > 0\&\&\beta + \beta 0 == 0))\|$
$(((\alpha \geq 0\&\&\alpha 0 > 0)\|(\alpha < 0\&\&\alpha + \alpha 0 > 0))\&\&(\beta < 0\&\&\beta + \beta 0 == 0))\|$
$((\alpha > 0\&\&\alpha 0 == 0)\&\&(\beta < 0\&\&\beta + \beta 0 == 0))\|$
$((\alpha < 0\&\&\alpha + \alpha 0 == 0)\&\&(\beta > 0\&\&\beta + \beta 0 == 0))$ },

(*10*) { {Point[{{1,1}}]},
$(((\alpha \geq 0\&\&\alpha 0 > 0)\|(\alpha < 0\&\&\alpha + \alpha 0 > 0))\&\&((\beta \geq 0\&\&\beta 0 > 0)\|$
$(\beta < 0\&\&\beta + \beta 0 > 0)))\|(((\alpha \geq 0\&\&\alpha 0 > 0)\|$
$(\alpha < 0\&\&\alpha + \alpha 0 > 0))\&\&(\beta > 0\&\&\beta 0 == 0))\|(((\alpha \geq 0\&\&\alpha 0 > 0)\|$
$(\alpha < 0\&\&\alpha + \alpha 0 > 0))\&\&(\beta > 0\&\&\beta 0 < 0\&\&\beta + \beta 0 > 0))\|$
$((\alpha > 0\&\&\alpha 0 == 0)\&\&((\beta \geq 0\&\&\beta 0 > 0)\|(\beta < 0\&\&\beta + \beta 0 > 0)))\|$
$((\alpha > 0\&\&\alpha 0 < 0\&\&\alpha + \alpha 0 > 0)\&\&((\beta \geq 0\&\&\beta 0 > 0)\|$
$(\beta < 0\&\&\beta + \beta 0 > 0)))$ },

(*11*) { {Point[{{1,0}}]},
$(((\alpha \geq 0\&\&\alpha 0 > 0)\|(\alpha < 0\&\&\alpha + \alpha 0 > 0))\&\&((\beta \leq 0\&\&\beta 0 < 0)\|$
$(\beta > 0\&\&\beta + \beta 0 < 0)))\|(((\alpha \geq 0\&\&\alpha 0 > 0)\|$

```
(α < 0&&α + α0 > 0))&&(β < 0&&β0 == 0))||(((α ≥ 0&&α0 > 0)||
(α < 0&&α + α0 > 0))&&(β < 0&&β0 > 0&&β + β0 < 0))||
((α < 0&&α + α0 == 0)&&((β ≤ 0&&β0 < 0)||(β > 0&&β + β0 < 0)))||
((α < 0&&α0 > 0&&α + α0 < 0)&&((β ≤ 0&&β0 < 0)||
(β > 0&&β + β0 < 0))) },

(*12*) { {Point[{{0,0},{0,1}}], Line[{{0,0},{0,1}}]},
    (((α ≤ 0&&α0 < 0)||
(α > 0&&α + α0 < 0))&&(β == 0&&β0 == 0))||
(((α ≤ 0&&α0 < 0)||(α > 0&&α + α0 < 0))&&(β > 0&&β0 == 0))||
(((α ≤ 0&&α0 < 0)||(α > 0&&α + α0 < 0))&&(β < 0&&β0 == 0))||
((α > 0&&α + α0 == 0)&&(β < 0&&β0 == 0))||
((α < 0&&α0 == 0)&&(β > 0&&β0 == 0)) },

(*13*) { {Point[{0,1}]},
    (((α ≤ 0&&α0 < 0)||(α > 0&&α + α0 < 0))&&((β ≥ 0&&β0 > 0)||
(β < 0&&β + β0 > 0)))||(((α ≤ 0&&α0 < 0)||
(α > 0&&α + α0 < 0))&&(β < 0&&β0 > 0&&β + β0 < 0))||
(((α ≤ 0&&α0 < 0)||(α > 0&&α + α0 < 0))&&(β < 0&&β + β0 == 0))||
((α < 0&&α0 == 0)&&((β ≥ 0&&β0 > 0)||(β < 0&&β + β0 > 0)))||
((α < 0&&α0 > 0&&α + α0 < 0)&&((β ≥ 0&&β0 > 0)||
(β < 0&&β + β0 > 0))) },

(*14*) { {Point[{0,0}]},
    (((α ≤ 0&&α0 < 0)||
(α > 0&&α + α0 < 0))&&(β > 0&&β + β0 == 0))||
(((α ≤ 0&&α0 < 0)||(α > 0&&α + α0 < 0))&&((β ≤ 0&&β0 < 0)||
(β > 0&&β + β0 < 0)))||(((α ≤ 0&&α0 < 0)||
(α > 0&&α + α0 < 0))&&(β > 0&&β0 < 0&&β + β0 > 0))||
((α > 0&&α0 < 0&&α + α0 > 0)&&((β ≤ 0&&β0 < 0)||
(β > 0&&β + β0 < 0)))||((α > 0&&α + α0 == 0)&&((β ≤ 0&&β0 < 0)||
(β > 0&&β + β0 < 0))) },

(*15*) { {Point[{{0,0},{1,1}}]},
    ((α > 0&&α0 == 0)&&(β > 0&&β0 == 0))||
((α > 0&&α + α0 == 0)&&(β > 0&&β + β0 == 0)) },

(*16*) { {Point[{{0,0},{-β0/β, 0}, {1, 1}}], Line[{{0, 0}, {-β0/β, 0}}]},
    (α > 0&&α0 == 0)&&(β > 0&&β0 < 0&&β + β0 > 0) },

(*17*) { {Point[{{-β0/β, 0}, {1, 0}}], Line[{{-β0/β, 0}, {1, 0}}]},
    (α > 0&&α0 == 0)&&(β < 0&&β0 > 0&&β + β0 < 0) },

(*18*) { {Point[g1[α, α0][[1]]], g1[α, α0]},
    ((α > 0&&α0 < 0&&α + α0 > 0)&&(β == 0&&β0 == 0))||
((α < 0&&α0 > 0&&α + α0 < 0)&&(β == 0&&β0 == 0)) },

(*19*) { {Point[{{0,0},{0,-α0/α},{1,1}}], Line[{{0,0},{0,-α0/α}}]},
    (α > 0&&α0 < 0&&α + α0 > 0)&&(β > 0&&β0 == 0) },

(*20*) { {Point[{{0,0},{-β0/β, -α0/α},{1,1}}]},
    (α > 0&&α0 < 0&&α + α0 > 0)&&
(β > 0&&β0 < 0&&β + β0 > 0) },
```

(*21*) $\{\{\text{Point}[\{\{0,0\},\{1,-\frac{\alpha0}{\alpha}\},\{1,1\}\}],\text{Line}[\{\{1,-\frac{\alpha0}{\alpha}\},\{1,1\}\}]\},$
$(\alpha > 0\&\&\alpha0 < 0\&\&\alpha + \alpha0 > 0)\&\&(\beta > 0\&\&\beta + \beta0 == 0)\},$

(*22*) $\{\{\text{Point}[\{\{0,0\},\{0,-\frac{\alpha0}{\alpha}\}\}],\text{Line}[\{\{0,0\},\{0,-\frac{\alpha0}{\alpha}\}\}]\},$
$(\alpha > 0\&\&\alpha0 < 0\&\&\alpha + \alpha0 > 0)\&\&(\beta < 0\&\&\beta0 == 0)\},$

(*23*) $\{\{\text{Point}[\{\{-\frac{\beta0}{\beta}, -\frac{\alpha0}{\alpha}\}\}]\},$
$((\alpha > 0\&\&\alpha0 < 0\&\&\alpha + \alpha0 > 0)\&\&$
$(\beta < 0\&\&\beta0 > 0\&\&\beta + \beta0 < 0))\|$
$((\alpha < 0\&\&\alpha0 > 0\&\&\alpha + \alpha0 < 0)\&\&$
$(\beta > 0\&\&\beta0 < 0\&\&\beta + \beta0 > 0))\},$

(*24*) $\{\{\text{Point}[\{\{1,-\frac{\alpha0}{\alpha}\},\{1,1\}\}],\text{Line}[\{\{1,-\frac{\alpha0}{\alpha}\},\{1,1\}\}]\},$
$(\alpha > 0\&\&\alpha0 < 0\&\&\alpha + \alpha0 > 0)\&\&(\beta < 0\&\&\beta + \beta0 == 0)\},$

(*25*) $\{\{\text{Point}[\{\{0,0\},\{-\frac{\beta0}{\beta},1\},\{1,1\}\}],\text{Line}[\{\{-\frac{\beta0}{\beta},1\},\{1,1\}\}]\},$
$(\alpha > 0\&\&\alpha + \alpha0 == 0)\&\&(\beta > 0\&\&\beta0 < 0\&\&\beta + \beta0 > 0)\},$

(*26*) $\{\{\text{Point}[\{\{0,1\},\{-\frac{\beta0}{\beta},1\}\}],\text{Line}[\{\{0,1\},\{-\frac{\beta0}{\beta},1\}\}]\},$
$(\alpha > 0\&\&\alpha + \alpha0 == 0)\&\&(\beta < 0\&\&\beta0 > 0\&\&\beta + \beta0 < 0)\},$

(*27*) $\{\{\text{Point}[\{\{0,0\},\{-\frac{\beta0}{\beta},0\}\}],\text{Line}[\{\{0,0\},\{-\frac{\beta0}{\beta},0\}\}]\},$
$(\alpha < 0\&\&\alpha0 == 0)\&\&(\beta > 0\&\&\beta0 < 0\&\&\beta + \beta0 > 0)\},$

(*28*) $\{\{\text{Point}[\{\{-\frac{\beta0}{\beta},0\}, \{1,0\},\{0,1\}\}],\text{Line}[\{\{-\frac{\beta0}{\beta},0\}, \{1,0\}\}]\},$
$(\alpha < 0\&\&\alpha0 == 0)\&\&(\beta < 0\&\&\beta0 > 0\&\&\beta + \beta0 < 0)\},$

(*29*) $\{\{\text{Point}[\{\{1,0\},\{0,1\}\}]\},$
$((\alpha < 0\&\&\alpha0 == 0)\&\&(\beta < 0\&\&\beta + \beta0 == 0))\|$
$((\alpha < 0\&\&\alpha + \alpha0 == 0)\&\&(\beta < 0\&\&\beta0 == 0))\},$

(*30*) $\{\{\text{Point}[\{\{0,-\frac{\alpha0}{\alpha}\},\{0,1\}\}],\text{Line}[\{\{0,-\frac{\alpha0}{\alpha}\},\{0,1\}\}]\},$
$(\alpha < 0\&\&\alpha0 > 0\&\&\alpha + \alpha0 < 0)\&\&(\beta > 0\&\&\beta0 == 0)\},$

(*31*) $\{\{\text{Point}[\{\{1,-\frac{\alpha0}{\alpha}\},\{1,0\}\}],\text{Line}[\{\{1,-\frac{\alpha0}{\alpha}\},\{1,0\}\}]\},$
$(\alpha < 0\&\&\alpha0 > 0\&\&\alpha + \alpha0 < 0)\&\&(\beta > 0\&\&\beta + \beta0 == 0)\},$

(*32*) $\{\{\text{Point}[\{\{0,-\frac{\alpha0}{\alpha}\},\{0,1\},\{1,0\}\}],\text{Line}[\{\{0,-\frac{\alpha0}{\alpha}\},\{0,1\}\}]\},$
$(\alpha < 0\&\&\alpha0 > 0\&\&\alpha + \alpha0 < 0)\&\&(\beta < 0\&\&\beta0 == 0)\},$

(*33*) $\{\{\text{Point}[\{\{0,1\},\{-\frac{\beta0}{\beta}, -\frac{\alpha0}{\alpha}\},\{1,0\}\}]\},$
$(\alpha < 0\&\&\alpha0 > 0\&\&\alpha + \alpha0 < 0)\&\&$
$(\beta < 0\&\&\beta0 > 0\&\&\beta + \beta0 < 0)\},$

(*34*) $\{\{\text{Point}[\{\{1,-\frac{\alpha0}{\alpha}\},\{1,0\},\{0,1\}\}],\text{Line}[\{\{1,-\frac{\alpha0}{\alpha}\},\{1,0\}\}]\},$
$(\alpha < 0\&\&\alpha0 > 0\&\&\alpha + \alpha0 < 0)\&\&(\beta < 0\&\&\beta + \beta0 == 0)\},$

(*35*) $\{\{\text{Point}[\{\{-\frac{\beta0}{\beta},1\},\{1,1\}\}],\text{Line}[\{\{-\frac{\beta0}{\beta},1\},\{1,1\}\}]\},$
$(\alpha < 0\&\&\alpha + \alpha0 == 0)\&\&(\beta > 0\&\&\beta0 < 0\&\&\beta + \beta0 > 0)\},$

(*36*) $\{\{\text{Point}[\{\{0,1\},\{-\frac{\beta0}{\beta},1\},\{1,0\}\}],\text{Line}[\{\{0,1\},\{-\frac{\beta0}{\beta},1\}\}]\},$
$(\alpha < 0\&\&\alpha + \alpha0 == 0)\&\&(\beta < 0\&\&\beta0 > 0\&\&\beta + \beta0 < 0)\}$

}]

```
Manipulate[
Grid[{{Graphics[{Thick,
Blue,g1[a11-a21+a22-a12,a12-a22],
Green,g2[b11-b12+b22-b21,b21-b22],
Red,PointSize[Large],
nes[a11-a12-a21+a22,a12-a22,b11-b12-b21+b22,b21-b22]},
PlotRange → {{0,1},{0,1}},Axes → True,AxesLabel → {"x1","y1"},
ImageSize → {400,400} ]},{" "},
{Text@Style["Reference Nash Equilibria",Bold]},
{Text@Style[nes[a11-a12-a21+a22,a12-a22,
b11-b12-b21+b22,b21-b22][[1,1]],
Bold]}},ItemSize → {Automatic,{10,1,1,3}},
Alignment → {Center,Top}], Style["Matrix A",Bold],
{{a11,10,"a11"},-10,10,1,Appearance → "Labeled",ImageSize → Tiny},
{{a12, 1,"a12"},-10,10,1,Appearance → "Labeled",ImageSize → Tiny},
{{a21,-2,"a21"},-10,10,1,Appearance → "Labeled",ImageSize → Tiny},
{{a22,-4,"a22"},-10,10,1,Appearance → "Labeled",ImageSize → Tiny},
Delimiter,{{NonAntagonistic,True},{True,False}},
Delimiter,Style["Matrix B",Bold],
{{b11, 4,"b11"},-10,10,1,Enabled → NonAntagonistic,
Appearance → "Labeled",ImageSize → Tiny},
{{b12,-3,"b12"},-10,10,1,Enabled → NonAntagonistic,
Appearance → "Labeled",ImageSize → Tiny},
{{b21,-1,"b21"},-10,10,1,Enabled → NonAntagonistic,
Appearance → "Labeled",ImageSize → Tiny},
{{b22,-4,"b22"},-10,10,1,Enabled → NonAntagonistic,
Appearance → "Labeled",ImageSize → Tiny},
Delimiter, Style["Matrices A and B",Bold],
Dynamic[TableForm[{{ToString[a11]<>" , "<>
ToString[If[NonAntagonistic,b11,b11=-a11]],ToString[a12]<>" , "<>
ToString[If[NonAntagonistic,b12,b12=-a12]]},{ToString[a21]<>" , "<>
ToString[If[NonAntagonistic,b21,b21=-a21]],ToString[a22]<>" , "<>
ToString[If[NonAntagonistic,b22,b22=-a22]]}},
TableHeadings → {{"1","2"},{" 1"," 2"}}, TableSpacing →{2,2}]],
SaveDefinitions → True]
```

As the enumerated 36 cases include all the 81 abstractly possible cases of the results of graph intersections, the proof is complete. □

Even though we establish the 36 different pieces/forms/cases needed to define exhaustively the Nash equilibrium set function, we can summarize them and can conclude additionally that a Nash equilibrium set may be in a particular game:

1. **a border point** as one of the vertices of the square (cases: 10, 11, 13, 14),
2. **an interior point** of the square (cases: 5, 23),
3. **two border points** as two opposite vertices of the square (cases: 15, 29),

4. **a unit border segment** as one of the sides of the square (cases: 2, 3, 9, 12),
5. **two unit border segments** as two connected sides of the square (one vertical and one horizontal) (cases: 4, 6, 7, 8),
6. **a union of one point and one non-unit segment** as one vertex of the square and one non-unit segment on opposite side of the square (cases: 16, 19, 21, 25, 28, 32, 34, 36),
7. **one non-unit segment** as a segment on one of the sides of the square (cases: 17, 22, 24, 26, 27, 30, 31, 35),
8. **a graph** of one of the players as a union of three connected segments (case 18),
9. **three distinct points** as two corner opposite vertices of the square and one interior point (cases: 20, 33),
10. **a unit square** (case 1).

Corollary 6.1 *Nash equilibrium set in dyadic mixed strategy game may be formed by*

1. *a point,*
2. *two points,*
3. *three points,*
4. *a segment,*
5. *two connected segments,*
6. *three connected segments,*
7. *union of non-connected one point and one segment,*
8. *the unit square.*

6.4 Conclusions

For the dyadic mixed strategy games we developed an analytic method for Nash equilibrium set computing as the value of the NES function $nes[\alpha, \alpha_0, \beta, \beta_0]$. The function $nes[\alpha, \alpha_0, \beta, \beta_0]$ is defined in the proof of Theorem 6.1 as a Wolfram language function.

The corollary summarizes all the results in a very simple and useful conceptual statement.

An earlier version algorithm was realized in the Wolfram language too. It was published on the Wolfram Demonstrations Project [7, 17]. That code may be freely viewed, verified and downloaded from the address [7]. A preliminary test version of the exposed in this chapter results where presented in [18].

References

1. Wolfram, S. 2016. *An Elementary Introduction to the Wolfram Language*. Champaign: Wolfram Media, Inc., XV+324 pp.

2. Hastings, C., K. Mischo, and M. Morrison. 2015. *Hands-on Start to Wolfram Mathematica and Programming with Wolfram Language*. Champaign: Wolfram Media, Inc., X+470 pp.
3. Vorob'ev, N.N. 1984. *Foundations of Game Theory: Noncooperative Games*. Moscow: Nauka (in Russian); Translated by Boas, R.P. 1994. Basel-Boston: Birkhäuser, 497 pp.
4. Vorob'ev, N.N. 1985. *Game Theory: Lectures for Economists and Systems Scientists*, Moscow: Nauka (in Russian); Translated and supplemented by Kotz, S. 1985. New York: Springer, 193 pp.
5. Gonzalez-Diaz, J., I. Garcia-Jurado, and M. Fiestras-Janeiro. 2010. *An Introductory Course on Mathematical Game Theory*. American Mathematical Society, XIV+324 pp.
6. Sagaidac, M., and V. Ungureanu. 2004. *Operational Research*. Chişinău: CEP USM, 296 pp. (in Romanian).
7. Ungureanu, V. 2007. Set of Nash Equilibria in 2 × 2 Mixed Extended Games, from the Wolfram Demonstrations Project. http://demonstrations.wolfram.com/SetOfNashEquilibriaIn2x2MixedExtendedGames/. Accessed 17 Nov 2007.
8. Stahl, S. 1999. *A Gentle Introduction to Game Theory*. American Mathematical Society, XII+176 pp.
9. Barron, E.N. 2013. *Game Theory: An Introduction*, 2nd ed. Hoboken: Wiley, XVIII+555 pp.
10. Gintis, H. 2009. *Game Theory Evolving: A Problem-Centered Introduction to Modeling Strategic Interaction*, 2nd ed. Princeton and Oxford: Princeton University Press, XVIII+390 pp.
11. Dickhaut, J., and T. Kaplan. 1993. A program for finding nash equilibria. In *Economic and Financial Modeling with Mathematica®*, ed. H.R. Varian, 148–166. New York: TELOS and Springer.
12. Rapoport, A. 1966. *Two-Person Game Theory: The Essential Ideas*. University of Michigan Press, 229 pp.
13. Rapoport, A. 1970. *N-Person Game Theory: Concepts and Applications*. Mineola: Dover Publications, 331 pp.
14. Friedman, J.W. 1990. *Game Theory with Applications to Economics*, 2nd ed. Oxford: Oxford University Press, XIX+322 pp.
15. Kreps, D.M. 1990. *A Course in Microeconomic Theory*. Princeton: Princeton University Press, XVIII+839 pp.
16. Harsanyi, J.C., and R. Selten. 1988. *General Theory of Equilibrium Selection in Games*. Cambridge: The MIT Press, XVI+378 pp.
17. Muñoz, M.G., and J.C. Rodríguez. Recursos informáticos para la docencia en Matemáticas y Finanzas: The Wolfram Demonstrations Project, XIX Jornadas ASEPUMA? VII Encuentro Internacional, Anales de ASEPUMA, nr. 19: 0407, 1–14. (in Spain).
18. Ungureanu, V., and M. Cirnat. 2014. *Wolfram Mathematica applications for computing sets of Nash equilibria in dyadic mixed-strategy games*, Mathematical Modelling, Optimization and Information Technologies, International Conference Proceedings, ATIC, March 25–28, 4th Edition. Chişinău: Evrica, 66–76. (in Romanian).

Chapter 7
Stackelberg Equilibrium Sets in Polymatrix Mixed-Strategy Generalized Stackelberg Games

Abstract By modifying the principle of simultaneity in strategy/simultaneous games to a hierarchical/sequential principle according to which players select their strategies in a known order, we obtain other class of games called Generalised Stackelberg Games or simply Stackelberg Games. Such games are called sequential games to especially highlight a sequential process of decision making. At every stage of Stackelberg games one player selects his strategy. To ensure choosing of his optimal strategy he solves an optimization problem. For such games a Stackelberg equilibrium concept is considered as a solution concept (Von Stackelberg in Marktform und Gleichgewicht (Market Structure and Equilibrium). Springer, Vienna, XIV+134, 1934, [1], Chenet al., in IEEE Transactions on Automatic Control AC-17, 791–798, 1972, [2], Simaan and Cruz in Journal of Optimization Theory and Applications 11, 613–626, 1973, [3], Simaan and Cruz in Journal of Optimization Theory and Applications 11, 533–555, 1973, [4], Leitmann in Journal of Optimization Theory and Applications 26, 637–648, 1978, [5], Blaquière in Une géneralisation du concept d'optimalité et des certains notions geometriques qui's rattachent, No. 1–2, Bruxelles, 49–61, 1976, [6], Başar and Olsder in Dynamic noncooperative game theory. SIAM, Philadelphia, [7], Ungureanu in ROMAI Journal 4(1), 225–242, 2008, [8], Ungureanu in Mathematical Modelling, Optimization and Information Technologies, International Conference Proceedings, ATIC, 181–189, Evrica, Chişinău, [9], Peters in Game theory: A multi-leveled approach, p. XVII+494, Springer, Berlin, 2015, [10], Ungureanu and Lozan in Mathematical Modelling, Optimization and Information Technologies, International Conference Proceedings, ATIC, 370–382, Evrica, Chişinău, 2016, [11], Korzhyk et al. in Journal of Artificial Intelligence Research, 41, 297–327, 2011, [12]). The set of all Stackelberg equilibria is described/investigated as a set of optimal solutions of an optimization problem. The last problem is obtained by considering results of solving a sequence of optimization problems that reduce all at once the graph of best response mapping of the last player to the Stackelberg equilibrium set. Namely the problem of Stackelberg equilibrium set computing in bimatrix and polymatrix finite mixed-strategy games is considered in this chapter. A method for Stackelberg equilibrium set computing is exposed.

V. Ungureanu, *Pareto-Nash-Stackelberg Game and Control Theory*,
Smart Innovation, Systems and Technologies 89,
https://doi.org/10.1007/978-3-319-75151-1_7

7.1 Introduction

A Stackelberg equilibrium set can be determined by reducing the graph of best response mapping of the last player to the Stackelberg equilibrium set via a series of optimization problems solving. This idea serves as a basis for the method of Stackelberg equilibrium set computing in finite mixed-strategy n-player hierarchical Stackelberg games. The method was initially considered in the papers [8, 9], as an extension of the intersection method, developed earlier in a series of works (e.g. [8, 9, 13–15]) and exposed in this monograph in precedent sections.

In this chapter, for convenience, we apply bold face notation for sets, vectors, and matrices.

Consider the finite strategic form game

$$\Gamma = \langle \mathbf{N}, \{\mathbf{S_p}\}_{p\in\mathbf{N}}, \{u_p(\mathbf{s})\}_{p\in\mathbf{N}}\rangle,$$

where

- $\mathbf{N} = \{1, 2, \ldots, n\}$ is a set of players;
- n is a number of players;
- $\mathbf{S_p} = \{1, 2, \ldots, m_p\}$ is a set of strategies of the player $p \in \mathbf{N}$;
- m_p is a number of strategies of the player p, where $m_p < +\infty$, $p \in \mathbf{N}$;
- $u_p(\mathbf{s})$ is a utility (payoff, cost) function of the player $p \in \mathbf{N}$, where the function $u_p(\mathbf{s})$ is defined on the Cartesian product $\mathbf{S} = \underset{p\in N}{\times} \mathbf{S_p}$ which is called a profile set;
- $\mathbf{s} = (s_1, s_2, \ldots, s_n) \in S = \times_{p\in N} \mathbf{S_p}$ is an element of the profile set $\mathbf{S}$.

Let us associate with the payoff/utility function $u_p(\mathbf{s})$, $p \in \mathbf{N}$, its matrix representation

$$u_p(\mathbf{s}) = \mathbf{A}^p_s = \left[a^p_{s_1 s_2 \ldots s_n}\right]_\mathbf{s} \in \mathbf{S} \in \mathbb{R}^\mathbf{m},$$

where

$$\mathbf{m} = m_1 \times m_2 \times \cdots \times m_n.$$

The pure-strategy game Γ imposes in an evident manner a mixed-strategy game

$$\widetilde{\Gamma} = \langle \mathbf{N}, \{\mathbf{X_p}\}_{p\in\mathbf{N}}, \{f_p(\mathbf{x})\}_{p\in\mathbf{N}}\rangle.$$

where

- $\mathbf{X}_p = \{x^p \in \mathbb{R}^{m_p}_{\geq} : x^p_1 + x^p_2 + \cdots + x^p_{m_p} = 1\}$ is a set of mixed strategies of the player $p \in \mathbf{N}$;
- $f_p(\mathbf{x})$ is a utility function of the player $p \in \mathbf{N}$ defined on the Cartesian product $X = \times_{p\in N} X_p$ and

$$f_p(\mathbf{x}) = \sum_{s_1=1}^{m_1} \sum_{s_2=1}^{m_2} \cdots \sum_{s_n=1}^{m_n} a^p_{s_1 s_2 \ldots s_n} \prod_{p=1}^{n} x^p_{s_p}.$$

In a **h**ierarchical/generalised Stackelberg game it is supposed the players make their moves sequentially (consecutively, hierarchically):

1. the first player chooses his strategy $\mathbf{x}^1 \in X_1$ and informs the second player about his choice,
2. the second player chooses his strategy $\mathbf{x}^2 \in X_2$ and informs the third player about the choices $\mathbf{x}^1, x^2$, and so on,

n. at last, the nth player selects his strategy $\mathbf{x}^n \in X_n$ after knowing the choices $\mathbf{x}^1, \ldots, x^{n-1}$, of the preceding players.

On the resulting profile $\mathbf{x} = (x^1, \ldots, x^n) \in X$, every player computes his payoff as the value of his utility function $f_p(\mathbf{x})$, $p = 1, \ldots, n$.

When the player $p \in$ **N moves, the players** $1, 2, \ldots, p-1$ **are leaders or predecessors of the player** p **and the players** $p+1, \ldots, n$ **are followers or successors of the player** p. **Players have full information about predecessors choices and doesn't have information about successors choices, but the** pth **player** $(p < n)$ **has full information about strategy sets and cost functions of the players** $p, p+1, \ldots, n$. **Without loss of generality let us suppose that all the players maximize values of their utility/payoff functions.**

The Stackelberg equilibrium definition essentially needs induction on reversed sequence of players. By backward induction, the player n computes his *best move mapping* and the players n-1, n-2, ..., 2, 1 compute their *(best move) Stackelberg mappings*; the first player computes the *set of his best Stackelberg moves*, which is the searched Stackelberg equilibrium set:

$$\mathbf{Br_n}(\mathbf{x^1}, \ldots, \mathbf{x^{n-1}}) = \operatorname*{Arg\,max}_{\mathbf{y^n} \in \mathbf{X_n}} f_n\left(\mathbf{x^1}, \ldots, \mathbf{x^{n-1}}, \mathbf{y^n}\right),$$

$$\mathbf{Br}_p(\mathbf{x^1}, \ldots, \mathbf{x^{p-1}}) = \operatorname*{Arg\,max}_{\substack{\mathbf{y^p}, \ldots, \mathbf{y^n}: \\ (\mathbf{x^1}, \ldots, \mathbf{x^{p-1}}, \mathbf{y^p}, \ldots, \mathbf{y^n}) \in \mathbf{Gr_{p+1}}}} f_p\left(\mathbf{x}^1, \ldots, x^{p-1}, y^p, \ldots, y^n\right),$$

for $p = n-1, n-2, \ldots, 2,$

$$\hat{\mathbf{S}} = \operatorname*{Arg\,max}_{(\mathbf{y^1}, \ldots, \mathbf{y^n}) \in \mathbf{Gr_2}} f_1\left(\mathbf{y}^1, \ldots, y^n\right),$$

where

$$\mathbf{Gr_n} = \left\{\mathbf{x} = (\mathbf{x^{-n}}, \mathbf{x^n}) \in \mathbf{X} : \mathbf{x^{-n}} \in \mathbf{X_{-n}}, \mathbf{x^n} \in \mathbf{Br_n}(\mathbf{x^{-n}})\right\}$$

$$= \bigcup_{\mathbf{x^{-n}} \in \mathbf{X_{-n}}} \mathbf{x^{-n}} \times \mathbf{Br_n}(\mathbf{x^{-n}}),$$

$$\mathbf{Gr_p} = \left\{ \mathbf{x} \in \mathbf{Gr_{p+1}} : \begin{matrix} \mathbf{x^1} \in \mathbf{X_1} \\ \vdots \\ \mathbf{x^{p-1}} \in \mathbf{X_{p-1}} \\ (\mathbf{x^p}, \dots, \mathbf{x^n}) \in \mathbf{Br_p}(\mathbf{x^1}, \dots, \mathbf{x^{p-1}}) \end{matrix} \right\}$$

$$= \bigcup_{(\mathbf{x^1}, \mathbf{x^2}, \dots, \mathbf{x^{p-1}}) \in \mathbf{X_1} \times \mathbf{X_2} \times \cdots \times \mathbf{X_{p-1}}} (\mathbf{x^1}, \mathbf{x^2}, \dots, \mathbf{x^{p-1}}) \times \mathbf{Br_p}(\mathbf{x^1}, \dots, \mathbf{x^{p-1}}),$$

for $p = n - 1, n - 2, \dots, 2$.

Proposition 7.1 *For every game $\widetilde{\Gamma}$ a following sequence of inclusions holds*

$$\hat{\mathbf{S}} \subseteq \mathbf{Gr_2} \subseteq \mathbf{Gr_3} \subseteq \cdots \subseteq \mathbf{Gr_n}.$$

Proof It is sufficient to go forward or backward on the above series of inclusions considering and arguing them one by one. □

Proposition 7.2 *The graph $\mathbf{Gr_n}$ represents a union of convex polyhedra*

$$\mathbf{Gr_n} = \mathbf{X}_{-n} \times \bigcup_{\mathbf{x^{-n}} \in \mathbf{X_{-n}}} \mathbf{Br_n}(\mathbf{x^{-n}}).$$

Proof Let us recall the definition of the graph of best response set-valued mapping $\mathbf{Br_n}(\mathbf{x^{-n}})$. For every $\mathbf{x^{-n}} \in \mathbf{X_{-n}}$, the value $\mathbf{Br_n}(\mathbf{x^-n})$ represents a set of optimal solutions of a linear programming problem, i.e. it is a convex polyhedron. Consequently, it follows that

$$\mathbf{Gr_n} = \bigcup_{\mathbf{x^{-n}} \in \mathbf{X_{-n}}} (\mathbf{x^{-n}} \times \mathbf{Br_n}(\mathbf{x^{-n}})) = \mathbf{X_{-n}} \times \bigcup_{\mathbf{x^{-n}} \in \mathbf{X_{-n}}} \mathbf{Br_n}(\mathbf{x^{-n}}).$$

Thus, the graph $\mathbf{Gr_n}$ is really a union of convex polyhedra. □

Remark 7.1 Generally, the non-finite union of compact sets is not always a compact set. It is sufficient to refer the union of all points (compact sets) of the unit interval (0; 1) that is performed on an open set. In precedent Propositions 7.1 and 7.2 the union is performed on the Cartesian product X_{-n} of compact strategy sets of players $1, 2, \dots, n - 1$.

Taking into account Lemma 3.1 the following statement becomes evident.

Lemma 7.1 *The graph* $\mathbf{Gr}_n$ *is a compact set.*

Remark 7.2 Let us observe once again that the graph of the best response mapping of the last player is a compact set formed of a finite union of compact subsets (Lemma 3.1), but we can refer in this context as well both functional analysis foundations [16–26] and set-valued analysis foundations [27–31].

Now, we can define the first concept of an unsafe Stackelberg equilibrium.

Definition 7.1 Any profile $\hat{\mathbf{x}} \in \hat{S}$ of the game $\widetilde{\Gamma}$ is called an unsafe Stackelberg equilibrium.

Theorem 7.1 *For every game* $\widetilde{\Gamma}$ *the set* $\hat{\mathbf{S}}$ *of unsafe Stackelberg equilibria is non empty.*

Proof First, let us recall that the graph $\mathbf{G}r_n$ is a compact set (see Lemma 7.1), i.e. it is a bounded and closed set. It consists of a finite union of compact subsets (see Lemma 3.1). The player $n - 1$ maximize his payoff linear function on the graph $\mathbf{G}r_n$, i.e. on the compact set. So, the optimal value is attained and the set of optimal solutions is compact (see Theorems 5.2.2 and 5.2.3 from [16]).

By taking into account theoretical foundations from [16, 22, 23] we can assert that the players $n-2, n-3, \ldots, 1$, maximize subsequently their payoff multilinear functions on the compact sets $\mathbf{G}r_{n-1}, Gr_{n-2}, \ldots, Gr_2$.

The set $\hat{\mathbf{S}}$, obtained at the end stage, is non empty as the set of solutions of a problem of multilinear (continuous) function maximization on a compact set. So, the set of unsafe Stackelberg equilibria is non empty (Theorems 5.2.2 and 5.2.3 from [16]). □

Remark 7.3 The proof of Theorem 7.1 is essentially based on notions of compactness and continuity which are important subjects of calculus, functional analysis, and general topology [32, 33].

The Arzelà-Ascoli Theorem [34–36] may be regarded both as an incipient and fundamental result in the context of Theorem 7.1. It represents a starting point from which a lot of other results where developed. So, there exists various results which extend and generalise The Arzelà-Ascoli Theorem (see [37], p. 382). More general formulations of The Arzelà-Ascoli Theorem may be found in [33, 38, 39] too.

As our research purpose does not relate directly to subjects of topology, we do not stop to detail on these topics, but have to mention their importance in game theory studies. It is sufficient to recall that the fundamental result of game theory — the Nash Theorem [40], is traditionally proved by applying Kakutani Fixed Point Theorem [41], an extension of Brouwer's Fixed Point Theorem [42]. Both theorems are only two among hundreds of topology fixed point theorems [43–49].

Unfortunately, the practical achievement of unsafe equilibria is not ensured when successor cost functions are surjective, i.e. they can attain the same value for different particular strategies. In order to exceed such confusing and unwanted situations, the notion of a safe Stackelberg equilibrium is introduced [5, 8].

By backward induction, the player n computes his optimal response mapping; the players $n-1, n-2, \ldots, 2$ compute their max-min mappings on the graphs of max-min mappings of the precedent players; the first player computes the set of his max-min moves on the graph $\mathbf{Gr_2}$:

$$\mathbf{Br_n}(\mathbf{x}^1, \ldots, \mathbf{x}^{n-1}) = \underset{\mathbf{y}^n \in \mathbf{X_n}}{\text{Arg max}}\ f_n\left(\mathbf{x}^1, \ldots, \mathbf{x}^{n-1}, \mathbf{y}^n\right),$$

$$\tilde{\mathbf{Br}}_\mathbf{p}(\mathbf{x}^1, \ldots, \mathbf{x}^{p-1}) =$$

$$= \underset{(\mathbf{x}^1, \ldots, \mathbf{x}^{p-1}, \mathbf{y}^p, \mathbf{y}^{p+1}, \ldots, \mathbf{y}^n) \in \tilde{\mathbf{Gr}}_{p+1}}{\underset{\mathbf{y}^p}{\text{Arg max}}\ \underset{(\mathbf{y}^{p+1}, \ldots, \mathbf{y}^n)}{\min}}\ f_p\left(\mathbf{x}^1, \ldots, \mathbf{x}^{p-1}, \mathbf{y}^p, \mathbf{y}^{p+1}, \ldots, \mathbf{y}^n\right),$$

$$\tilde{\mathbf{S}} = \underset{(\mathbf{y}^1, \ldots, \mathbf{y}^n) \in \tilde{\mathbf{Gr}}_2}{\underset{\mathbf{y}^1}{\text{Arg max}}\ \underset{(\mathbf{y}^2, \ldots, \mathbf{y}^n)}{\min}}\ f_1\left(\mathbf{y}^1, \ldots, \mathbf{y}^n\right)$$

where

$$\mathbf{Gr_n} = \left\{\mathbf{x} \in \mathbf{X} : \mathbf{x}^{-n} \in \mathbf{X}_{-n},\ \mathbf{x}^n \in \mathbf{Br_n}(\mathbf{x}^{-n})\right\}$$

$$= \bigcup_{\mathbf{x}^{-n} \in \mathbf{X}_{-n}} \mathbf{x}^{-}\mathbf{n} \times \mathbf{Br_n}(\mathbf{x}^{-n}),$$

$$\tilde{\mathbf{Gr}}_\mathbf{p} = \left\{\mathbf{x} \in \mathbf{Gr_{p+1}} : \begin{array}{l} \mathbf{x}^1 \in \mathbf{X}_1, \ldots, \mathbf{x}^{p-1} \in X_{p-1}, \\ (\mathbf{x}^p, \ldots, \mathbf{x}^n) \in \tilde{\mathbf{Br}}_\mathbf{p}(\mathbf{x}^1, \ldots, \mathbf{x}^{p-1}) \end{array}\right\}$$

$$= \bigcup_{(\mathbf{x}^1, \mathbf{x}^2, \ldots, \mathbf{x}^{p-1}) \in \mathbf{X}_1 \times \mathbf{X}_2 \times \cdots \times \mathbf{X}_{p-1}} (\mathbf{x}^1, \mathbf{x}^2, \ldots, \mathbf{x}^{p-1}) \times \tilde{\mathbf{Br}}_\mathbf{p}(\mathbf{x}^1, \ldots, \mathbf{x}^{p-1}),$$

for $p = n-1, n-2, \ldots, 2$.
Evidently, $\tilde{\mathbf{Gr}}_2 \subseteq \tilde{\mathbf{Gr}}_3 \subseteq \cdots \subseteq \tilde{\mathbf{Gr}}_{\mathbf{n-1}} \subseteq \mathbf{Gr_n}$, too.

Definition 7.2 Any profile $\tilde{\mathbf{x}} \in \tilde{\mathbf{S}}$ of the game $\tilde{\Gamma}$ is called a safe Stackelberg equilibrium.

Theorem 7.2 *For every game $\tilde{\Gamma}$, the set $\tilde{\mathbf{S}}$ of safe Stackelberg equilibria is non empty.*

Proof The proof is based on a multi-linear property of payoff functions and is governed by the same arguments as those that are applied in the proof of Theorem 7.1. □

In the context of individual features of a safe equilibrium concept, some remarks may be highlighted.

Proposition 7.3 *If the functions*

$$f_p\left(\mathbf{x^1}, \ldots, \mathbf{x^p}, \mathbf{x^{p+1}}, \ \ldots, \ \mathbf{x^n}\right), p = 1, 2, \ldots, n,$$

are injective in $\mathbf{X}_{p+1}, \ldots, X_p$ *for any fixed profile of strategies* $\mathbf{x^1}, \ldots, \mathbf{x^p}$, *then* $\hat{\mathbf{S}} = \tilde{\mathbf{S}}$.

Proof It is sufficient to observe that for injective (one to one) functions

$$\begin{aligned} &\mathop{\mathop{\mathrm{Arg\,max}}_{\mathbf{y^p}} \ \min_{(\mathbf{y^{p+1}}, \ldots, \mathbf{y^n})}}_{(\mathbf{x^1}, \ldots, \mathbf{x^{p-1}}, \mathbf{y^p}, \mathbf{y^{p+1}}, \ldots, \mathbf{y^n}) \in \tilde{\mathbf{G}}\mathbf{r_{p+1}}} f_p\left(\mathbf{x^1}, \ldots, \mathbf{x^{p-1}}, \mathbf{y^p}, \mathbf{y^{p+1}}, \ \ldots, \ \mathbf{y^n}\right) \\ = &\mathop{\mathrm{Arg} \max_{(\mathbf{y^p}, \ldots, \mathbf{y^n})}}_{(\mathbf{x^1}, \ldots, \mathbf{x^{p-1}}, \mathbf{y^p}, \ldots, \mathbf{y^n}) \in \tilde{\mathbf{G}}\mathbf{r_{p+1}}} f_p\left(\mathbf{x^1}, \ldots, \mathbf{x^{p-1}}, \ \mathbf{y^p}, \ldots, \ \mathbf{y^n}\right). \end{aligned}$$

So, safe Stackelberg equilibria and unsafe ones are simply identical for such (injective) games. □

Proposition 7.4 *If the payoff functions of the players are not injective, then the relation* $\hat{\mathbf{S}} \neq \tilde{\mathbf{S}}$ *is possible.*

Proof It is sufficient to recall an example, e.g., from Sect. 9.2 or from paper [8]. □

Remark 7.4 Even for dyadic mixed-strategy games an unsafe Stackelberg equilibrium is not always a safe one.

Remark 7.5 The problem of Stackelberg equilibrium computing in the game $\widetilde{\Gamma}$ with two players has polynomial complexity, but the problem with more than two players is NP-hard [50, 51]. Evidently, a problem of computing the whole equilibrium set is not easier.

Remark 7.6 The concept of an ε-mixt solution for weak Stackelberg problems, corresponding to two-player nonzero-sum noncooperative games, are considered by Marhfour in the [52].

Further and first of all, let us consider in next Sect. 7.2 the bimatrix mixed-strategy Stackelberg game $\widetilde{\Gamma}$ [11] and let us construct a polynomial algorithm for the set of Stackelberg equilibria computing in Sect. 7.3. After that, let us sketch an exponential algorithm for the set of Stackelberg equilibria computing in polymatrix mixed-strategy Stackelberg games in Sect. 7.4.

7.2 Stackelberg Equilibrium Sets in Bimatrix Mixed-Strategy Stackelberg Games

Consider payoff matrices

$$\mathbf{A} = (a_{ij}), \mathbf{B} = (b_{ij}), i = 1, \ldots, m, j = 1, \ldots, n.$$

Let $\mathbf{a^j}$, $j = 1, \ldots, n$, be the columns of the matrix $\mathbf{A}$,
$\mathbf{b^j}$, $j = 1, \ldots, n$, be the columns of the matrix $\mathbf{B}$.

Assume that

$$\mathbf{X} = \{\mathbf{x} \in \mathbb{R}^{\mathbf{m}}_{\geq} : x_1 + x_2 + \cdots + x_m = 1\},$$

$$\mathbf{Y} = \{\mathbf{y} \in \mathbb{R}^{\mathbf{n}}_{\geq} : y_1 + y_2 + \cdots + y_n = 1\},$$

denote sets of mixed strategies of the players, and

$$\mathbb{R}^{\mathbf{m}}_{\geq} = \left\{\mathbf{x} \in \mathbb{R}^{\mathbf{m}} : x_1 \geq 0, x_2 \geq 0, \ldots, x_m \geq 0\right\},$$

$$\mathbb{R}^{\mathbf{n}}_{\geq} = \left\{\mathbf{y} \in \mathbb{R}^{\mathbf{n}} : y_1 \geq 0, y_2 \geq 0, \ldots, y_n \geq 0\right\}$$

denote closed non-negative orthants.

Consider a mixed-strategy bimatrix hierarchical/Stackelberg game

$$\widetilde{\Gamma} = \langle \mathbf{X}, \mathbf{Y}; f_1, f_2 \rangle$$

with payoff functions:

$$f_1(\mathbf{x}, \mathbf{y}) = (\mathbf{x}^T\mathbf{a}^1)y_1 + (\mathbf{x}^T\mathbf{a}^2)y_2 + \cdots + (\mathbf{x}^T\mathbf{a}^\mathbf{n})y_n,$$

$$f_2(\mathbf{x}, \mathbf{y}) = (\mathbf{x}^T\mathbf{b}^1)y_1 + (\mathbf{x}^T\mathbf{b}^2)y_2 + \cdots + (\mathbf{x}^T\mathbf{b}^\mathbf{n})y_n.$$

The procedure of Stackelberg equilibrium set computing consists of two stages. It begins with the construction of the graph of optimal response mapping of the second player at the first stage and finalises with the Stackelberg equilibrium set computing as the solution of a constrained optimization problem at the second stage. The optimization problem solved at the second stage has as an objective function the payoff function of the first player and as an admissible set the graph of optimal response mapping of the second player. Let us expose this procedure.

Stage 1. If the strategy $\mathbf{x}$ of the first player is considered in the quality of a parameter, then the second player has to solve a linear programming parametric problem:

$$f_2(\mathbf{x}, \mathbf{y}) \to \max, \ \mathbf{y} \in \mathbf{Y}, \tag{7.1}$$

with the parameter-vector $\mathbf{x} \in \mathbf{X}$ [13, 53].

For an exposition convenience let us introduce the following notation

$$\mathbf{ex}^T = (1, \dots, 1) \in \mathbb{R}^{\mathbf{m}}, \quad \mathbf{ey}^T = (1, \dots, 1) \in \mathbb{R}^{\mathbf{n}}.$$

It is well known that solutions of linear programming problems are realized on vertices of feasible sets [22, 53, 54]. In problem (7.1), the set $\mathbf{Y}$ has n vertices — the y_j-axis unit vectors $\mathbf{e}^{y_j} \in \mathbb{R}^{\mathbf{n}}$, $j = 1, \dots, n$. Thus, in accordance with the simplex method and its optimality criterion [22, 53, 54], in parametric problem (7.1) the parameter set $\mathbf{X}$ is partitioned into such n subsets

$$\mathbf{X}^j = \left\{ \mathbf{x} \in \mathbb{R}^{\mathbf{m}} : \begin{array}{l} (\mathbf{b}^{\mathbf{k}} - \mathbf{b}^{\mathbf{j}})\mathbf{x} \leq 0, \ k = 1, \dots, n, \\ \mathbf{ex}^T \mathbf{x} = 1, \\ \mathbf{x} \geq \mathbf{0}, \end{array} \right\}, \ j = 1, \dots, n,$$

for which one of the optimal solutions of linear programming problem (7.1) is $\mathbf{e}^{y_j}$ - the corresponding unit vector of the axis y_j.

Let us introduce the notation

$$\mathbf{V} = \left\{ j \in \{1, 2, \dots, n\} : \mathbf{X}^j \neq \emptyset \right\}.$$

By the optimality criterion of the simplex method, for all $j \in \mathbf{V}$ and for all $\mathbf{J} \in 2^{\mathbf{V}} \setminus \{j\}$ all the points of

$$\Delta^{j\mathbf{J}} = \mathbf{Conv}\{\mathbf{e}^{y_k}, k \in \mathbf{J} \cup \{j\}\} = \left\{ \mathbf{y} \in \mathbb{R}^{\mathbf{n}} : \begin{array}{r} \mathbf{ey}^T \mathbf{y} = 1, \\ \mathbf{y} \geq 0, \\ y_k = 0, k \notin \mathbf{J} \cup \{j\} \end{array} \right\}$$

are optimal for

$$\mathbf{x} \in \mathbf{X}^{j\mathbf{J}} = \left\{ \mathbf{x} \in \mathbb{R}^{\mathbf{m}} : \begin{array}{l} (\mathbf{b}^{\mathbf{k}} - \mathbf{b}^{\mathbf{j}})\mathbf{x} = 0, k \in \mathbf{J}, \\ (\mathbf{b}^{\mathbf{k}} - \mathbf{b}^{\mathbf{j}})\mathbf{x} \leq 0, k \notin \mathbf{J} \cup \{j\}, \\ \mathbf{ex}^T \mathbf{x} = 1, \\ \mathbf{x} \geq \mathbf{0} \end{array} \right\}.$$

Evidently $\mathbf{X}^{j\emptyset} = \mathbf{X}^j$. Hence,

$$\mathbf{Gr}_2 = \bigcup_{j \in \mathbf{V}, \mathbf{J} \in 2^{\mathbf{V}} \setminus \{j\}} \mathbf{X}^{j\mathbf{J}} \times \Delta^{j\mathbf{J}} = \bigcup_{j \in \mathbf{V}, \mathbf{J} \in 2^{\mathbf{V}} \setminus \{j\}} \mathbf{XY}^{j\mathbf{J}},$$

where

$$\mathbf{XY}^{j\mathbf{J}} = \left\{ (\mathbf{x}, \mathbf{y}) \in \mathbb{R}^{\mathbf{m}} \times \mathbb{R}^{\mathbf{n}} : \begin{array}{l} (\mathbf{b^k} - \mathbf{b^j})\mathbf{x} = 0, k \in \mathbf{J}, \\ (\mathbf{b^k} - \mathbf{b^j})\mathbf{x} \le 0, k \notin \mathbf{J} \cup \{j\}, \\ \mathbf{e}\mathbf{x}^T\mathbf{x} = 1, \mathbf{x} \ge \mathbf{0}, \\ \mathbf{e}\mathbf{y}^T\mathbf{y} = 1, \mathbf{y} \ge \mathbf{0}, \\ y_k = 0, k \notin \mathbf{J} \cup \{j\} \end{array} \right\}.$$

Stage 2. The first player computes the set of his best moves on the graph of best response mapping of the second player if the problem of unsafe equilibria is considered, and the set of the maximin moves on the graph of best response mapping of the second player if the problem of safe equilibria is considered. **He determines the optimal values on each non-empty component $\mathbf{XY}^{j\mathbf{J}}$, comparing them and the best one defines the equilibrium.**

The precedent highlighted statement may be exposed briefly in the following notation

$$\mu_{jJ} = \max_{\mathbf{x} \in \mathbf{X}^{j\mathbf{J}}, \mathbf{y} \in \Delta^{j\mathbf{J}}} \sum_{k \in \mathbf{J} \cup \{j\}} (\mathbf{x}^T \mathbf{a^k}) y_k,$$

$$\nu_{jJ} = \max_{\mathbf{x} \in \mathbf{X}^{j\mathbf{J}}} \min_{\mathbf{y} \in \Delta^{j\mathbf{J}}} \sum_{k \in \mathbf{J} \cup \{j\}} (\mathbf{x}^T \mathbf{a^k}) y_k,$$

$$\mu = \max_{jJ} \mu_{jJ} \quad \text{and} \quad \nu = \max_{jJ} \nu_{jJ}.$$

Finally, the safe Stackelberg equilibrium set is

$$\mathbf{USES}(\widetilde{\Gamma}) = \operatorname*{Arg\,max}_{(\mathbf{x},\mathbf{y}) \in \mathbf{Gr_2}} \sum_{k=1}^{n} (\mathbf{x}^T \mathbf{a^k}) y_k,$$

and the unsafe Stackelberg equilibrium set is

$$\mathbf{SSES}(\widetilde{\Gamma}) = \operatorname*{Arg\,max}_{\substack{\mathbf{x} \\ (\mathbf{x},\mathbf{y}) \in \mathbf{Gr_2}}} \min_{\mathbf{y}} \sum_{k=1}^{n} (\mathbf{x}^T \mathbf{a^k}) y_k.$$

The above procedure is applied to prove the following theorem.

Theorem 7.3 *In the game $\widetilde{\Gamma}$, the unsafe Stackelberg equilibrium set is not empty and it is compact, i.e.* $\mathbf{USES}(\widetilde{\Gamma}) \neq \emptyset$ *and* $\mathbf{USES}(\widetilde{\Gamma})$ *is compact.*

Proof The proof follows from the fact that the graph $\mathbf{Gr_2}$ is constructed on the bases of problem (7.1), which is a parametric linear programming problem on a unit simplex. Evidently, such a problem has a solution for any value of the parameter $\mathbf{x} \in \mathbf{X}$, i.e. for any strategy of the first player. More the more, the graph $\mathbf{Gr_2}$ is formed by a finite union of convex closed polyhedra which is a compact set accordingly to [27] (Theorem 4, p. 69).

The execution of the second stage of the above procedure needs a solution of an optimization problem with a continuous objective function defined on a compact admissible set. By well known Weierstrass theorem such a problem has global solution. So, the set of unsafe Stackeberg equilibria is non empty, i.e. $\mathbf{USES}(\widetilde{\Gamma}) \neq \emptyset$, and it is bounded. By applying property III from [22, 55] (p. 9), we can conclude that the set **USES** is closed, as a finite union of convex polyhedral closed sets. It results that the set **USES** is compact.

The same property of the set **USES** may be proved alternatively in other way. According to the method described above, the function $f_1(\mathbf{x}, \mathbf{y})$ achieves its maximal value on each compact set $\mathbf{XY}^{j\mathbf{J}}$. The respective optimal set $\mathbf{XY}^{j\mathbf{J}*}$ is bounded as a facet, edge or vertex of the polyhedron $\mathbf{XY}^{j\mathbf{J}}$. According to [22, 55], the set $\mathbf{XY}^{j\mathbf{J}*}$ is a closed convex polyhedral set. So, the set $\mathbf{XY}^{j\mathbf{J}*}$ is compact.

By comparing the maximal values on all $\mathbf{XY}^{j\mathbf{J}}$, for all $j \in \mathbf{V}$, and for all $\mathbf{J} \in 2^{\mathbf{V}\setminus\{j\}}$, the record will define μ. Evidently, if μ is achieved on more than one component of the type $\mathbf{XY}^{j\mathbf{J}}$, the set $\mathbf{USES}(\widetilde{\Gamma})$ is not empty; it is represented as a finite union of compact sets and the set **USES** is compact [27]. □

A similar theorem is valid for a safe Stackelberg equilibrium set.

Theorem 7.4 *In the game $\widetilde{\Gamma}$, the safe Stackelberg equilibrium set is not empty and it is compact, i.e.* $\mathbf{SSES}(\widetilde{\Gamma}) \neq \emptyset$ *and* $\mathbf{SSES}(\widetilde{\Gamma})$ *is compact.*

Proof As in the precedent theorem, it is argued that $\mathbf{Gr_2}$ is compact.

The function $\varphi(\mathbf{x}) = \min\limits_{\mathbf{y} \in \Delta^{j\mathbf{J}}} \sum\limits_{k \in \mathbf{J} \cup \{j\}} (\mathbf{x}^T \mathbf{a}^{\mathbf{k}}) y_k$ is continuous and concave on the convex compact polyhedron $\mathbf{X}^{j\mathbf{J}}$ (see e.g. [55], p. 233). According to Weierstrass theorem, the function $\varphi(\mathbf{x})$ achieves its maximal value on the compact $\mathbf{X}^{j\mathbf{J}}$. The optimal set $\mathbf{X}^{j\mathbf{J}*}$ is bounded as a facet, edge or vertex of the polyhedron $\mathbf{X}^{j\mathbf{J}}$. In accordance with [22, 55], the set $\mathbf{X}^{j\mathbf{J}*}$ is a closed convex polyhedral set. So, $\mathbf{X}^{j\mathbf{J}*}$ is compact.

By comparing maximal values on all $\mathbf{X}^{j\mathbf{J}}$, for all $j \in \mathbf{V}$ and for all $\mathbf{J} \in 2^{\mathbf{V}\setminus\{j\}}$, the record will define ν. Evidently, if the record ν is achieved on more than one component of the type $\mathbf{X}^{j\mathbf{J}}$, the set **SSES** is formed by a finite union of compact sets and the set **SSES** is compact [27]. □

Summing it up, an algorithm for the set **USES** computing follows.

Algorithm 7.2.1

$V = \{j \in \{1, 2, \dots, n\} : X^j \neq \emptyset\}; \quad k = V[1];$

$\mu = \mu_k; \quad \mathbf{USES} = \text{Arg} \max\limits_{\mathbf{x} \in \mathbf{X}^k} \mathbf{x}^T \mathbf{a}^\mathbf{k} \times \mathbf{e}^{y_k};$

for $j \in V$ **do**

{

for $J \in 2^{V \setminus \{j\}}$ **do**

case $(XY^{jJ} \neq \emptyset)$ **and** $(\mu_{jJ} > \mu)$ **then** $\mu = \mu_{jJ}$,
$\mathbf{USES} = \text{Arg} \max\limits_{(\mathbf{x},\mathbf{y}) \in XY^{jJ}} \sum\limits_{k \in \mathbf{J} \cup \{j\}} (\mathbf{x}^T \mathbf{a}^\mathbf{k}) y_k;$

case $(XY^{jJ} \neq \emptyset)$ **and** $(\mu_{jJ} = \mu)$ **then**
$\mathbf{USES} = \mathbf{USES} \bigcup \text{Arg} \max\limits_{(\mathbf{x},\mathbf{y}) \in XY^{jJ}} \sum\limits_{k \in \mathbf{J} \cup \{j\}} (\mathbf{x}^T \mathbf{a}^\mathbf{k}) y_k;$

}

Analogically, an algorithm for the set **SSES** determining follows.

Algorithm 7.2.2

$V = \{j \in \{1, 2, \dots, n\} : X^j \neq \emptyset\}; k = V[1];$

$\nu = \nu_k; \quad \mathbf{SSES} = \text{Arg} \max\limits_{\mathbf{x} \in \mathbf{X}^k} \mathbf{x}^T \mathbf{a}^\mathbf{k} \times \mathbf{e}^{y_k};$

for $j \in V$ **do**

{

for $J \in 2^{V \setminus \{j\}}$ **do**

case $(XY^{jJ} \neq \emptyset)$ **and** $(\nu_{jJ} > \nu)$ **then** $\nu = \nu_{jJ}$,
$\mathbf{SSES} = \text{Arg} \max\limits_{\substack{\mathbf{x} \\ (\mathbf{x},\mathbf{y}) \in XY^{jJ}}} \min\limits_{\mathbf{y}} \sum\limits_{k \in \mathbf{J} \cup \{j\}} (\mathbf{x}^T \mathbf{a}^\mathbf{k}) y_k;$

case $(XY^{jJ} \neq \emptyset)$ **and** $(\nu_{jJ} = \nu)$ **then**
$\mathbf{SSES} = \mathbf{SSES} \bigcup \text{Arg} \max\limits_{\substack{\mathbf{x} \\ (\mathbf{x},\mathbf{y}) \in XY^{jJ}}} \min\limits_{\mathbf{y}} \sum\limits_{k \in \mathbf{J} \cup \{j\}} (\mathbf{x}^T \mathbf{a}^\mathbf{k}) y_k;$

}

It can be easily set out that the above algorithms execute the statement **case** no more then

$$2^{n-1} + 2^{n-2} + \cdots + 2^1 + 2^0 = 2^n - 1$$

times.

Theorem 7.5 *The sets* $\mathbf{USES}(\widetilde{\Gamma})$ *and* $\mathbf{SSES}(\widetilde{\Gamma})$ *consist of no more than* $2^n - 1$ *components of the* $\mathbf{XY}^{j\mathbf{J}*}$ *type, where* $\mathbf{XY}^{j\mathbf{J}*}$ *is defined as in the proofs of the above theorems.*

Proof It is enough to refer the above algorithms and the structure of the graph $\mathbf{Gr_2}$. □

Remark that Theorem 7.5 estimates the number of the convex components of the Stackelberg equilibrium sets that are represented by the systems of linear equations and inequalities. If the problem of a convex component representation by its vertices is posed, then the above algorithms must be supplemented by a procedure that calculates the vertices of polyhedra.

The computational complexity of such algorithms may be estimated by identifying the number of equations and inequalities of every convex polyhedron, i.e. the number of equations and inequalities which defines the component $\mathbf{XY}^{j\mathbf{J}}$ type. It's easy to set out that the component is defined by a system with $m + 2$ equations and $m + n$ inequalities, i.e. by a system with $2m + n + 2$ constraints. A component of the type $\mathbf{XY}^{j\mathbf{J}}$ may have no more than C^{m+n}_{2m+n+2} vertices. A polyhedron vertex may be identified as the solution of a system with $m + n$ equations in $m + n$ variables and additional $m + 2$ constraints. We can suppose that the complexity of an algorithm for solving a system of $m + n$ equations in $m + n$ variables is of order $O((m + n)^3)$ [56, 57]. So, the following theorem becomes evident.

Theorem 7.6 *The computational complexity of the algorithms with vertex computation is approximately* $O\left(C^{m+n}_{2m+n+2}(m + n)^3(2^n - 1)\right)$.

For practical reasons the presented method may be improved by identifying equivalent, dominant and dominated strategies in pure-strategy game Γ [58], with the following pure and mixed-strategy games simplification.

Example 7.1 The game matrices are:

$$\mathbf{A} = \begin{bmatrix} 2 & 6 & 4 \\ 3 & 3 & 9 \end{bmatrix}, \quad \mathbf{B} = \begin{bmatrix} 1 & 2 & 3 \\ 6 & 5 & 3 \end{bmatrix}.$$

Let us determine the safe and unsafe Stackelberg equilibrium sets.

The cost functions in the mixed strategy game are

$$f_1(\mathbf{x}, \mathbf{y}) = (2x_1 + 3x_2)y_1 + (6x_1 + 3x_2)y_2 + (4x_1 + 9x_2)y_3,$$

$$f_2(\mathbf{x}, \mathbf{y}) = (x_1 + 6x_2)y_1 + (2x_1 + 5x_2)y_2 + (3x_1 + 3x_2)y_3.$$

By applying exposed method, all the components $\mathbf{XY}^{j\mathbf{J}}$ must be examined. Let us begin with the value $j = 1$.

The set

$$\mathbf{XY}^{1\emptyset} = \left\{ (\mathbf{x}, \mathbf{y}) \in \mathbb{R}^2 \times \mathbb{R}^3 : \begin{array}{l} x_1 - x_2 \leq 0, \\ 2x_1 - 3x_2 \leq 0, \\ x_1 + x_2 = 1, x_1 \geq 0, x_2 \geq 0, \\ y_1 + y_2 + y_3 = 1, \\ y_1 \geq 0, y_2 = 0, y_3 = 0, \end{array} \right\}$$

is reduced to

$$\mathbf{XY}^{1\emptyset} = \begin{pmatrix} 0 \leq x_1 \leq \frac{1}{2} \\ 1 - x_1 \end{pmatrix} \times \begin{pmatrix} 1 \\ 0 \\ 0 \end{pmatrix},$$

and

$$\nu = \nu_{1\emptyset} = \max_{\substack{\mathbf{x} \\ (\mathbf{x},\mathbf{y}) \in \mathbf{XY}^{1\emptyset}}} \min_{\mathbf{y}} (2x_1 + 3x_2) = 3, \ \mathbf{SSES} = \left\{ \begin{pmatrix} 0 \\ 1 \end{pmatrix} \times \begin{pmatrix} 1 \\ 0 \\ 0 \end{pmatrix} \right\};$$

$$\mu = \mu_{1\emptyset} = \max_{(\mathbf{x},\mathbf{y}) \in \mathbf{XY}^{1\emptyset}} (2x_1 + 3x_2) = 3, \ \mathbf{USES} = \left\{ \begin{pmatrix} 0 \\ 1 \end{pmatrix} \times \begin{pmatrix} 1 \\ 0 \\ 0 \end{pmatrix} \right\}.$$

By analogy, we investigate other components.

$$\mathbf{XY}^{1\{2\}} = \left\{ (\mathbf{x}, \mathbf{y}) \in \mathbb{R}^2 \times \mathbb{R}^3 : \begin{array}{l} x_1 - x_2 = 0, \\ 2x_1 - 3x_2 \leq 0, \\ x_1 + x_2 = 1, x_1 \geq 0, x_2 \geq 0, \\ y_1 + y_2 + y_3 = 1, \\ y_1 \geq 0, y_2 \geq 0, y_3 = 0, \end{array} \right\}$$

$$= \begin{pmatrix} \frac{1}{2} \\ \frac{1}{2} \end{pmatrix} \times \begin{pmatrix} 0 \leq y_1 \leq 1 \\ 1 - y_1 \\ 0 \end{pmatrix},$$

$$\nu_{1\{2\}} = \max_{\substack{\mathbf{x} \\ (\mathbf{x},\mathbf{y}) \in \mathbf{XY}^{1\{2\}}}} \min_{\mathbf{y}} \left(\frac{5}{2} y_1 + \frac{9}{2} y_2 \right) = \frac{5}{2}.$$

As $\nu > \frac{5}{2}$, the set **SSES** remains unmodified.

$$\mu_{1\{2\}} = \max_{(\mathbf{x},\mathbf{y})\in\mathbf{XY}^{1\{2\}}} \left(\frac{5}{2}y_1 + \frac{9}{2}y_2\right) = \frac{9}{2}.$$

Since $\mu_{1\{2\}} > 3$, we put $\mu = \frac{9}{2}$ and $\mathbf{USES} = \left\{ \begin{pmatrix} \frac{1}{2} \\ \frac{1}{2} \end{pmatrix} \times \begin{pmatrix} 0 \\ 1 \\ 0 \end{pmatrix} \right\}$.

$$\mathbf{XY}^{1\{3\}} = \left\{ (\mathbf{x},\mathbf{y}) \in \mathbb{R}^2 \times \mathbb{R}^3 : \begin{array}{l} x_1 - x_2 \le 0, \\ 2x_1 - 3x_2 = 0, \\ x_1 + x_2 = 1, x_1 \ge 0, x_2 \ge 0, \\ y_1 + y_2 + y_3 = 1, \\ y_1 \ge 0, y_2 = 0, y_3 \ge 0. \end{array} \right\} = \emptyset.$$

$$\mathbf{XY}^{1\{2,3\}} = \left\{ (\mathbf{x},\mathbf{y}) \in \mathbb{R}^2 \times \mathbb{R}^3 : \begin{array}{l} x_1 - x_2 = 0, \\ 2x_1 - 3x_2 = 0, \\ x_1 + x_2 = 1, x_1 \ge 0, x_2 \ge 0, \\ y_1 + y_2 + y_3 = 1, \\ y_1 \ge 0, y_2 \ge 0, y_3 \ge 0. \end{array} \right\} = \emptyset.$$

All the components $\mathbf{XY}^{1\mathbf{J}}$ are examined.
Let us continue with $j = 2$, i.e. with components $\mathbf{XY}^{2\mathbf{J}}$.

$$\mathbf{XY}^{2\emptyset} = \left\{ (\mathbf{x},\mathbf{y}) \in \mathbb{R}^2 \times \mathbb{R}^3 : \begin{array}{l} -x_1 + x_2 \le 0, \\ x_1 - 2x_2 \le 0, \\ x_1 + x_2 = 1, x_1 \ge 0, x_2 \ge 0, \\ y_1 + y_2 + y_3 = 1, \\ y_1 = 0, y_2 \ge 0, y_3 = 0. \end{array} \right\}$$

$$= \begin{pmatrix} \frac{1}{2} \le x_1 \le \frac{2}{3} \\ 1 - x_1 \end{pmatrix} \times \begin{pmatrix} 0 \\ 1 \\ 0 \end{pmatrix},$$

$$\nu_{2\emptyset} = \max_{\mathbf{x}} \min_{\mathbf{y}} (6x_1 + 3x_2) = 5.$$
$$(\mathbf{x},\mathbf{y})\in\mathbf{XY}^{2\emptyset}$$

As $\nu_{2\emptyset} > 3$, we assign $\nu = 5$ and

$$\mathbf{SSES} = \left\{ \begin{pmatrix} \frac{2}{3} \\ \frac{1}{3} \end{pmatrix} \times \begin{pmatrix} 0 \\ 1 \\ 0 \end{pmatrix} \right\}.$$

$$\mu_{2\emptyset} = \max_{(\mathbf{x},\mathbf{y}) \in \mathbf{XY}^{2\emptyset}} (6x_1 + 3x_2) = 5.$$

As $\mu_{2\emptyset} > \frac{9}{2}$, it follows that $\mu = 5$ and

$$\mathbf{USES} = \left\{ \begin{pmatrix} \frac{2}{3} \\ \frac{1}{3} \end{pmatrix} \times \begin{pmatrix} 0 \\ 1 \\ 0 \end{pmatrix} \right\}.$$

$$\mathbf{XY}^{2\{3\}} = \left\{ (\mathbf{x}, \mathbf{y}) \in \mathbb{R}^2 \times \mathbb{R}^3 : \begin{array}{l} -x_1 + x_2 \le 0, \\ x_1 - 2x_2 = 0, \\ x_1 + x_2 = 1, x_1 \ge 0, x_2 \ge 0, \\ y_1 + y_2 + y_3 = 1, \\ y_1 = 0, y_2 \ge 0, y_3 \ge 0. \end{array} \right\}$$

$$= \begin{pmatrix} \frac{2}{3} \\ \frac{1}{3} \end{pmatrix} \times \begin{pmatrix} 0 \\ 0 \le y_2 \le 1 \\ 1 - y_2 \end{pmatrix},$$

$$\nu_{2\{3\}} = \max_{\mathbf{x}} \min_{\mathbf{y}} \underset{(\mathbf{x},\mathbf{y}) \in \mathbf{XY}^{2\{3\}}}{} \left(5y_2 + \frac{17}{3} y_3 \right) = 5, \; \nu = 5,$$

$$\mathbf{SSES} = \left\{ \begin{pmatrix} \frac{2}{3} \\ \frac{1}{3} \end{pmatrix} \times \begin{pmatrix} 0 \\ 1 \\ 0 \end{pmatrix} \right\};$$

$$\mu_{2\{3\}} = \max_{(\mathbf{x},\mathbf{y}) \in \mathbf{XY}^{2\{3\}}} \left(5y_2 + \frac{17}{3} y_3 \right) = \frac{17}{3}.$$

Since $\mu_{2\{3\}} > 5$, it follows that $\mu = \frac{17}{3}$ and

$$\mathbf{USES} = \left\{ \begin{pmatrix} \frac{2}{3} \\ \frac{1}{3} \end{pmatrix} \times \begin{pmatrix} 0 \\ 0 \\ 1 \end{pmatrix} \right\}.$$

$$\mathbf{XY}^{3\emptyset} = \left\{ (\mathbf{x}, \mathbf{y}) \in \mathbb{R}^2 \times \mathbb{R}^3 : \begin{array}{l} -2x_1 + 3x_2 \le 0, \\ -x_1 + 2x_2 \le 0, \\ x_1 + x_2 = 1, x_1 \ge 0, x_2 \ge 0, \\ y_1 + y_2 + y_3 = 1, \\ y_1 = 0, y_2 = 0, y_3 \ge 0. \end{array} \right\}$$

$$= \begin{pmatrix} \frac{2}{3} \le x_1 \le 1 \\ 1 - x_1 \end{pmatrix} \times \begin{pmatrix} 0 \\ 0 \\ 1 \end{pmatrix},$$

$$\nu_{3\emptyset} = \max_{\substack{\mathbf{x} \\ (\mathbf{x},\mathbf{y}) \in \mathbf{XY}^{3\emptyset}}} \min_{\mathbf{y}} (4x_1 + 9x_2) = \frac{17}{3},$$

As $\nu_{3\emptyset} > 5$, we assign $\nu = \frac{17}{3}$ and $\mathbf{SSES} = \left\{ \begin{pmatrix} \frac{2}{3} \\ \frac{1}{3} \end{pmatrix} \times \begin{pmatrix} 0 \\ 0 \\ 1 \end{pmatrix} \right\}$;

$$\mu_{3\emptyset} = \max_{(\mathbf{x},\mathbf{y}) \in \mathbf{XY}^{3\emptyset}} (4x_1 + 9x_2) = \frac{17}{3}, \ \mu = \frac{17}{3},$$

$$\mathbf{USES} = \left\{ \begin{pmatrix} \frac{2}{3} \\ \frac{1}{3} \end{pmatrix} \times \begin{pmatrix} 0 \\ 0 \\ 1 \end{pmatrix} \right\}.$$

Finally, the sets of Stackelberg equilibria are

$$\mathbf{SSES} = \mathbf{USES} = \left\{ \begin{pmatrix} \frac{2}{3} \\ \frac{1}{3} \end{pmatrix} \times \begin{pmatrix} 0 \\ 0 \\ 1 \end{pmatrix} \right\}$$

with the payoffs $\left(\frac{17}{3}, 3\right)$.

Remark that the pure strategy game Γ has a pure strategy Stackelberg equilibrium $(1, 3)$, or $(1, 0) \times (0, 0, 1)$ in mixed-strategy notation, with the payoffs $(4, 3)$, correspondingly. □

Example 7.1 establishes the truth of the following propositions concerning relations between pure and mixed-strategy Stackelberg equilibrium sets.

Proposition 7.5 *A pure-strategy Stackelberg equilibrium in Γ is not necessarily a mixed-strategy Stackelberg equilibrium in $\widetilde{\Gamma}$.*

Proposition 7.6 *The mixed-strategy Stackelberg equilibrium set in $\widetilde{\Gamma}$ does not necessarily include the pure-strategy Stackelberg equilibrium set from Γ.*

In Example 7.1, the solution sets are equivalent and consist of one element. By a small changing of the initial data, e.g. of elements of the matrix $\mathbf{A}$, the sets of solutions are enlarged drastically.

Example 7.2 The game matrices are:

$$\mathbf{A} = \begin{bmatrix} 3 & 4 & 4 \\ 1 & 4 & 4 \end{bmatrix}, \quad \mathbf{B} = \begin{bmatrix} 1 & 2 & 3 \\ 6 & 5 & 3 \end{bmatrix}.$$

Let us determine the safe and unsafe Stackelberg equilibrium sets.

The cost functions in the mixed strategy game are

$$f_1(\mathbf{x}, \mathbf{y}) = (3x_1 + x_2)y_1 + (4x_1 + 4x_2)y_2 + (4x_1 + 4x_2)y_3,$$

$$f_2(\mathbf{x}, \mathbf{y}) = (x_1 + 6x_2)y_1 + (2x_1 + 5x_2)y_2 + (3x_1 + 3x_2)y_3.$$

In conformity with the exposed method, we must examine all the components of the type $\mathbf{XY}^{j\mathbf{J}}$. Let us begin with the value $j = 1$.

$$\mathbf{XY}^{1\emptyset} = \left\{ (\mathbf{x}, \mathbf{y}) \in \mathbb{R}^2 \times \mathbb{R}^3 : \begin{array}{l} x_1 - x_2 \le 0, \\ 2x_1 - 3x_2 \le 0, \\ x_1 + x_2 = 1, x_1 \ge 0, x_2 \ge 0, \\ y_1 + y_2 + y_3 = 1, \\ y_1 \ge 0, y_2 = 0, y_3 = 0, \end{array} \right\}$$

$$= \begin{pmatrix} 0 \le x_1 \le \frac{1}{2} \\ 1 - x_1 \end{pmatrix} \times \begin{pmatrix} 1 \\ 0 \\ 0 \end{pmatrix},$$

$$\nu = \nu_{1\emptyset} = \max_{\mathbf{x}} \min_{\substack{\mathbf{y} \\ (\mathbf{x},\mathbf{y}) \in \mathbf{XY}^{1\emptyset}}} (3x_1 + x_2) = 2, \ \mathbf{SSES} = \left\{ \begin{pmatrix} \frac{1}{2} \\ \frac{1}{2} \end{pmatrix} \times \begin{pmatrix} 1 \\ 0 \\ 0 \end{pmatrix} \right\};$$

$$\mu = \mu_{1\emptyset} = \max_{(\mathbf{x},\mathbf{y}) \in \mathbf{XY}^{1\emptyset}} (3x_1 + x_2) = 2, \ \mathbf{USES} = \left\{ \begin{pmatrix} \frac{1}{2} \\ \frac{1}{2} \end{pmatrix} \times \begin{pmatrix} 1 \\ 0 \\ 0 \end{pmatrix} \right\}.$$

$$\mathbf{XY}^{1\{2\}} = \left\{ (\mathbf{x}, \mathbf{y}) \in \mathbb{R}^2 \times \mathbb{R}^3 : \begin{array}{l} x_1 - x_2 = 0, \\ 2x_1 - 3x_2 \le 0, \\ x_1 + x_2 = 1, \\ x_1 \ge 0, x_2 \ge 0, \\ y_1 + y_2 + y_3 = 1, \\ y_1 \ge 0, y_2 \ge 0, y_3 = 0. \end{array} \right\}$$

$$= \begin{pmatrix} \frac{1}{2} \\ \frac{1}{2} \end{pmatrix} \times \begin{pmatrix} 0 \le y_1 \le 1 \\ 1 - y_1 \\ 0 \end{pmatrix},$$

$$\nu_{1\{2\}} = \max_{\substack{\mathbf{x} \\ (\mathbf{x},\mathbf{y})\in\mathbf{XY}^{1\{2\}}}} \min_{\mathbf{y}} (2y_1 + 4y_2) = 2.$$

Whereas $\nu = 2$, the set **SSES** remains unchanged.
$\mu_{1\{2\}} = \max_{(\mathbf{x},\mathbf{y})\in\mathbf{XY}^{1\{2\}}} (2y_1 + 4y_2) = 4.$

Since $\mu_{1\{2\}} > 2$, we assign $\mu = 4$ and $\mathbf{USES} = \left\{ \begin{pmatrix} \frac{1}{2} \\ \frac{1}{2} \end{pmatrix} \times \begin{pmatrix} 0 \\ 1 \\ 0 \end{pmatrix} \right\}.$

$\mathbf{XY}^{1\{3\}} = \emptyset$, $\mathbf{XY}^{1\{2,3\}} = \emptyset$.
The components $\mathbf{XY}^{1\mathbf{J}}$ are examined and let us continue with the value $j = 2$.

$$\mathbf{XY}^{2\emptyset} = \left\{ (\mathbf{x},\mathbf{y}) \in \mathbb{R}^2 \times \mathbb{R}^3 : \begin{array}{l} -x_1 + x_2 \le 0, \\ x_1 - 2x_2 \le 0, \\ x_1 + x_2 = 1, x_1 \ge 0, x_2 \ge 0, \\ y_1 + y_2 + y_3 = 1, \\ y_1 = 0,\ y_2 \ge 0,\ y_3 = 0. \end{array} \right\}$$
$$= \begin{pmatrix} \frac{1}{2} \le x_1 \le \frac{2}{3} \\ 1 - x_1 \end{pmatrix} \times \begin{pmatrix} 0 \\ 1 \\ 0 \end{pmatrix},$$

$$\nu_{2\emptyset} = \max_{\substack{\mathbf{x} \\ (\mathbf{x},\mathbf{y})\in\mathbf{XY}^{2\emptyset}}} \min_{\mathbf{y}} (4x_1 + 4x_2) = 4.$$

Because $\nu_{2\emptyset} > 2$, it follows that $\nu = 4$ and

$$\mathbf{SSES} = \left\{ \begin{pmatrix} \frac{1}{2} \le x_1 \le \frac{2}{3} \\ 1 - x_1 \end{pmatrix} \times \begin{pmatrix} 0 \\ 1 \\ 0 \end{pmatrix} \right\}.$$

$$\mu_{2\emptyset} = \max_{(\mathbf{x},\mathbf{y})\in\mathbf{XY}^{2\emptyset}} (4x_1 + 4x_2) = 4,\ \mu = 4,$$

and $\mathbf{USES} = \left\{ \begin{pmatrix} \frac{1}{2} \le x_1 \le \frac{2}{3} \\ 1 - x_1 \end{pmatrix} \times \begin{pmatrix} 0 \\ 1 \\ 0 \end{pmatrix} \right\}.$

$$\mathbf{XY}^{2\{3\}} = \left\{ (\mathbf{x},\mathbf{y}) \in \mathbb{R}^2 \times \mathbb{R}^3 : \begin{array}{l} -x_1 + x_2 \le 0, \\ x_1 - 2x_2 = 0, \\ x_1 + x_2 = 1, x_1 \ge 0, x_2 \ge 0, \\ y_1 + y_2 + y_3 = 1, \\ y_1 = 0,\ y_2 \ge 0,\ y_3 \ge 0. \end{array} \right\}$$

$$= \begin{pmatrix} \frac{2}{3} \\ \frac{1}{3} \end{pmatrix} \times \begin{pmatrix} 0 \\ 0 \le y_2 \le 1 \\ 1 - y_2 \end{pmatrix},$$

$$\nu_{2\{3\}} = \max_{\mathbf{x}} \min_{\mathbf{y}}_{(\mathbf{x},\mathbf{y}) \in \mathbf{XY}^{2\{3\}}} (4y_2 + 4y_3) = 4, \ \nu = 4,$$

$$\mathbf{SSES} = \left\{ \begin{pmatrix} \frac{2}{3} \\ \frac{1}{3} \end{pmatrix} \times \begin{pmatrix} 0 \\ 0 \le y_2 \le 1 \\ 1 - y_2 \end{pmatrix} \right\} \bigcup \left\{ \begin{pmatrix} \frac{1}{2} \le x_1 \le \frac{2}{3} \\ 1 - x_1 \end{pmatrix} \times \begin{pmatrix} 0 \\ 1 \\ 0 \end{pmatrix} \right\};$$

$$\mu_{2\{3\}} = \max_{(\mathbf{x},\mathbf{y}) \in \mathbf{XY}^{2\{3\}}} (4y_2 + 4y_3) = 4, \ \mu = 4,$$

$$\mathbf{USES} = \left\{ \begin{pmatrix} \frac{2}{3} \\ \frac{1}{3} \end{pmatrix} \times \begin{pmatrix} 0 \\ 0 \le y_2 \le 1 \\ 1 - y_2 \end{pmatrix} \right\} \bigcup \left\{ \begin{pmatrix} \frac{1}{2} \le x_1 \le \frac{2}{3} \\ 1 - x_1 \end{pmatrix} \times \begin{pmatrix} 0 \\ 1 \\ 0 \end{pmatrix} \right\}.$$

$$\mathbf{XY}^{3\emptyset} = \left\{ (\mathbf{x}, \mathbf{y}) \in \mathbb{R}^2 \times \mathbb{R}^3 : \begin{array}{l} -2x_1 + 3x_2 \le 0, \\ -x_1 + 2x_2 \le 0, \\ x_1 + x_2 = 1, x_1 \ge 0, x_2 \ge 0, \\ y_1 + y_2 + y_3 = 1, \\ y_1 = 0, y_2 = 0, y_3 \ge 0. \end{array} \right\}$$

$$= \begin{pmatrix} \frac{2}{3} \le x_1 \le 1 \\ 1 - x_1 \end{pmatrix} \times \begin{pmatrix} 0 \\ 0 \\ 1 \end{pmatrix},$$

$$\nu_{3\emptyset} = \max_{\mathbf{x}} \min_{\mathbf{y}}_{(\mathbf{x},\mathbf{y}) \in \mathbf{XY}^{3\emptyset}} (4x_1 + 4x_2) = 4, \ \nu = 4,$$

$$\mathbf{SSES} = \left\{ \begin{pmatrix} \frac{2}{3} \\ \frac{1}{3} \end{pmatrix} \times \begin{pmatrix} 0 \\ 0 \le y_2 \le 1 \\ 1 - y_2 \end{pmatrix} \right\} \bigcup$$

$$\bigcup \left\{ \begin{pmatrix} \frac{1}{2} \le x_1 \le \frac{2}{3} \\ 1 - x_1 \end{pmatrix} \times \begin{pmatrix} 0 \\ 1 \\ 0 \end{pmatrix} \right\} \bigcup \left\{ \begin{pmatrix} \frac{2}{3} \le x_1 \le 1 \\ 1 - x_1 \end{pmatrix} \times \begin{pmatrix} 0 \\ 0 \\ 1 \end{pmatrix} \right\}.$$

$$\mu_{3\emptyset} = \max_{(\mathbf{x},\mathbf{y}) \in \mathbf{XY}^{3\emptyset}} (4x_1 + 4x_2) = 4, \ \mu = 4,$$

$$\mathbf{USES} = \left\{ \begin{pmatrix} \frac{2}{3} \\ \frac{1}{3} \end{pmatrix} \times \begin{pmatrix} 0 \\ 0 \le y_2 \le 1 \\ 1 - y_2 \end{pmatrix} \right\} \bigcup$$

$$\bigcup \left\{ \begin{pmatrix} \frac{1}{2} \le x_1 \le \frac{2}{3} \\ 1 - x_1 \end{pmatrix} \times \begin{pmatrix} 0 \\ 1 \\ 0 \end{pmatrix} \right\} \bigcup \left\{ \begin{pmatrix} \frac{2}{3} \le x_1 \le 1 \\ 1 - x_1 \end{pmatrix} \times \begin{pmatrix} 0 \\ 0 \\ 1 \end{pmatrix} \right\}.$$

Finally,

$$\mathbf{SSES} = \mathbf{USES} = \left\{ \begin{pmatrix} \frac{2}{3} \\ \frac{1}{3} \end{pmatrix} \times \begin{pmatrix} 0 \\ 0 \le y_2 \le 1 \\ 1 - y_2 \end{pmatrix} \right\} \bigcup$$

$$\bigcup \left\{ \begin{pmatrix} \frac{1}{2} \le x_1 \le \frac{2}{3} \\ 1 - x_1 \end{pmatrix} \times \begin{pmatrix} 0 \\ 1 \\ 0 \end{pmatrix} \right\} \bigcup \left\{ \begin{pmatrix} \frac{2}{3} \le x_1 \le 1 \\ 1 - x_1 \end{pmatrix} \times \begin{pmatrix} 0 \\ 0 \\ 1 \end{pmatrix} \right\}.$$

□

Next example has to illustrate the above method for 3×3 games.

Example 7.3 The game matrices are:

$$\mathbf{A} = \begin{bmatrix} 2 & 1 & 3 \\ 3 & 2 & 6 \\ 2 & 2 & 4 \end{bmatrix}, \quad \mathbf{B} = \begin{bmatrix} 1 & 0 & 3 \\ -1 & 1 & -2 \\ 2 & -1 & 2 \end{bmatrix}.$$

Let us determine the safe and unsafe Stackelberg equilibrium sets.

The cost functions in the mixed strategy game are

$$f_1(\mathbf{x}, \mathbf{y}) = (2x_1 + 3x_2 + 2x_3)y_1 + (x_1 + 2x_2 + 2x_3)y_2 + (3x_1 + 6x_2 + 4x_3)y_3,$$

$$f_2(\mathbf{x}, \mathbf{y}) = (x_1 - x_2 + 2x_3)y_1 + (x_2 - x_3)y_2 + (3x_1 - 2x_2 + 2x_3)y_3.$$

By applying the above method, seven components $\mathbf{XY}^{j\mathbf{J}}$ are investigated. The equilibrium sets are: $\mathbf{SSES} = \mathbf{USES} = \left\{ \begin{pmatrix} \frac{1}{4} \\ \frac{1}{2} \\ \frac{1}{4} \end{pmatrix} \times \begin{pmatrix} 0 \\ 0 \\ 1 \end{pmatrix} \right\}$ with the payoffs $(\frac{19}{4}, \frac{1}{4})$.

□

7.3 Polynomial Algorithm for a Single Stackelberg Equilibrium Computing in Bimatrix Mixed-Strategy Games

In the context of problems [50, 51] related to the complexity of Stackelberg equilibrium determining, a certain dubiety about the importance of Algorithms 7.2.1, 7.2.2 may appear because of their exponential complexity. It's rather a false and superficial first impression whereas we are interested there to find the whole set of Stackelberg equilibria, while paper [50] considers the problem of a single Stackelberg equilibrium determining.

The complexity of the algorithm exposed in paper [50] is polynomial in the number of strategies. The algorithm becomes exponential if it is modified to determine the whole set of Stackelberg equilibria which has, e.g. about $C^m_{m+n} \geq \left(1+\dfrac{n}{m}\right)^m$ vertices [59]. More the more, the algorithm doesn't guarantee the whole set of equilibria determining in all possible cases.

Algorithms 7.2.1, 7.2.2 may be simplified to determine only a single Stackelberg equilibrium and the computational complexity of such algorithms is not exponential but polynomial. Indeed, the first player strategy set $\mathbf{X}$ is partitioned into n subsets $\mathbf{X}^j$, $j \in \{1, 2, \ldots, n\}$, on which the first player optimizes his utility function. The record value from these optimal values defines the Stackelberg equilibrium. So, we can expose a procedure with polynomial complexity to determine a sample of unsafe Stackelberg equilibria.

Stage 1. The set $\mathbf{X}$ is partitioned into n subsets

$$\mathbf{X}^j = \left\{ \mathbf{x} \in \mathbb{R}^{\mathbf{m}} : \begin{array}{l} (\mathbf{b^k} - \mathbf{b^j})\mathbf{x} \leq 0, k = 1, \ldots, n, \\ \mathbf{e}\mathbf{x}^T\mathbf{x} = 1, \\ \mathbf{x} \geq \mathbf{0}, \end{array} \right\}, \quad j = 1, \ldots, n,$$

for which one of the optimal solutions of problem (7.1) is $\mathbf{e}^{y_j}$ — the corresponding unit vector of the y_j-axis, $j \in \{1, 2, \ldots, n\}$.

Stage 2. The first player computes his best strategy on every non-empty component $\mathbf{X}^j$: $\mu_j = \max\limits_{\mathbf{x} \in \mathbf{X}^j} \mathbf{x}^T\mathbf{a^j}$. By comparing them and by selecting the best value $\mu = \max\limits_j \mu_j$, the corresponding to μ optimal solution is the searched sample of unsafe Stackelberg equilibria.

Alternatively, the procedure may be exposed in the form of the following algorithm.

Algorithm 7.3.1

$$
\begin{array}{l}
\mu = -\infty; \\
\textbf{for } \ j \in \{1, 2, \ldots, n\} \ \textbf{do} \\
\{ \\
\qquad \mu_j = \max\limits_{\mathbf{x} \in \mathbf{X}^j} \mathbf{x}^T \mathbf{a}^{\mathbf{j}}; \\
\qquad \textbf{if } \ (\mu_j > \mu) \ \textbf{ then } \ \mu = \mu_j, k = j; \\
\} \\
\textbf{print } \ \arg\max\limits_{\mathbf{x} \in \mathbf{X}^k} \mathbf{x}^T \mathbf{a}^{\mathbf{k}} \times \mathbf{e}^{y_k}
\end{array}
$$

Apparently, Algorithm 7.3.1 may be modified to determine the set of unsafe Stackelberg equilibria:

Algorithm 7.3.2

$$
\begin{array}{l}
\mu = -\infty; \quad \textbf{USES} = \emptyset; \\
\textbf{for } \ j \in \{1, 2, \ldots, n\} \ \textbf{do} \\
\{ \\
\qquad \mu_j = \max\limits_{\mathbf{x} \in \mathbf{X}^j} (\mathbf{x}^T \mathbf{a}^{\mathbf{j}}) y_j; \\
\qquad \textbf{case } \ (X^j \neq \emptyset) \ \textbf{ and } \ (\mu_j > \mu) \ \textbf{ then } \ \mu = \mu_j, \\
\qquad\qquad \textbf{USES} = \mathrm{Arg}\max\limits_{\mathbf{x} \in X^j} \mathbf{x}^T \mathbf{a}^{\mathbf{j}} \times \mathbf{e}^{y_j}; \\
\qquad \textbf{case } \ (X^j \neq \emptyset) \ \textbf{ and } \ (\mu_j = \mu) \ \textbf{ then } \\
\qquad\qquad \textbf{USES} = \textbf{USES} \bigcup \mathrm{Arg}\max\limits_{\mathbf{x} \in X^j} \mathbf{x}^T \mathbf{a}^{\mathbf{j}} \times \mathbf{e}^{y_j}; \\
\}
\end{array}
$$

Unfortunately, Algorithm 7.3.2 doesn't ensure that it determines the whole set of equilibria. Example 7.2 illustrates that for an entire set of unsafe Stackelberg equilibria determining it's not sufficient to find the solutions on all components $\mathbf{X}^j$, $j \in \{1, 2, \ldots, n\}$, because, e.g., the component $\left\{ \begin{pmatrix} \frac{2}{3} \\ \frac{1}{3} \end{pmatrix} \times \begin{pmatrix} 0 \\ 0 \leq y_2 \leq 1 \\ 1 - y_2 \end{pmatrix} \right\}$ of **USES** in the example is not included by Algorithm 7.3.2 in the set of unsafe Stackelberg equilibria.

The systems which defines the components $\mathbf{X}^j$ have $2m + 1$ constraints in m variables. A maximal number of vertices for a component of the type $\mathbf{X}^j$ is approximately $C_{2m+1}^m \geq \left(2 + \dfrac{1}{m}\right)^m$ (see Proposition 1.4 in [59]). A polyhedron vertex may be computed as a solution of a system of m equations and $m + 1$ inequalities in m variables. It is well known that there are algorithms for solving systems of m equations in m variables which needs $O(m^3)$ time. To solve a linear programming problem in m variables, with the length of input equals to L, Karmarkar's algorithm [60] requires $O(m^{3.5} L)$ time.

Theorem 7.7 *Algorithm 7.3.1, for determining a single unsafe Stackelberg equilibrium, needs* $O\left(nm^{3.5}L\right)$ *time.*

Proof We must recall that Algorithm 7.3.1 needs to solve n linear programming problems and any of them needs $O(m^{3.5}L)$ time. □

Based on appropriate arguments, the following theorems becomes truth.

Theorem 7.8 *Algorithm 7.3.2 needs* $O\left(nm^{3.5}L + C_{m+n}^{m}m^{3}n\right)$ *time.*

Nevertheless, in the context of polynomial Algorithm 7.3.1 and the polynomial algorithm from [50], for determining a single equilibrium, a question about polynomial characteristics arises in the case when, e.g., $m = 2^n$. Are the algorithms steel polynomial or they are pseudo-polynomial?

Remark 7.7 Let us observe that Algorithms 7.3.1 and 7.3.2 are aimed to find only unsafe Stackelberg equilibria. They may be modified to find safe Stackelberg equilibria too, by taking into account the theoretical results of previous section.

7.4 Stackelberg Equilibrium Sets in Polymatrix Mixed-Strategy Games

Consider the n-matrix $m_1 \times m_2 \times \cdots \times m_n$ mixed-strategy game

$$\widetilde{\Gamma} = \langle \mathbf{N}, \{\mathbf{X_p}\}_{p\in\mathbf{N}}, \{f_p(\mathbf{x})\}_{p\in\mathbf{N}} \rangle,$$

formulated in Sect. 7.1. The utility function of the player p is linear when the strategies of the rest players are fixed

$$f_p(\mathbf{x}) = \sum_{k=1}^{m_p} \left(\sum_{s_{-p}\in S_{-p}} a^p_{k\|s_{-p}} \prod_{q=1,\ldots,n,q\neq p} x^q_{s_q} \right) x^p_k .$$

Theorem 7.9 *For every Stackelberg game* $\widetilde{\Gamma}$ *the set* **USES** *is not empty.*

Proof Let us apply backward induction to analyse the Stackelberg solution set in the game $\widetilde{\Gamma}$.

The player n has to solve a linear programming parametric problem with the vector-parameter $\mathbf{x^{-n}} \in \mathbf{X_{-n}}$:

$$f_n(\mathbf{x^n}, \mathbf{x^{-n}}) \to max, \ \mathbf{x^n} \in \mathbf{X_n}. \tag{7.2}$$

The solution of this problem is located on the polytope $\mathbf{X_n}$ vertices — x_i^n axis unit vectors $\mathbf{e}^{x_i^n} \in \mathbb{R}^{\mathbf{m_n}}$, $i = 1, \ldots, m_n$. In accordance with the simplex method and its optimality criterion, the parameter set $\mathbf{X_{-n}}$ is partitioned into the such m_n subsets $\mathbf{X_{-n}(i_n)}$:

$$\begin{cases} \sum\limits_{s_{-n}\in S_{-n}} (a^n_{k\|s_{-n}} - a^n_{i_n\|s_{-n}}) \prod\limits_{q=\overline{1,n-1}} x^q_{s_q} \leq 0,\ k = 1,\ldots,m_n, \\ x^q_1 + x^q_2 + \cdots + x^q_{m_q} = 1,\ q = 1,\ldots,n-1, \\ \mathbf{x^{-n}} \geq \mathbf{0}, \end{cases}$$

$\mathbf{x^{-n}} \in \mathbb{R}^{\mathbf{m_1\times\cdots\times m_{n-1}}}$, $i_n = 1,\ldots,m_n$, for which one of the optimal solution of linear programming problem (7.2) is $\mathbf{e}^{x^n_i}$.

Consider notation

$$\mathbf{U_n} = \{i_n \in \{1,2,\ldots,m_n\} : \mathbf{X_{-n}(i_n)} \neq \emptyset\},\ \mathbf{ep}^T = (1,\ldots,1) \in \mathbb{R}^{\mathbf{m_n}}.$$

According to the simplex method optimality criterion, for all $i_n \in \mathbf{U_n}$ and for all $\mathbf{I_n} \in 2^{\mathbf{U_n}\setminus\{i_n\}}$, all the points of

$$\begin{aligned} \Delta(i_n I_n) &= \mathbf{Conv}\{\mathbf{e}^{x^n_k}, k \in \mathbf{I_n} \cup \{i_n\}\} \\ &= \left\{ \mathbf{x} \in \mathbb{R}^{\mathbf{m_n}} : \begin{array}{c} \mathbf{ep}^T\mathbf{x^n} = 1, \\ \mathbf{x^n} \geq 0, \\ x^n_k = 0, k \notin \mathbf{I_n} \cup \{i_n\} \end{array} \right\} \end{aligned}$$

are optimal for $\mathbf{x^{-n}} \in \mathbf{X_{-n}(i_n I_n)} \subset \mathbb{R}^{\mathbf{m_1\times\cdots\times m_{n-1}}}$, where $\mathbf{X_{-n}(i_n I_n)}$ is the solution set of the system:

$$\begin{cases} \sum\limits_{s_{-n}\in S_{-n}} (a^n_{k\|s_{-n}} - a^n_{i_n\|s_{-n}}) \prod\limits_{q=1,\ldots,n-1} x^q_{s_q} = 0, k \in \mathbf{I_n}, \\ \sum\limits_{s_{-n}\in S_{-n}} (a^n_{k\|s_{-n}} - a^n_{i_n\|s_{-n}}) \prod\limits_{q=1,\ldots,n-1} x^q_{s_q} \leq 0, k \notin \mathbf{I_n} \cup \{i_n\}, \\ \mathbf{er}^T\mathbf{x^r} = 1, r = 1,\ldots,n-1, \\ \mathbf{x^r} \geq \mathbf{0}, r = 1,\ldots,n-1. \end{cases}$$

Evidently, $\mathbf{X_{-n}}(i_n\emptyset) = \mathbf{X_{-n}(i_n)}$. Hence,

$$\begin{aligned} \mathbf{Gr_n} &= \bigcup_{\mathbf{i_n}\in\mathbf{U_n},\ \mathbf{I_n}\in 2^{\mathbf{U_n}\setminus\{\mathbf{i_n}\}}} \Delta(i_n I_n) \times \mathbf{X_n(i_n I_n)} \\ &= \bigcup_{\mathbf{i_n}\in\mathbf{U_n},\ \mathbf{I_n}\in 2^{\mathbf{U_n}\Delta\{\mathbf{i_n}\}}} \mathbf{X(i_n I_n)}, \end{aligned} \tag{7.3}$$

where $\mathbf{X(i_n I_n)}$ is the solution set of the system:

$$\begin{cases} \sum\limits_{s_{-n} \in S_{-n}} (a^n_{k\|s_{-n}} - a^n_{i_n\|s_{-n}}) \prod\limits_{q=\overline{1,n-1}} x^q_{s_q} = 0, k \in \mathbf{I_n}, \\ \sum\limits_{s_{-n} \in S_{-n}} (a^n_{k\|s_{-n}} - a^n_{i_n\|s_{-n}}) \prod\limits_{q=\overline{1,n-1}} x^q_{s_q} \leq 0, k \notin \mathbf{I_n} \cup \{i_n\}, \\ \mathbf{er}^T \mathbf{x^r} = 1,\ \mathbf{x^r} \geq \mathbf{0}, r = 1, \ldots, n-1, \\ \mathbf{ep}^T \mathbf{x^n} = 1, \mathbf{x^n} \geq \mathbf{0}, \\ x^n_k = 0, k \notin \mathbf{I_n} \cup \{i_n\}, \end{cases} \tag{7.4}$$

which is an $n-1$-linear system of equalities and inequalities.

Taking into account the indices i_n and $\mathbf{I_n}$, the graph $\mathbf{Gr_n}$ may by represented in the following form too:

$$\left[\begin{array}{l} \begin{cases} \sum\limits_{s_{-n} \in S_{-n}} (a^n_{k\|s_{-n}} - a^n_{1\|s_{-n}}) \prod\limits_{q=1,\ldots,n-1} x^q_{s_q} = 0, k \in \mathbf{I_1}, \\ \sum\limits_{s_{-n} \in S_{-n}} (a^n_{k\|s_{-n}} - a^n_{1\|s_{-n}}) \prod\limits_{q=1,\ldots,n-1} x^q_{s_q} \leq 0, k \notin \mathbf{I_1} \cup \{1\}, \\ \mathbf{er}^T \mathbf{x^r} = 1,\ \mathbf{x^r} \geq \mathbf{0}, r = 1, \ldots, n-1, \\ \mathbf{ep}^T \mathbf{x^n} = 1, \mathbf{x^n} \geq \mathbf{0}, \\ x^n_k = 0, k \notin \mathbf{I_1} \cup \{1\}. \end{cases} \\ \ldots \\ \begin{cases} \sum\limits_{s_{-n} \in S_{-n}} (a^n_{k\|s_{-n}} - a^n_{m_n-1\|s_{-n}}) \prod\limits_{q=1,\ldots,n-1} x^q_{s_q} = 0, k \in \mathbf{I_{m_n-1}}, \\ \sum\limits_{s_{-n} \in S_{-n}} (a^n_{k\|s_{-n}} - a^n_{m_n-1\|s_{-n}}) \prod\limits_{q=1,\ldots,n-1} x^q_{s_q} \leq 0, \\ k \notin \mathbf{I_{m_n-1}} \cup \{m_n - 1\}, \\ \mathbf{er}^T \mathbf{x^r} = 1,\ \mathbf{x^r} \geq \mathbf{0}, r = 1, \ldots, n-1, \\ \mathbf{ep}^T \mathbf{x^n} = 1, \mathbf{x^n} \geq \mathbf{0}, \\ x^n_k = 0, k \notin \mathbf{I_{m_n-1}} \cup \{m_n - 1\}. \end{cases} \\ \begin{cases} \sum\limits_{s_{-n} \in S_{-n}} (a^n_{k\|s_{-n}} - a^n_{m_n\|s_{-n}}) \prod\limits_{q=1,\ldots,n-1} x^q_{s_q} \leq 0, k = 1, \ldots, m_n, \\ \mathbf{er}^T \mathbf{x^r} = 1,\ \mathbf{x^r} \geq \mathbf{0}, r = 1, \ldots, n-1, \\ \mathbf{ep}^T \mathbf{x^n} = 1, \\ x^n_{m_n} \geq 0, x^n_k = 0, k = 1, \ldots, (m_n - 1). \end{cases} \end{array}\right.$$

The player $n-1$ solves a bilinear parametric optimization problem on the graph $\mathbf{Gr_n}$:

$$\begin{array}{l} f_{n-1}(\mathbf{x^1}, \ldots, \mathbf{x^{n-2}}, \mathbf{y^{n-1}}, \mathbf{y^n}) \rightarrow max, \\ (\mathbf{x^1}, \ldots, \mathbf{x^{n-2}}, \mathbf{y^{n-1}}, \mathbf{y^n}) \in \mathbf{Gr_n}. \end{array}$$

Evidently, this problem is equivalent to the series of optimization problems of the form:

$$f_{n-1}(\mathbf{x^1}, \ldots, \mathbf{x^{n-2}}, \mathbf{y^{n-1}}, \mathbf{y^n}) \to max,$$

$$(\mathbf{x^1}, \ldots, \mathbf{x^{n-2}}, \mathbf{y^{n-1}}, \mathbf{y^n}) \in \mathbf{X(i_n I_n)},$$

where $\mathbf{X(i_n I_n)}$ denotes component (7.4) of graph (7.3).

When the player $n-1$ maximizes his payoff function on each non-empty component $\mathbf{X(i_n I_n)}$, the simplex corresponding to $\mathbf{x^1}, \ldots, \mathbf{x^{n-2}}$ is divided into unknown prior k_{n-1} parts. The values of the function

$$f_{n-1}(\mathbf{x^1}, \ldots, \mathbf{x^{n-2}}, \mathbf{y^{n-1}}, \mathbf{y^n})$$

on all the parts of $\mathbf{X(i_n I_n)}$ are compared and the optimal solutions are saved. The graph $\mathbf{Gr_{n-1}}$ consists of all the parts for which the best values of the function $f_{n-1}(\mathbf{x^1}, \ldots, \mathbf{x^{n-2}}, \mathbf{y^{n-1}}, \mathbf{y^n})$ are attained and the best strategies of the players $n-1$ and n are the vectors $\mathbf{y^{n-1}}$ and $\mathbf{y^n}$ which correspond to the saved solutions.

The process is repeated for players $n-2, n-3, \ldots, 3$.

The second player maximizes his gain function $f_2(\mathbf{x^1}, \mathbf{y^2}, \ldots, \mathbf{y^n})$ on each component of the graph $\mathbf{Gr_3}$. The simplex corresponding to $\mathbf{x^1}$ is split into unknown prior k_2 parts. The optimal values of the function $f_2(\mathbf{x^1}, \mathbf{y^2}, \ldots, \mathbf{y^n})$ on all the parts are compared and the best is saved. The graph $\mathbf{Gr_2}$ consists of the parts for which the best value of the function $f_2(\mathbf{x^1}, \mathbf{y^2}, \ldots, \mathbf{y^n})$ is achieved and the best strategies of the players $2, 3, \ldots, n$: $\mathbf{y^2}, \mathbf{y^3}, \ldots, \mathbf{y^n}$, corresponding to the saved solutions.

First player calculates his best moves on each components of the $\mathbf{Gr_2}$ and determines the set of unsafe Stackelberg equilibria:

$$\mathbf{USES} = \operatorname*{Arg\,max}_{(\mathbf{y^1}, \ldots, \mathbf{y^n}) \in \mathbf{Gr_2}} f_1\left(\mathbf{y^1}, \ldots, \mathbf{y^n}\right)$$

The multi-linear property of the payoff functions, the properties of components (7.4) and the entire above exposition proves the theorem. □

Theorem 7.10 *The set* **USES** *of unsafe Stackelberg equilibria has no more than* $2^{m_1+m_2+\cdots+m_n} - 1$ *components described as a system of equalities and inequalities.*

Proof It is enough to refer expression (7.3) which defines the structure of the graph $\mathbf{Gr_n}$, expression (7.4) that defines components of the graph, the number n of stages and the numbers $m_1, m_2, \ldots, m_n$ of pure strategies at stages $1, 2, \ldots, n$. □

Example 7.4 Let us consider a three-player mixed-strategy $2 \times 2 \times 2$ game with matrices:

$$\mathbf{a_{1**}} = \begin{bmatrix} 9 & 0 \\ 0 & 3 \end{bmatrix}, \quad a_{2**} = \begin{bmatrix} 0 & 3 \\ 9 & 0 \end{bmatrix},$$

$$\mathbf{b}_{*1*} = \begin{bmatrix} 8 & 0 \\ 0 & 4 \end{bmatrix}, \quad b_{*2*} = \begin{bmatrix} 0 & 4 \\ 8 & 0 \end{bmatrix},$$

$$\mathbf{c}_{**1} = \begin{bmatrix} 12 & 0 \\ 0 & 2 \end{bmatrix}, \quad c_{**2} = \begin{bmatrix} 0 & 6 \\ 4 & 0 \end{bmatrix}.$$

and the payoff functions:

$$f_1(\mathbf{x}, \mathbf{y}, \mathbf{z}) = (9y_1z_1 + 3y_2z_2)x_1 + (9y_2z_1 + 3y_1z_2)x_2,$$

$$f_2(\mathbf{x}, \mathbf{y}, \mathbf{z}) = (8x_1z_1 + 4x_2z_2)y_1 + (8x_2z_1 + 4x_1z_2)y_2,$$

$$f_3(\mathbf{x}, \mathbf{y}, \mathbf{z}) = (12x_1y_1 + 2x_2y_2)z_1 + (4x_2y_1 + 6x_1y_2)z_2.$$

First, the graph of best response mapping of the third player is determined. The component $\mathbf{X}(\mathbf{1}\emptyset)$ is defined by all the solutions of the following system

$$\begin{cases} 4x_2y_1 + 6x_1y_2 - 12x_1y_1 - 2x_2y_2 \leq 0, \\ x_1 + x_2 = 1, x_1 \geq 0, x_2 \geq 0, \\ y_1 + y_2 = 1, y_1 \geq 0, y_2 \geq 0, \\ z_1 + z_2 = 1, z_1 \geq 0, z_2 = 0. \end{cases}$$

So,

$$\mathbf{X}(\mathbf{1}\emptyset) = \left\{ \begin{pmatrix} 0 \leq x_1 < \frac{1}{4} \\ 1 - x_1 \end{pmatrix} \times \begin{pmatrix} 0 \leq y_1 \leq \frac{1}{3} \\ 1 - y_1 \end{pmatrix} \times \begin{pmatrix} 1 \\ 0 \end{pmatrix} \right.$$
$$\cup \begin{pmatrix} \frac{1}{4} \\ \frac{3}{4} \end{pmatrix} \times \begin{pmatrix} 0 \leq y_1 \leq 1 \\ 1 - y_1 \end{pmatrix} \times \begin{pmatrix} 1 \\ 0 \end{pmatrix}$$
$$\left. \cup \begin{pmatrix} \frac{1}{4} < x_1 \leq 1 \\ 1 - x_1 \end{pmatrix} \times \begin{pmatrix} \frac{1}{3} \leq y_1 \leq 1 \\ 1 - y_1 \end{pmatrix} \times \begin{pmatrix} 1 \\ 0 \end{pmatrix} \right\},$$

For $\mathbf{X}(\mathbf{1}\{\mathbf{2}\})$ the system is

$$\begin{cases} 4x_2y_1 + 6x_1y_2 - 12x_1y_1 - 2x_2y_2 = 0, \\ x_1 + x_2 = 1, x_1 \geq 0, x_2 \geq 0, \\ y_1 + y_2 = 1, y_1 \geq 0, y_2 \geq 0, \\ z_1 + z_2 = 1, z_1 \geq 0, z_2 \geq 0, \end{cases}$$

and

$$\mathbf{X}(\mathbf{1}\{\mathbf{2}\}) = \left\{ \begin{pmatrix} 0 \le x_1 < \frac{1}{4} \\ 1 - x_1 \end{pmatrix} \times \begin{pmatrix} \frac{1}{3} \\ \frac{2}{3} \end{pmatrix} \times \begin{pmatrix} 0 \le z_1 \le 1 \\ 1 - z_1 \end{pmatrix} \right.$$
$$\bigcup \begin{pmatrix} \frac{1}{4} \\ \frac{3}{4} \end{pmatrix} \times \begin{pmatrix} 0 \le y_1 \le 1 \\ 1 - y_1 \end{pmatrix} \times \begin{pmatrix} 0 \le z_1 \le 1 \\ 1 - z_1 \end{pmatrix}$$
$$\left. \bigcup \begin{pmatrix} \frac{1}{4} < x_1 \le 1 \\ 1 - x_1 \end{pmatrix} \times \begin{pmatrix} \frac{1}{3} \\ \frac{2}{3} \end{pmatrix} \times \begin{pmatrix} 0 \le z_1 \le 1 \\ 1 - z_1 \end{pmatrix} \right\}.$$

For $\mathbf{X}(\mathbf{2}\emptyset)$ the system is

$$\begin{cases} -4x_2y_1 - 6x_1y_2 + 12x_1y_1 + 2x_2y_2 \le 0, \\ x_1 + x_2 = 1, x_1 \ge 0, x_2 \ge 0, \\ y_1 + y_2 = 1, y_1 \ge 0, y_2 \ge 0, \\ z_1 + z_2 = 1, z_1 = 0, z_2 \ge 0, \end{cases}$$

and

$$\mathbf{X}(\mathbf{2}\emptyset) = \left\{ \begin{pmatrix} 0 \le x_1 < \frac{1}{4} \\ 1 - x_1 \end{pmatrix} \times \begin{pmatrix} \frac{1}{3} \le y_1 \le 1 \\ 1 - y_1 \end{pmatrix} \times \begin{pmatrix} 0 \\ 1 \end{pmatrix} \right.$$
$$\bigcup \begin{pmatrix} \frac{1}{4} \\ \frac{3}{4} \end{pmatrix} \times \begin{pmatrix} 0 \le y_1 \le 1 \\ 1 - y_1 \end{pmatrix} \times \begin{pmatrix} 0 \\ 1 \end{pmatrix}$$
$$\left. \bigcup \begin{pmatrix} \frac{1}{4} < x_1 \le 1 \\ 1 - x_1 \end{pmatrix} \times \begin{pmatrix} 0 \le y_1 \le \frac{1}{3} \\ 1 - y_1 \end{pmatrix} \times \begin{pmatrix} 0 \\ 1 \end{pmatrix} \right\}.$$

So, $\mathbf{Gr_3} = \mathbf{X}(\mathbf{1}\emptyset) \bigcup \mathbf{X}(\mathbf{1}\{\mathbf{2}\}) \bigcup \mathbf{X}(\mathbf{2}\emptyset)$.

Let us consider the problem of determining the set of optimal points of the second player utility function on the graph of best response mapping of the third player. The maximum value of the second player utility function on the each component of the graph Gr_3 is determined.

By applying substitutions: $x_1 = x$ and $x_2 = 1 - x$, $y_1 = y$ and $y_2 = 1 - y$, $z_1 = z$ and $z_2 = 1 - z$, we obtain:

$$f_2(x, y, z) = 24xyz - 8xy - 12xz - 12yz + 4x + 4y + 8z.$$

Let us determine the maximum of this function on the each component of $\mathbf{Gr_3}$.

For $\mathbf{X}(\mathbf{1}\emptyset)$ the system is

$$\begin{cases} 8 - 8x, & x \in \left[0, \frac{1}{4}\right), \ y = 0, \ z = 1, \\ 6, & x = \frac{1}{4}, \ y = 0, \ z = 1, \\ \frac{1}{3}(16 - 8x), & x \in \left(\frac{1}{4}, \frac{1}{2}\right), \ y = \frac{1}{3}, \ z = 1, \\ 4, & x\frac{1}{2}, \ y = \frac{2}{3}, \ z = 1, \\ 8x, & x \in \left(\frac{1}{2}, 1\right], \ y = 1, \ z = 1. \end{cases}$$

For $\mathbf{X}(\mathbf{1}\{\mathbf{2}\})$ the system is

$$\begin{cases} \frac{1}{3}(16-8x), \; x \in \left[0, \frac{1}{4}\right), \; y = \frac{1}{3}, \; z = 1, \\ 6, \qquad\qquad x = \frac{1}{4}, \; y = 0, \; z = 1, \\ \frac{1}{3}(16-8x), \; x \in \left(\frac{1}{4}, 1\right], \; y = \frac{1}{3}, \; z = 1. \end{cases}$$

For $\mathbf{X}(\mathbf{2}\emptyset)$ the system is

$$\begin{cases} 4-4x, \qquad x \in \left[0, \frac{1}{4}\right), \; y = 1, \; z = 0, \\ 3, \qquad\qquad x = \frac{1}{4}, \; y = 1, \; z = 0, \\ \frac{1}{3}(4+4x), \; x \in \left(\frac{1}{4}, \frac{1}{2}\right), \; y = \frac{1}{3}, \; z = 0, \\ 2, \qquad\qquad x\frac{1}{2}, \; y = \frac{1}{6}, \; z = 0, \\ 4x, \qquad\quad x \in \left(\frac{1}{2}, 1\right], \; y = 0, \; z = 0. \end{cases}$$

As a result, the $\mathbf{Gr_2}$ may be represented as

$$\begin{cases} \begin{pmatrix} 0 \le x \le \frac{1}{4} \\ 1-x \end{pmatrix} \times \begin{pmatrix} 0 \\ 1 \end{pmatrix} \times \begin{pmatrix} 1 \\ 0 \end{pmatrix}, \\ \begin{pmatrix} \frac{1}{4} < x \le \frac{1}{2} \\ 1-x \end{pmatrix} \times \begin{pmatrix} \frac{1}{3} \\ \frac{2}{3} \end{pmatrix} \times \begin{pmatrix} 1 \\ 0 \end{pmatrix}, \\ \begin{pmatrix} \frac{1}{2} < x \le 1 \\ 1-x \end{pmatrix} \times \begin{pmatrix} 1 \\ 0 \end{pmatrix} \times \begin{pmatrix} 1 \\ 0 \end{pmatrix}. \end{cases}$$

Let us determine the Stackelberg equilibrium set.
The gain function of the first player is:

$$f_1(x, y, z) = 24xyz - 6xy - 12xz - 12yz + 3x + 3y + 9z.$$

So, the components of Gr_2 may be used to determine the Stackelberg equilibrium set.

For,

$$\mathbf{Gr_2}(1) : x \in \left[0, \tfrac{1}{4}\right], \; y = 0, \; z = 1,$$

$$\hat{S} = \hat{S}_1 = \max_{(x,y,z) \in \mathbf{Gr_2}(1)} f_1(x, y, z) = \max_{(x,y,z) \in \mathbf{Gr_2}(1)} (-9x + 9) = 9,$$

$$\mathbf{USES} = \left\{ \begin{pmatrix} 0 \\ 1 \end{pmatrix} \times \begin{pmatrix} 0 \\ 1 \end{pmatrix} \times \begin{pmatrix} 1 \\ 0 \end{pmatrix} \right\}.$$

For

$$\mathbf{Gr_2}(2) : x \in \left(\tfrac{1}{4}, \tfrac{1}{2}\right], \ y = \tfrac{1}{3}, \ z = 1,$$

$$\hat{S}_2 = \max_{(x,y,z)\in \mathbf{Gr_2}(2)} f_1(x, y, z) = \max_{(x,y,z)\in \mathbf{Gr_2}(2)} (-3x + 6) = \frac{21}{4},$$

in the point $\left(\frac{1}{4}, \frac{1}{3}, 1\right)$, but $\frac{1}{4} \notin \left(\frac{1}{4}, \frac{1}{2}\right]$, anyway, $\hat{S} > \frac{21}{4}$ and the set **USES** remains unchanged.

For

$$\mathbf{Gr_2}(3) : x \in \left(\tfrac{1}{2}, 1\right], \ y = 1, \ z = 1,$$

$$\hat{S}_3 = \max_{(x,y,z)\in \mathbf{Gr_2}(3)} f_1(x, y, z) = \max_{(x,y,z)\in \mathbf{Gr_2}(3)} 9x = 9, \ \hat{S} = 9,$$

$$\mathbf{USES} = \left\{ \begin{pmatrix} 0 \\ 1 \end{pmatrix} \times \begin{pmatrix} 0 \\ 1 \end{pmatrix} \times \begin{pmatrix} 1 \\ 0 \end{pmatrix} \cup \begin{pmatrix} 1 \\ 0 \end{pmatrix} \times \begin{pmatrix} 1 \\ 0 \end{pmatrix} \times \begin{pmatrix} 1 \\ 0 \end{pmatrix} \right\}.$$

So,

$$\mathbf{USES} = \left\{ \begin{pmatrix} 0 \\ 1 \end{pmatrix} \times \begin{pmatrix} 0 \\ 1 \end{pmatrix} \times \begin{pmatrix} 1 \\ 0 \end{pmatrix} \cup \begin{pmatrix} 1 \\ 0 \end{pmatrix} \times \begin{pmatrix} 1 \\ 0 \end{pmatrix} \times \begin{pmatrix} 1 \\ 0 \end{pmatrix} \right\}$$

with the gains (9, 8, 12) and (9, 8, 4). □

Both equilibria in the precedent example are Stackelberg equilibria in the both, pure and mixed strategy, games. By modifying the example, a continuum set of equilibria may be obtained.

Example 7.5 Consider a three-player mixed-strategy $2 \times 2 \times 2$ game with matrices

$$\mathbf{a}_{1**} = \begin{bmatrix} 2 & 5 \\ 5 & 3 \end{bmatrix}, \quad a_{2**} = \begin{bmatrix} 3 & 5 \\ 5 & 2 \end{bmatrix},$$

$$\mathbf{b}_{*1*} = \begin{bmatrix} 6 & 6 \\ 0 & 6 \end{bmatrix}, \quad b_{*2*} = \begin{bmatrix} 0 & 4 \\ 4 & 4 \end{bmatrix},$$

$$\mathbf{c}_{**1} = \begin{bmatrix} 12 & 0 \\ 0 & 2 \end{bmatrix}, \quad c_{**2} = \begin{bmatrix} 0 & 6 \\ 4 & 0 \end{bmatrix},$$

and the gain functions

$$\begin{aligned} f_1(\mathbf{x}, \mathbf{y}, \mathbf{z}) &= (2y_1z_1 + 5y_1z_2 + 5y_2z_1 + 3y_2z_2)x_1 \\ &\quad +(3y_1z_1 + 5y_1z_2 + 5y_2z_1 + 2y_2z_2)x_2, \\ f_2(\mathbf{x}, \mathbf{y}, \mathbf{z}) &= (6x_1z_1 + 6x_1z_2 + 6x_2z_2)y_1 \\ &\quad +(4x_1z_2 + 4x_2z_1 + 4x_2z_2)y_2, \\ f_3(\mathbf{x}, \mathbf{y}, \mathbf{z}) &= (12x_1y_1 + 2x_2y_2)z_1 + (6x_1y_2 + 4x_2y_1)z_2. \end{aligned}$$

First, let us determine the graph of best response mapping of the third player. The component $\mathbf{X}(1\emptyset)$ is defined by all the solutions of the system

$$\begin{cases} 4x_2y_1 + 6x_1y_2 - 12x_1y_1 - 2x_2y_2 \le 0, \\ x_1 + x_2 = 1,\, x_1 \ge 0,\, x_2 \ge 0, \\ y_1 + y_2 = 1,\, y_1 \ge 0,\, y_2 \ge 0, \\ z_1 + z_2 = 1,\, z_1 \ge 0,\, z_2 = 0, \end{cases}$$

so

$$\begin{aligned} \mathbf{X}(1\emptyset) = & \left\{ \begin{pmatrix} 0 \le x_1 < \frac{1}{4} \\ 1 - x_1 \end{pmatrix} \times \begin{pmatrix} 0 \le y_1 \le \frac{1}{3} \\ 1 - y_1 \end{pmatrix} \times \begin{pmatrix} 1 \\ 0 \end{pmatrix} \right. \\ & \cup \begin{pmatrix} \frac{1}{4} \\ \frac{3}{4} \end{pmatrix} \times \begin{pmatrix} 0 \le y_1 \le 1 \\ 1 - y_1 \end{pmatrix} \times \begin{pmatrix} 1 \\ 0 \end{pmatrix} \\ & \left. \cup \begin{pmatrix} \frac{1}{4} < x_1 \le 1 \\ 1 - x_1 \end{pmatrix} \times \begin{pmatrix} \frac{1}{3} \le y_1 \le 1 \\ 1 - y_1 \end{pmatrix} \times \begin{pmatrix} 1 \\ 0 \end{pmatrix} \right\}, \end{aligned}$$

For $\mathbf{X}(1\{2\})$ the system is

$$\begin{cases} 4x_2y_1 + 6x_1y_2 - 12x_1y_1 - 2x_2y_2 = 0, \\ x_1 + x_2 = 1,\, x_1 \ge 0,\, x_2 \ge 0, \\ y_1 + y_2 = 1,\, y_1 \ge 0,\, y_2 \ge 0, \\ z_1 + z_2 = 1,\, z_1 \ge 0,\, z_2 \ge 0, \end{cases}$$

and

$$\begin{aligned} \mathbf{X}(1\{2\}) = & \left\{ \begin{pmatrix} 0 \le x_1 < \frac{1}{4} \\ 1 - x_1 \end{pmatrix} \times \begin{pmatrix} \frac{1}{3} \\ \frac{2}{3} \end{pmatrix} \times \begin{pmatrix} 0 \le z_1 \le 1 \\ 1 - z_1 \end{pmatrix} \right. \\ & \cup \begin{pmatrix} \frac{1}{4} \\ \frac{3}{4} \end{pmatrix} \times \begin{pmatrix} 0 \le y_1 \le 1 \\ 1 - y_1 \end{pmatrix} \times \begin{pmatrix} 0 \le z_1 \le 1 \\ 1 - z_1 \end{pmatrix} \\ & \left. \cup \begin{pmatrix} \frac{1}{4} < x_1 \le 1 \\ 1 - x_1 \end{pmatrix} \times \begin{pmatrix} \frac{1}{3} \\ \frac{2}{3} \end{pmatrix} \times \begin{pmatrix} 0 \le z_1 \le 1 \\ 1 - z_1 \end{pmatrix} \right\}, \end{aligned}$$

For $\mathbf{X}(2\emptyset)$ the system is

$$\begin{cases} -4x_2y_1 - 6x_1y_2 + 12x_1y_1 + 2x_2y_2 \le 0, \\ x_1 + x_2 = 1,\, x_1 \ge 0,\, x_2 \ge 0, \\ y_1 + y_2 = 1,\, y_1 \ge 0,\, y_2 \ge 0, \\ z_1 + z_2 = 1,\, z_1 = 0,\, z_2 \ge 0, \end{cases}$$

and

$$\mathbf{X}(\mathbf{2}\emptyset) = \left\{ \begin{pmatrix} 0 \le x_1 < \frac{1}{4} \\ 1 - x_1 \end{pmatrix} \times \begin{pmatrix} \frac{1}{3} \le y_1 \le 1 \\ 1 - y_1 \end{pmatrix} \times \begin{pmatrix} 0 \\ 1 \end{pmatrix} \right.$$
$$\bigcup \begin{pmatrix} \frac{1}{4} \\ \frac{3}{4} \end{pmatrix} \times \begin{pmatrix} 0 \le y_1 \le 1 \\ 1 - y_1 \end{pmatrix} \times \begin{pmatrix} 0 \\ 1 \end{pmatrix}$$
$$\left. \bigcup \begin{pmatrix} \frac{1}{4} < x_1 \le 1 \\ 1 - x_1 \end{pmatrix} \times \begin{pmatrix} 0 \le y_1 \le \frac{1}{3} \\ 1 - y_1 \end{pmatrix} \times \begin{pmatrix} 0 \\ 1 \end{pmatrix} \right\}.$$

Finally, $\mathbf{Gr_3} = \mathbf{X}(\mathbf{1}\emptyset) \bigcup \mathbf{X}(\mathbf{1}\{\mathbf{2}\}) \bigcup \mathbf{X}(\mathbf{2}\emptyset)$.

Let us determine the set of optimal points of the second player utility function constrained on the graph of best response mapping of the third player. The maximum of the second player utility function on each component of the graph is determined.

By applying substitutions: $x_1 = x$ and $x_2 = 1-x$, $y_1 = y$ and $y_2 = 1-y$, $z_1 = z$ and $z_2 = 1 - z$, it follows:

$$f_2(x, y, z) = 10xyz + 2y - 4xz - 6yz + 4.$$

The maximum of the second player utility function on the $\mathbf{Gr_3}$ is determined.

For $\mathbf{X}(\mathbf{1}\emptyset)$:

$$\begin{cases} 4 - 4x, & x \in \left[0, \frac{1}{4}\right),\ y = 0,\ z = 1, \\ 3, & x = \frac{1}{4},\ y = 0,\ z = 1, \\ \frac{1}{3}(8 - 2x), & x \in \left(\frac{1}{4}, \frac{2}{5}\right),\ y = \frac{1}{3},\ z = 1, \\ \frac{12}{5}, & x = \frac{2}{5},\ y = \frac{2}{3},\ z = 1, \\ 6x, & x \in \left(\frac{2}{5}, 1\right],\ y = 1,\ z = 1. \end{cases}$$

For $\mathbf{X}(\mathbf{1}\{\mathbf{2}\})$:

$$\begin{cases} \frac{14}{3}, x \in \left[0, \frac{1}{4}\right),\ y = \frac{1}{3},\ z = 0, \\ 6,\ \ x = \frac{1}{4},\ y = 1,\ z = 0, \\ \frac{14}{3}, x \in \left(\frac{1}{4}, 1\right],\ y = \frac{1}{3},\ z = 0. \end{cases}$$

For $\mathbf{X}(\mathbf{2}\emptyset)$:

$$\begin{cases} 6,\ \ x \in \left[0, \frac{1}{4}\right),\ y = 1,\ z = 0, \\ 6,\ \ x = \frac{1}{4},\ y = 1,\ z = 0, \\ \frac{14}{3}, x \in \left(\frac{1}{4}, 1\right],\ y = \frac{1}{3},\ z = 0. \end{cases}$$

The $\mathbf{Gr_2}$:

$$\left\{\begin{array}{l}\begin{pmatrix}0 \le x \le \frac{1}{4}\\ 1-x\end{pmatrix} \times \begin{pmatrix}1\\0\end{pmatrix} \times \begin{pmatrix}0\\1\end{pmatrix},\\ \begin{pmatrix}\frac{1}{4} < x \le \frac{7}{9}\\ 1-x\end{pmatrix} \times \begin{pmatrix}\frac{1}{3}\\\frac{2}{3}\end{pmatrix} \times \begin{pmatrix}0\\1\end{pmatrix},\\ \begin{pmatrix}\frac{7}{9} \le x \le 1\\ 1-x\end{pmatrix} \times \begin{pmatrix}1\\0\end{pmatrix} \times \begin{pmatrix}1\\0\end{pmatrix}.\end{array}\right.$$

So, the **SES** is determined.

$$f_1(x, y, z) = -xy - xz - 5yz + x + 3y + 3z + 2.$$

$$\mathbf{Gr_2}(1) : x \in \left[0, \tfrac{1}{4}\right], \ y = 1, \ z = 0,$$

$$\hat{S} = \hat{S}_1 = \max_{(x,y,z)\in\mathbf{Gr_2}(1)} f_1(x, y, z) = \max_{(x,y,z)\in\mathbf{Gr_2}(1)} 5 = 5,$$

$$\mathbf{SES} = \left\{\begin{pmatrix}0 \le x \le \frac{1}{4}\\ 1-x\end{pmatrix} \times \begin{pmatrix}1\\0\end{pmatrix} \times \begin{pmatrix}0\\1\end{pmatrix}\right\}.$$

$$\mathbf{Gr_2}(2) : x \in \left(\tfrac{1}{4}, \tfrac{7}{9}\right], \ y = \tfrac{1}{3}, \ z = 0,$$

$$\hat{S}_2 = \max_{(x,y,z)\in\mathbf{Gr_2}(2)} f_1(x, y, z) = \max_{(x,y,z)\in\mathbf{Gr_2}(2)} \left(\frac{2}{3}x + 3\right) = \frac{95}{27}.$$

$\hat{S} > \frac{95}{27}$, the **SES** remain unchanged.

$$\mathbf{Gr_2}(3) : x \in \left[\tfrac{7}{9}, 1\right], \ y = 1, \ z = 1,$$

$$\hat{S}_3 = \max_{(x,y,z)\in\mathbf{Gr_2}(3)} f_1(x, y, z) = \max_{(x,y,z)\in\mathbf{Gr_2}(3)} (-x + 3) = \frac{20}{9}.$$

$\hat{S} > \frac{20}{9}$, the **SES** remain unchanged.

Finally, $\mathbf{SES} = \left\{\begin{pmatrix}0 \le x \le \frac{1}{4}\\ 1-x\end{pmatrix} \times \begin{pmatrix}1\\0\end{pmatrix} \times \begin{pmatrix}0\\1\end{pmatrix}\right\}.$ □

7.5 Conclusions

The idea to consider the Stackelberg equilibrium set as a set of optimal points of the first player utility function constrained on the graph of best response mapping of the second player yields to a method of Stackelberg equilibrium set computing in a Stackelberg two-player mixed-strategy game.

The Stackelberg equilibrium set in bimatrix mixed-strategy games may be partitioned into finite number of polytopes $\mathbf{XY}^{jJ}$, no more than $(2^n - 1)$; the safe and unsafe Stackelberg equilibra are calculated on each polytopes by comparing optimal values of first player's payoff function on convex components of the graph of best response mapping of the second player. The components' points on which the record is attained form a Stackelberg equilibrium set.

References

1. Von Stackelberg, H. 1934. *Marktform und Gleichgewicht (Market Structure and Equilibrium)*. Vienna: Springer, XIV+134 pp. (in German).
2. Chen, C.I., and J.B. Cruz, Jr. 1972. Stackelberg solution for two-person games with biased information patterns. *IEEE Transactions on Automatic Control* AC-17: 791–798.
3. Simaan, M., and J.B. Cruz Jr. 1973. Additional aspects of the Stackelberg strategy in nonzero sum games. *Journal of Optimization Theory and Applications* 11: 613–626.
4. Simaan, M., and J.B. Cruz Jr. 1973. On the Stackelberg strategy in nonzero sum games. *Journal of Optimization Theory and Applications* 11: 533–555.
5. Leitmann, G. 1978. On generalized Stackelberg strategies. *Journal of Optimization Theory and Applications* 26: 637–648.
6. Blaquière, A. 1976. Une géneralisation du concept d'optimalité et des certains notions geometriques qui's rattachent, Institute des Hautes Etudes de Belgique, Cahiers du centre d'études de recherche operationelle, Vol. 18, No. 1–2, Bruxelles, 49–61.
7. Başar, T., and G.J. Olsder. 1999. *Dynamic noncooperative game theory*. Society for Industrial and Applied Mathematics, vol. 536. Philadelphia: SIAM.
8. Ungureanu, V. 2008. Solution principles for simultaneous and sequential games mixture. *ROMAI Journal* 4 (1): 225–242.
9. Ungureanu, V. 2008. Solution principles for generalized Stackelberg games. In *Mathematical Modelling, Optimization and Information Technologies, International Conference Proceedings, ATIC, March 19–21*. Chişinău. *Evrica*, 181–189.
10. Peters, H. 2015. *Game theory: A multi-leveled approach*, 2nd edn. Berlin: Springer, XVII+494 pp.
11. Ungureanu, V., and V. Lozan, 2016. Stackelberg equilibrium sets in bimatrix mixed-strategy games. In *Mathematical Modelling, Optimization and Information Technologies, International Conference Proceedings, ATIC, March 22–25, 2016*, Chişinău: Evrica, 370–382 (in Romanian).
12. Korzhyk, D., Z. Yin, C. Kiekintveld, V. Conitzer, and M. Tambe. 2011. Stackelberg versus Nash in security games: An extended investigation of interchangeability, equivalence, and uniqueness. *Journal of Artificial Intelligence Research* 41: 297–327.
13. Ungureanu, V. 2006. Nash equilibrium set computing in finite extended games. *Computer Science Journal of Moldova*, 14, No. 3 (42): 345–365.
14. Ungureanu, V. 2013. Linear discrete-time Pareto-Nash-Stackelberg control problem and principles for its solving. *Computer Science Journal of Moldova* 21, No. 1 (61): 65–85.

15. Ungureanu, V., and V. Lozan, 2013. Stackelberg equilibria set in multi-matrix mixed strategy games. In *Proceedings IIS: International Conference on Intelligent Information Systems*, Chişinău: Institute of Mathematics and Computer Science, August 20–23, 2013, 114–117.
16. Arkhangel'skii, A.V., and V.V. Fedorchuk. 1990. The basic concepts and constructions of general topology. In *General topology I: Basic concepts and constructions*, vol. 17, ed. A.V. Arkhangel'skii, and L.S. Pontrjagin, 1–90. Berlin: Dimension Theory, Encyclopaedia of the Mathematical Sciences Springer.
17. Rockafellar, R.T., and R.J.-B. Wets, 2009. *Variational analysis*, 3rd edn. Berlin: Springer, XII+726 pp.
18. Kolmogorov, A.N., and S.V. Fomin. 1957. *Elements of the theory of functions and functional analysis: Metric and normed spaces*, vol. 1, 141. Rochester, New York: Graylock Press.
19. Kolmogorov, A.N., and S.V. Fomin. 1961. *Elements of the theory of functions and functional analysis: Measure, the lebesgue integral, hilbert space*, vol. 2, 139. Rochester, New York: Graylock Press.
20. Kolmogorov , A.N., and S.V. Fomin, 1989. *Elements of the Theory of Functions and Functional Analysis (Elementy teorii funczii i funczionalnogo analiza) Moskva: Nauka*, 6th edn. 632 pp. (in Russian).
21. Kantorovich, L.V., and G.P. Akilov. 1977. *Functional analysis*, 742. Moskva: Nauka.
22. Dantzig, G.B., and M.N. Thapa. 2003. *Linear programming 2: Theory and extensions*, 475. New York: Springer.
23. Rockafellar, T. 1970. *Convex analysis*, 468. Princeton: Princeton University Press.
24. Curtain, R.F., and A.J. Pritchard, 1977. *Functional analysis in modern applied mathematics*. London: Academic Press Inc., IX+339 pp.
25. Collatz, L. 1966. *Functional analysis and numerical mathematics*. New York: Academic Press, X+473 pp.
26. Göpfert, A., Tammer, C., Riahi, H., and C. Zălinescu, 2003. *Variational methods in partially ordered spaces*. New York: Springer, XIV+350 pp.
27. Berge, C. 1963. *Topological Spaces: Including a Treatment of Multi-Valued Functions, Vector Spaces and Convexity*. Edinburgh and London: Oliver and Boyd Ltd, XIII+270 pp.
28. Aubin, J.-P., and H. Frankowska. 1990. *Set-valued analysis, systems and control series*, vol. 2, 480. Boston: Birkhäuser.
29. Burachik, R.S., and A.N. Iusem. 2008. *Set-valued Mappings and Enlargements of Monotone Operators*, 305. New York: Springer Science.
30. Zelinskii, J.B. 1993. *Multi-valued mappings in analysis*. Kiev: Naukova Dumka, 1993, 362 pp. (in Russian).
31. Chen, G., Huang, X., and Yang, X. *Vector optimization: set-valued and variational analysis*. Berlin: Springer, X+308 pp.
32. Bourbaki, N. 1995. *General topology*. Chapters 1–4, Elements of Mathematics, vol. 18, New York: Springer, VII+437 pp.
33. Bourbaki, N. 1998. *General topology*, Chapters 5–10, Elements of Mathematics. New York: Springer, IV+363 pp.
34. Arzelà, C. 1895. Sulle funzioni di linee. Mem. Accad. Sci. Ist. Bologna Cl. Sci. Fis. Mat., 5(5): 55–74 (in Italian).
35. Arzelà, C. 1882–1883. Un'osservazione intorno alle serie di funzioni. Rend. Dell' Accad. R. Delle Sci. Dell'Istituto di Bologna, 142–159 (in Italian).
36. Ascoli, G. 1882–1883. Le curve limiti di una variet data di curve. Atti della R. Accad. Dei Lincei Memorie della Cl. Sci. Fis. Mat. Nat., 18(3): 521–586 (in Italian).
37. Dunford, N., and J.T. Schwartz, 1958. *Linear operators*, vol. 1: General Theory, New York: Wiley Interscience Publishers, XIV+858 pp.
38. Kelley, J.L. 1991. *General topology*. New York: Springer-Verlag, XIV+298 pp.
39. Arkhangel'skii, A.V. 1995. Spaces of mappings and rings of continuous functions. In *General topology III: Paracompactness, Function Spaces, Descriptive Theory*, ed. A.V. Arkhangel'skii. Encyclopaedia of the Mathematical Sciences, vol. 51, 1–70. Berlin: Springer.
40. Nash, J. 1951. Noncooperative games. *Annals of Mathematics* 54 (2): 280–295.

41. Kakutani, S. 1941. A generalization of Brouwer's fixed point theorem. *Duke Mathematical Journal* 8: 457–459.
42. Brouwer, L.E.J. 1911. Über Abbildungen von Mannigfaltigkeiten. *Mathematische Annalen* 71: 97–115. (in German).
43. Berinde, V. 2007. *Iterative approximation of fixed points*, 2nd edn. Berlin: Springer, XV+322 pp.
44. Agarwal, R.P., M. Meehan, and D. O'Regan, 2001. *Fixed point theory and applications*. Cambridge: Cambridge University Press, X+170 pp.
45. Górniewicz, L. 2006. *Topological fixed point theory of multivalued mappings*. Dordrecht: Springer, XIII+538 pp.
46. Carl, S., and S. Heikkilä, 2011. *Fixed point theory in ordered sets and applications: from differential and integral equations to game theory*. New York: Springer Science + Business Media, XIV+477 pp.
47. Border, K.C. 1985. *Fixed point theorems with applications to economics and game theory*. Cambridge: Cambridge University Press, VIII+129 pp.
48. Brown, R.F., M. Furi, L. Górniewicz, and E. Jing, 2005. *Handbook of topological fixed point theory*, Dordrecht, Netherlands: Springer, IX+971 pp.
49. Granas, A., and J. Dugundji, 2003. *Fixed point theory*. New York: Springer, XVI+690 pp.
50. Conitzer, V., and T. Sandholm, 2006. *Computing the optimal strategy to commit to, Proceedings of the 7th ACM Conference on Electronic Commerce, Ann Arbor, MI, USA, June 11–15, 2006*, 82–90.
51. Letchford, J., V. Conitzer, and K. Munagala. 2009. Learning and approximating the optimal strategy to commit to, algorithmic game theory. *Lecture Notes in Computer Science* 5814: 250–262.
52. Marhfour, A. 2000. Mixed solutions for weak Stackelberg problems: existence and stability results. *Journal of Optimization Theory and Applications* 105 (2): 417–440.
53. Ungureanu, V. 2001. *Mathematical programming*. Chişinău: USM, 348 pp. (in Romanian).
54. Dantzig, G.B., and M.N. Thapa. 1997. *Linear programming 1: Introduction*, 474. New York: Springer.
55. Demyanov, V., and V. Malozemov, 1972. *Introduction in minimax*. Moscow: Nauka, 368 pp. (in Russian).
56. Murty, K.G. 2014. *Computational and algorithmic linear algebra and n-dimenshional geometry*, 480. Singapore: World Scientific Publishing Company.
57. Trefethen, L.N., and D. Bau, III, 1997. *Numerical linear algebra*. Philadelphia: Society for Industrial and Applied Mathematics, XII+361 pp.
58. Shoham, Y., and K. Leyton-Brown. 2009. *Multi-agent systems: Algorithmic, game-theoretic, and logical foundations, 532*. Cambridge: Cambridge University Press.
59. Jukna, S. 2011. *Extremal combinatorics: With applications in computer science*, 2nd edn. Berlin: Springer, XXIII+411 pp.
60. Karmarkar, N. 1984. New polynomial-time algorithm for linear programming. *Combinatorica* 4 (4): 373–395.

Chapter 8
Strategic Form Games on Digraphs

Abstract The chapter deals with strategic form games on digraphs, and examines maximin solution concepts based on different types of digraph substructures Ungureanu (Computer Science Journal of Moldova 6 3(18): 313–337, 1998, [1]), Ungureanu (ROMAI Journal 12(1): 133–161, 2016, [2]). Necessary and sufficient conditions for maximin solution existence in digraph matrix games with pure strategies are formulated and proved. Some particular games are considered. Algorithms for finding maximin substructures are suggested. Multi-player simultaneous games and dynamical/hierarchical games on digraphs are considered too.

8.1 Introduction

We regard games which can appear in real situations when several companies manage the activity of a big network. Decision-making persons may have antagonistic interests. In such circumstances, well-known extremal network/digraph problems [3, 4] and problems of constructing various structures on networks/digraphs [3–5] become single or multi criteria strategic network game problems [6–8]. Systems of human, information, hardware (servers, routers, etc.) or other types, controlled by different agents, involve their interactions [7–9]. As a consequence, many traditional network problems have to be treated from the perspective of game theory [6–10], including problems of routing [11], load balancing [12–14], facility location [15], flow control [16, 17], network design [18, 19], network security [20], etc. We have to mention that the related literature is impressively reach and the number and the types of considered problems constantly grow [21–28].

A series of related problems have been investigated and described in scientific literature [6–8, 29] in the context of cyclic games solving. That approach has used a special type of strategy definition [29]. This chapter is based on paper [1] which introduced some types of games on digraphs by defining originally the notions of pure strategies, outcome,indexoutcome and payoff functions.

The chapter is divided into six sections, including introduction and conclusions. Section 8.2 introduces the notion of zero-sum matrix games on digraphs. Some properties giving a general tool for matrix games investigations are proved. Section 8.3

V. Ungureanu, *Pareto-Nash-Stackelberg Game and Control Theory*,
Smart Innovation, Systems and Technologies 89,
https://doi.org/10.1007/978-3-319-75151-1_8

presents some particular solvable games. A special investigation is provided on flow games. It is proved that the problem of maximin cost flow finding is NP-hard. Section 8.4 generalises the notion of digraph matrix game for an arbitrary finite number of players. Section 8.5 introduces the notion of dynamic games.

8.2 Matrix Games on Digraphs

In this section we investigate three types of matrix games on directed graphs:

- basic (root) matrix games or simply **matrix games**,
- **matrix games with admissible strategies**,
- and **matrix games with feasible strategies and profiles**.

The games are defined by means of two related matrices: an outcome matrix, and a payoff matrix.

8.2.1 Concepts

Let us consider a digraph $G = (V, E)$, $|V| = n$, $|E| = m$, further called simply graph. Every directed edge $e \in E$ has the length (weight) $c(e) \in Z$. The vertex set V is partitioned into two disjoint subsets

$$V_1, V_2 \quad (V_1 \cup V_2 = V, \; V_1 \cap V_2 = \varnothing),$$

being positions of two players. The edge set E is partitioned into two disjoint subsets too, as

$$E_1 = \{(u, v) \in E | u \in V_1\}, \quad E_2 = \{(u, v) \in E | u \in V_2\}.$$

Any subset $S_1 \subseteq E_1$ or $S_2 \subseteq E_2$ is called a strategy of the corresponding player. The pair of strategies $(S_1, S_2) \in 2^{E_1} \times 2^{E_2}$ is called a game profile. Any game profile generates a subgraph $G_S = (V, S_1 \cup S_2)$, called the graph of the profile (S_1, S_2), where $S = S_1 \cup S_2$.

Let us introduce some notation.

- $2^G = \{G' = (V, E') \mid E' \subseteq E\}$ denotes the set of all subgraphs of the graph G;
- $\mathscr{D} = \{G' \in 2^G \mid \mathscr{P}\}$ denotes the set of all subgraphs of G, verifying a set of properties $\mathscr{P}$, that is, $\mathscr{D}$ is the set of feasible subgraphs;
- $M: 2^{E_1} \times 2^{E_2} \multimap \mathscr{D}$, $M(S_1, S_2) = 2^{G_S} \cap \mathscr{D}$ denotes the set-valued choice function which maps the graph of the profile $(S_1, S_2) \in 2^{E_1} \times 2^{E_2}$ into the set of all feasible subgraphs of G_S, that is into the subgraphs which verify the set of properties $\mathscr{P}$;
- $C: \mathscr{D} \rightarrow \mathscr{R}$ denotes the choice criterion.

Let

$$k(E_1, E_2, M) = \max_{(S_1,S_2)\in 2^{E_1}\times 2^{E_2}} |M(S_1, S_2)|.$$

be the cardinality of the choice function. There exists four alternatives, for given E_1, E_2, and M:

0^0. $k(E_1, E_2, M) = 0$;
1^0. $k(E_1, E_2, M) = 1$;
2^0. $k(E_1, E_2, M) > 1$ and for any $M(S_1, S_2) \neq \varnothing, G', G'' \in M(S_1, S_2)$, the equality $C(G') = C(G'')$ is true;
3^0. $k(E_1, E_2, M) > 1$ and there exists $M(S_1, S_2) \neq \varnothing$, and the subgraphs $G', G'' \in M(S_1, S_2)$, such that the relation $C(G') \neq C(G'')$ holds.

The case 0^0 doesn't make sense.

The matrix game can be defined when the choice function $M(S_1, S_2)$ verifies either the property 1^0 or 2^0. The case 3^0 (as well as the case 2^0) can be reduced to 1^0 by introducing the choice function

$$\overline{M}(S_1, S_2) = \underset{G'\in M(S_1,S_2)}{\arg\max}\ C(G').$$

It is a mapping assigning to each profile $(S_1, S_2) \in 2^{E_1} \times 2^{E_2}$ an element of $M(S_1, S_2)$, optimal by criterion C. The value $\overline{M}(S_1, S_2) \neq \varnothing$ is a feasible subgraph of the profile (S_1, S_2). We has to remark that the choice function $\overline{M}(S_1, S_2)$ reduces both the cases 2^0 and 3^0 to 1^0.

Now, we can define the (root) matrix game on digraph G via the means of two related matrices: the outcome matrix $\overline{M}(S_1, S_2)$ and the payoff matrix $\overline{C}(S_1, S_2)$.

The outcome matrix of the matrix game is defined by the function $\overline{M}(S_1, S_2)$ and has the same notation. Its elements are either feasible subgraphs or the empty set. The rows of the matrix are identified by the strategies $S_1 \in 2^{E_1}$ of the first player and the columns are identified by the strategies $S_2 \in 2^{E_2}$ of the second player.

The payoff matrix of the matrix game with the same dimensions as the dimensions of the outcome matrix is defined by the function

$$\overline{C}(S_1, S_2) = \begin{cases} C(\overline{M}(S_1, S_2)), & \text{if } \overline{M}(S_1, S_2) \neq \varnothing, \\ -\infty, & \text{if } \overline{M}(S_1, S_2') = \varnothing \ \text{ for all } S_2' \in 2^{E_2}, \\ +\infty, & \text{otherwise}, \end{cases}$$

and has the same notation.

Depending on the set of properties $\mathscr{P}$, various types of games may be investigated.

Remark 8.1 The set $\mathscr{P}$ of properties can induce in a particular game various feasible subgraphs: trees; paths between two fixed vertices v_s and v_t; flows between output vertex v_s and input vertex v_t; matchings; medians; cliques; cycles; Hamiltonian cycles, etc. Any feasible subgraph $\overline{M}(S_1, S_2)$ satisfies both the set of properties $\mathscr{P}$ and the optimality criterion C.

First, let us consider **a root strategic game**

$$\Gamma = \langle 2^{E_1}, 2^{E_2}, \overline{C}(S_1, S_2)\rangle$$

which is a (zero-sum) **matrix game** defined on the graph G. The first player has $2^{|E_1|}$ strategies, the second — $2^{|E_2|}$. The players choose their strategies simultaneously and independently. The first player chooses his strategy S_1 from E_1, the second — S_2 from E_2. Every profile $(S_1, S_2) \in 2^{E_1} \times 2^{E_2}$ has a numerical value $\overline{C}(S_1, S_2)$. For the first player it means the gain $\overline{C}(S_1, S_2)$ if $\overline{C}(S_1, S_2) > 0$ and the loss $\overline{C}(S_1, S_2)$ if $\overline{C}(S_1, S_2) < 0$. For the second player is valid vice versa — it means the loss $\overline{C}(S_1, S_2)$ if $\overline{C}(S_1, S_2) > 0$ and the gain $|\overline{C}(S_1, S_2)|$ if $\overline{C}(S_1, S_2) < 0$. Let us recall that in a zero-sum game the gain $\overline{C}(S_1, S_2)$ of one of the players means the loss $\overline{C}(S_1, S_2)$ of the other.

To introduce two other types of games we need some additional notation and concepts. The sets

$$\mathscr{B}_1 = \left\{S_1 \in 2^{E_1} | \exists S_2 \in 2^{E_2} : \overline{M}(S_1, S_2) \neq \varnothing\right\}, \tag{8.1}$$

$$\mathscr{B}_2 = \left\{S_2 \in 2^{E_2} | \exists S_1 \in 2^{E_1} : \overline{M}(S_1, S_2) \neq \varnothing\right\}, \tag{8.2}$$

are sets of admissible strategies. The sets

$$\mathscr{B}_1(S_2) = \left\{S_1 \in \mathscr{B}_1 | \overline{M}(S_1, S_2) \neq \varnothing\right\},$$

$$\mathscr{B}_2(S_1) = \left\{S_2 \in \mathscr{B}_2 | \overline{M}(S_1, S_2) \neq \varnothing\right\},$$

are sets of admissible strategies, connected with S_1 and S_2 correspondingly.

In such notation, we may consider the game

$$\Gamma^+ = \langle \mathscr{B}_1, \mathscr{B}_2, \overline{C}(S_1, S_2)\rangle$$

which is **a matrix game with admissible strategies**. All the profiles of the game Γ^+ are admissible.

Let us introduce a generic notation

$$\Gamma_*^+ = \langle \mathscr{B}_1(S_2), \mathscr{B}_2(S_1), C(\overline{M}(S_1, S_2))\rangle$$

for a game based on two Stackelberg games [30]: the game

$$\Gamma_1^+ = \langle \mathscr{B}_1, \mathscr{B}_2(S_1), C(\overline{M}(S_1, S_2))\rangle,$$

and the game

$$\Gamma_2^+ = \langle \mathscr{B}_2, \mathscr{B}_1(S_2), C(\overline{M}(S_1, S_2)) \rangle.$$

The players select their strategies consecutively on two stages in these three games.

1. In the game Γ_1^+ the first player moves at the first stage and the second player moves at the second stage.
2. In the game Γ_2^+ the second player moves at the first stage and the first player moves at the second stage.
3. In the game Γ_*^+ we distinguish two stages as well as for Γ_1^+ and Γ_2^+. At the first stage the first player selects his strategy as he plays the game Γ_1^+, and the second player selects his strategy as he plays the game Γ_2^+. At the second stage, one of them, chosen aleatory, may change his strategy, knowing the choice of his opponent at the first stage.

Remark 8.2 It is obvious that the Stackelberg games Γ_1^+ and Γ_2^+ are **matrix games with feasible strategies**. As all the profiles of the game Γ_*^+ are feasible, we will call it **matrix game with feasible profiles**. Clearly, when the game Γ_*^+ is referred, it means implicitly that the games Γ_1^+ and Γ_2^+ are referred too.

Remark 8.3 The game Γ_*^+ may be seen as a special "*matrix*" game, for which the outcome and payoff matrices are obtained from the matrices of the game Γ^+ by deleting the elements with non-finite payoff values. These "*special*" or pseudo matrices may be associated with two dimensional lists in the Wolfram Language [31, 32].

8.2.2 Properties of Digraph Matrix Games

The games Γ, Γ^+, and Γ_*^+, have some interesting and important properties. Let us investigate and highlight them.

Lemma 8.1 *In any games Γ and Γ^+ the following relations between lower and upper values of the games hold:*

$$\max_{S_1 \in 2^{E_1}} \min_{S_2 \in 2^{E_2}} \overline{C}(S_1, S_2) \leq \min_{S_2 \in 2^{E_2}} \max_{S_1 \in 2^{E_1}} \overline{C}(S_1, S_2),$$

$$\max_{S_1 \in \mathscr{B}_1} \min_{S_2 \in \mathscr{B}_2} \overline{C}(S_1, S_2) \leq \min_{S_2 \in \mathscr{B}_2} \max_{S_1 \in \mathscr{B}_1} \overline{C}(S_1, S_2).$$

Lemma 8.1 exposes a well-known property of the matrix games. The concept of upper and lower values of the games Γ and Γ^+ are imposed by the right and left members of the inequalities in Lemma 8.1.

Definition 8.1 The matrix game has a solution (is solvable) if its upper and lower values are equal. The corresponding profile is called an **equilibrium** (equilibrium solution, equilibrium profile, equilibrium outcome) of the game and its value is called the value of the game.

In the game with feasible profiles Γ_*^+, the opposite inequality may occur when the payoff function C satisfies some special properties, that is, it is possible that the value $\min\limits_{S_2 \in \mathscr{B}_2} \max\limits_{S_1 \in \mathscr{B}_1(S_2)} \overline{C}(S_1, S_2)$ of the game Γ_2^+ do not surpass the value

$$\max_{S_1 \in \mathscr{B}_1} \min_{S_2 \in \mathscr{B}_2(S_1)} \overline{C}(S_1, S_2)$$

of the game Γ_1^+.

Lemma 8.2 *If* $C(\overline{M}(S_1, S_2)) = C'(S_1) + C''(S_2)$ *and* $\mathscr{B}_1 \neq \varnothing$, *then*

$$\max_{S_1 \in \mathscr{B}_1} \min_{S_2 \in \mathscr{B}_2(S_1)} \overline{C}(S_1, S_2) \geq \min_{S_2 \in \mathscr{B}_2} \max_{S_1 \in \mathscr{B}_1(S_2)} \overline{C}(S_1, S_2)$$

Proof It is obvious that if $\mathscr{B}_1 \neq \varnothing$, then $\mathscr{B}_2 \neq \varnothing$ and vice versa. Thus, we have

$$\max_{S_1 \in \mathscr{B}_1} \min_{S_2 \in \mathscr{B}_2(S_1)} C(\overline{M}(S_1, S_2)) =$$

$$= \max_{S_1 \in \mathscr{B}_1} [C'(S_1) + \min_{S_2 \in \mathscr{B}_2(S_1)} C''(S_2)] \geq$$

$$\geq \max_{S_1 \in \mathscr{B}_1} C'(S_1) + \min_{S_2 \in \mathscr{B}_2} C''(S_2) \geq$$

$$\geq \min_{S_2 \in \mathscr{B}_2} [C''(S_2) + \max_{S_1 \in \mathscr{B}_1(S_2)} C'(S_1)] =$$

$$= \min_{S_2 \in \mathscr{B}_2} \max_{S_1 \in \mathscr{B}_1(S_2)} C(\overline{M}(S_1, S_2)).$$

The truth of the lemma follows from the above chain of the equalities and inequalities. □

Definition 8.2 The game with feasible profiles Γ_*^+ has an equilibrium if the values of the Stackelberg games Γ_1^+ and Γ_2^+ are equal.

This solution concept may have an integrative power for all the precedent ones. The following results have to prove this.

Lemma 8.3 *If* $\mathscr{B}_1 \neq \varnothing$, *then*

$$\max_{S_1 \in 2^{E_1}} \min_{S_2 \in 2^{E_2}} \overline{C}(S_1, S_2) = \max_{S_1 \in \mathscr{B}_1} \min_{S_2 \in \mathscr{B}_2} \overline{C}(S_1, S_2) =$$

$$= \max_{S_1 \in \mathscr{B}_1} \min_{S_2 \in \mathscr{B}_2(S_1)} C(\overline{M}(S_1, S_2)).$$

Proof The payoff function is so defined that $\min_{S_2 \in 2^{E_2}} \overline{C}(S_1, S_2) < \infty$ for any strategy $S_1 \in 2^{E_1}$. As $\mathscr{B}_1 \neq \varnothing$, then

$$-\infty < \max_{S_1 \in 2^{E_1}} \min_{S_2 \in 2^{E_2}} \overline{C}(S_1, S_2) < +\infty.$$

Therefore, the maximin profile is feasible and

$$\max_{S_1 \in 2^{E_1}} \min_{S_2 \in 2^{E_2}} \overline{C}(S_1, S_2) = \max_{S_1 \in \mathscr{B}_1} \min_{S_2 \in \mathscr{B}_2} \overline{C}(S_1, S_2). \tag{8.3}$$

For some admissible strategy $S_1 \in \mathscr{B}_1$ we have

$$\min_{S_2 \in \mathscr{B}_2} \overline{C}(S_1, S_2) = \min_{S_2 \in \mathscr{B}_2(S_1)} \overline{C}(S_1, S_2),$$

as $\overline{C}(S_1, S_2) = +\infty$ for any strategy $S_2 \notin \mathscr{B}_2(S_1)$. Then

$$\max_{S_1 \in \mathscr{B}_1} \min_{S_2 \in \mathscr{B}_2} \overline{C}(S_1, S_2) = \max_{S_1 \in \mathscr{B}_1} \min_{S_2 \in \mathscr{B}_2(S_1)} \overline{C}(S_1, S_2). \tag{8.4}$$

Relations (8.3) and (8.4) prove the lemma. □

Considering Lemma 8.3 and the equality $\max_{S_1 \in 2^{E_1}} \overline{C}(S_1, S_2) = +\infty$ for all $S_2 \notin \mathscr{B}_2$, the following theorem becomes obvious.

Theorem 8.1 *Let $\mathscr{B}_1 \neq \varnothing$. The upper (lower) values of the games Γ and Γ^+ are equal.*

For a sufficiently large set of edges E, an exhaustive search of equilibria in the games Γ, Γ^+, Γ^+_* is a hard task. Theorem 8.1 suggests how to narrow sets of admissible strategies by taking into account properties of G and the set of properties $\mathscr{P}$ of feasible subgraphs.

Supposition 8.2.1 *Further on, we assume that $\mathscr{B}_1 \neq \varnothing$.*

Supposition 8.2.2 *We may define/consider the sets of admissible strategies $\mathscr{B}_1$ and $\mathscr{B}_2$ being subsets of 2^{E_1} and 2^{E_2} less powerful than (8.1) and (8.2) too.*

Thus, we have for the fixed game Γ several possible games Γ^+, and Γ^+_*, by defining, e.g., the game Γ^+ constrained by the condition

$$|M(S_1, S_2)| \leq 1,$$

for all $(S_1, S_2) \in \mathscr{B}_1 \times \mathscr{B}_2$.

It is easy to observe that the payoff function is defined in such a way, that if there exists only one feasible profile in Γ^+_*, then for both players it is advantageous the same profile, namely the equilibrium profile. However, in some games Γ^+_* several feasible profiles exist, but the equilibrium profile itself does not exist in all the games Γ, Γ^+ and Γ_*^+.

The following example has to illustrate the above exposition.

Example 8.1 Let the games Γ, Γ^+, and Γ_*^+, be formulated on the next acyclic digraph $G = (V, E)$:

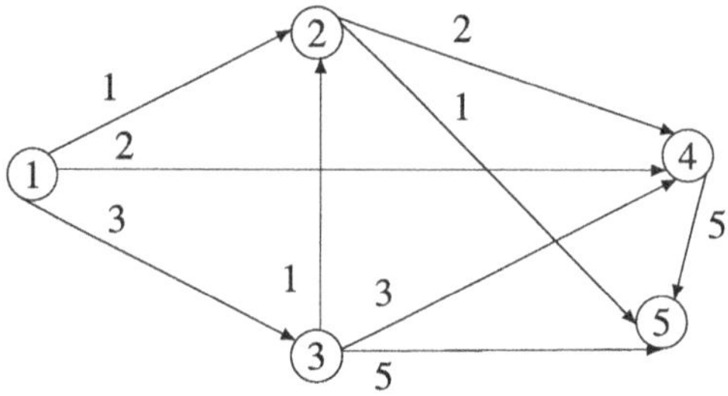

If $V_1 = \{1\}$ and $V_2 = \{2;\ 3;\ 4;\ 5\}$ are the positions of the players, then

$$E_1 = \{(1,2);\ (1,3);\ (1,4)\},$$
$$E_2 = \{(2,4);\ (2,5);\ (3,2);\ (3,4);\ (3,5);\ (4,5)\},$$

are their sets of edges.

We have a purpose to construct the matrix games based on three sets of paths from the input vertex $v_s = 1$ to the output vertex $v_t = 5$.

First, let us consider the set $\mathscr{D}_3$ of feasible graphs as the set of all paths from $v_s = 1$ to $v_t = 5$ that **contain exactly** 3 **edges**.

It is easy to observe that in the game Γ the first player has $2^3 = 8$ strategies and the second — $2^6 = 64$.

It is a difficult task to find the equilibrium by an exhaustive search because of 512 profiles of the game Γ. But taking into account the essence of the feasible graphs, admissible strategies are defined as the sets with cardinality $|E_1| = |V_1| = 1$ for the first player, and $|E_2| = |V_2| - 1 = 3$ for the second player, with an additional property that exactly one edge exits from every vertex, except $v_t = 5$.

As the graph G is acyclic, the subgraph G_S is a directed tree entering the vertex $v_t = 5$ for any profile S, that is, it has a path from $v_s = 1$ to $v_t = 5$, not obligatory with exactly 3 edges.

Observe that for $S_1 = \{(1,4)\}$ a $3-$path from $v_s = 1$ to $v_t = 5$ does not exist for any strategy of the second player. Besides that, for strategy $S_2 = \{(2,5);\ (3,5);\ (4,5)\}$ a $3-$path from $v_s = 1$ to $v_t = 5$ does not exist for any strategy of the first player.

If the payoff function C is defined as the length of the path from $v_s = 1$ to $v_t = 5$, and the first player has the purpose to maximise the length of the path (the second trying to minimise it), then it is easy to find the payoff matrices of the games Γ^+ and Γ_*^+ :

$\mathscr{B}_1$	$\mathscr{B}_2$							
	(3,2) (2,4) (4,5)	(2,4) (3,4) (4,5)	(2,4) (4,5) (3,5)	(3,2) (2,5) (4,5)	(2,5) (3,4) (4,5)	(2,5) (3,5) (4,5)	Γ^+: $\min_{S_2 \in \mathscr{B}_2}$	Γ_1^+: $\min_{S_2 \in \mathscr{B}_2(S_1)}$
(1,2)	8	8	8	$+\infty$	$+\infty$	$+\infty$	8	8
(1,3)	$+\infty$	11	$+\infty$	5	11	$+\infty$	5	5
(1,4)	$-\infty$	$-\infty$	$-\infty$	$-\infty$	$-\infty$	$-\infty$	$-\infty$	
Γ^+: $\max_{S_1 \in \mathscr{B}_1}$	$+\infty$	11	$+\infty$	$+\infty$	$+\infty$	$+\infty$	$11 \setminus 8$	
Γ_2^+: $\max_{S_1 \in \mathscr{B}_1(S_2)}$	8	11	8	5	11			$5 \setminus 8$

In the game Γ^+ (consequently in Γ)

$$\max_{S_1 \in \mathscr{B}_1} \min_{S_2 \in \mathscr{B}_2} \overline{C}(S_1, S_2) = 8 \leq \min_{S_2 \in \mathscr{B}_2} \max_{S_1 \in \mathscr{B}_1} \overline{C}(S_1, S_2) = 11.$$

But in the game Γ_*^+ we have

$$\max_{S_1 \in \mathscr{B}_1} \min_{S_2 \in \mathscr{B}_2(S_1)} \overline{C}(S_1, S_2) = 8 \geq \min_{S_2 \in \mathscr{B}_2} \max_{S_1 \in \mathscr{B}_1(S_2)} \overline{C}(S_1, S_2) = 5.$$

Second, let $\mathscr{D}_2$ be the set of all paths from $v_s = 1$ to $v_t = 5$, **having exactly** 2 **edges**.

As in the first case, the games Γ, Γ^+ and Γ_*^+, with their payoff matrices are considered:

$\mathscr{B}_1$	$\mathscr{B}_2$							
	(3,2) (2,4) (4,5)	(2,4) (3,4) (4,5)	(2,4) (4,5) (3,5)	(3,2) (2,5) (4,5)	(2,5) (3,4) (4,5)	(2,5) (3,5) (4,5)	Γ^+: $\min_{S_2 \in \mathscr{B}_2}$	Γ_1^+: $\min_{S_2 \in \mathscr{B}_2(S_1)}$
(1,2)	$+\infty$	$+\infty$	$+\infty$	2	2	2	2	2
(1,3)	$+\infty$	$+\infty$	8	$+\infty$	$+\infty$	8	8	8
(1,4)	7	7	7	7	7	7	7	7
Γ^+: $\max_{S_1 \in \mathscr{B}_1}$	$+\infty$	$+\infty$	$+\infty$	$+\infty$	11	8	$8 \setminus 8$	
Γ_2^+: $\max_{S_1 \in \mathscr{B}_1(S_2)}$	7	7	7	7	11	8		$7 \setminus 8$

In the game Γ^+ (consequently in Γ)

$$\max_{S_1 \in \mathscr{B}_1} \min_{S_2 \in \mathscr{B}_2} \overline{C}(S_1, S_2) = 8 = \min_{S_2 \in \mathscr{B}_2} \max_{S_1 \in \mathscr{B}_1} \overline{C}(S_1, S_2) = 8.$$

But in the game Γ_*^+

$$\max_{S_1 \in \mathscr{B}_1} \min_{S_2 \in \mathscr{B}_2(S_1)} \overline{C}(S_1, S_2) = 8 \geq \min_{S_2 \in \mathscr{B}_2} \max_{S_1 \in \mathscr{B}_1(S_2)} \overline{C}(S_1, S_2) = 7.$$

Remark, there exists an equilibrium only in the games Γ, Γ^+. The game Γ_*^+ doesn't have equilibrium.

Third, let us consider the set $\mathscr{D}$ of **all paths from** $v_s = 1$ **to** $v_t = 5$ as the set of feasible graphs.

$\mathscr{B}_1$	$\mathscr{B}_2$							
	(3,2) (2,4) (4,5)	(2,4) (3,4) (4,5)	(2,4) (4,5) (3,5)	(3,2) (2,5) (4,5)	(2,5) (3,4) (4,5)	(2,5) (3,5) (4,5)	Γ^+: $\min_{S_2 \in \mathscr{B}_2}$	Γ_1^+: $\min_{S_2 \in \mathscr{B}_2(S_1)}$
(1,2)	8	8	8	2	2	2	2	2
(1,3)	11	11	8	5	11	8	5	5
(1,4)	7	7	7	**[7]**	7	7	7	7
Γ^+: $\max_{S_1 \in \mathscr{B}_1}$	11	11	8	7	11	8	7\ 7	
Γ_2^+: $\max_{S_1 \in \mathscr{B}_1(S_2)}$	11	11	8	7	11	8		7\ 7

Let us remark, that all the games Γ, Γ^+, Γ_*^+ have the equilibrium

$$(S_1, S_2) = (\{(1,4)\},\ \{(3,2);\ (2,5);\ (4,5)\}),$$

with the 2-edge feasible path $P = \{(1,4); (4,5)\}$ and the length $C(P) = 7$. □

Theorem 8.2 *If*

$$C(\overline{M}(S_1, S_2)) = C'(S_1) + C''(S_2)$$

for all $\overline{M}(S_1, S_2) \neq \varnothing$ *and if all profiles in the game* Γ^+ *are feasible, then* Γ, Γ^+, Γ_*^+ *have an equilibrium, moreover, it is the same in all three games* Γ, Γ^+, Γ_*^+.

Proof All the profiles in the game Γ^+ are feasible. Then $\mathscr{B}_2(S_1) = \mathscr{B}_2$ for all $S_1 \in \mathscr{B}_1$, and $\mathscr{B}_1(S_2) = \mathscr{B}_1$ for all $S_2 \in \mathscr{B}_2$. Then, taking into account Lemma 8.1, we have

$$\max_{S_1 \in \mathscr{B}_1} \min_{S_2 \in \mathscr{B}_2(S_1)} C(\overline{M}(S_1, S_2)) = \max_{S_1 \in \mathscr{B}_1} \min_{S_2 \in \mathscr{B}_2} C(\overline{M}(S_1, S_2)) \leq$$

$$\leq \min_{S_2 \in \mathscr{B}_2} \max_{S_1 \in \mathscr{B}_1} C(\overline{M}(S_1, S_2)) = \min_{S_2 \in \mathscr{B}_2} \max_{S_1 \in \mathscr{B}_1(S_2)} C(\overline{M}(S_1, S_2)),$$

and, taking into account Lemma 8.2, we have

$$\max_{S_1\in\mathscr{B}_1}\min_{S_2\in\mathscr{B}_2} C(\overline{M}(S_1,S_2)) = \max_{S_1\in\mathscr{B}_1}\min_{S_2\in\mathscr{B}_2(S_1)} C(\overline{M}(S_1,S_2)) \geq$$

$$\geq \min_{S_2\in\mathscr{B}_2}\max_{S_1\in\mathscr{B}_1(S_2)} C(\overline{M}(S_1,S_2)) = \min_{S_2\in\mathscr{B}_2}\max_{S_1\in\mathscr{B}_1} C(\overline{M}(S_1,S_2)).$$

From these inequalities follows that

$$\max_{S_1\in\mathscr{B}_1}\min_{S_2\in\mathscr{B}_2} C(\overline{M}(S_1,S_2)) = \max_{S_1\in\mathscr{B}_1}\min_{S_2\in\mathscr{B}_2(S_1)} C(\overline{M}(S_1,S_2)) =$$

$$= \min_{S_2\in\mathscr{B}_2}\max_{S_1\in\mathscr{B}_1(S_2)} C(\overline{M}(S_1,S_2)) = \min_{S_2\in\mathscr{B}_2}\max_{S_1\in\mathscr{B}_1} C(\overline{M}(S_1,S_2)).$$

Therefore Γ^+ and Γ^+_* have the equilibrium profile. Finally, it follows from Theorem 8.1 that all the games Γ, Γ^+, Γ^+_* have the same equilibrium profile. □

Remark 8.4 Lemma 8.2 and Theorem 8.1 may be extended for other types of functions. One of such types is, for example: $C(\overline{M}(S_1,S_2)) = C'(S_1)\cdot C''(S_2)$, where $C'{:}\mathscr{B}_1 \to N^*$, $C''{:}\mathscr{B}_2 \to N^*$.

Remark 8.5 Example 8.1 shows that if in the game Γ^+ there is a profile that is not feasible, then the equilibrium profile may be absent from any game Γ, Γ^+, or Γ^+_*.

Remark 8.6 Example 8.1 illustrates also that the equilibrium profile may exist in the games Γ, Γ^+, but in the corresponding game Γ^+_* it may be absent. Inverse is possible: in the game Γ^+_* there is an equilibrium profile, but in the games Γ, Γ^+ it is absent.

Theorem 8.2 formulates only sufficient condition for the existence of the equilibrium profile in the games Γ, Γ^+, Γ^+_*. The following theorem formulates necessary and sufficient conditions.

Theorem 8.3 *Let*

$$C(\overline{M}(S_1,S_2)) = C'(S_1) + C''(S_2)$$

for all $\overline{M}(S_1,S_2) \neq \varnothing$. The profile (S_1^*, S_2^*) forms an equilibrium in all the games Γ, Γ^+, Γ^+_* if and only if the profile (S_1^*, S_2^*) is an equilibrium in the game Γ^+ and

$$\max_{S_1\in\mathscr{B}_1(S_2)} C(\overline{M}(S_1,S_2)) \geq C(\overline{M}(S_1^*,S_2^*)),$$

for all $S_2 \in \mathscr{B}_2$.

Proof Necessity is obvious.

Sufficiency follows from the following relations:

$$\begin{aligned} \min\max \Gamma^+ &\overset{\text{L. 7.2.1}}{\geq} \max\min \Gamma^+ = \\ \overset{\text{L. 7.2.3}}{=} \max\min \Gamma_1^+ &\overset{\text{L. 7.2.2}}{\geq} \min\max \Gamma_2^+, \end{aligned} \tag{8.5}$$

and Theorem 8.1. □

If the game Γ^+ has an equilibrium profile (S_1^*, S_2^*), then, taking into account Lemma 8.3 and Theorem 8.3, we deduce that $\min\max \Gamma^+ < +\infty$. This means that, for strategy $S_2^* \in \mathscr{B}_2$,

$$\overline{C}(S_1, S_2^*) = C(\overline{M}(S_1, S_2^*)) < +\infty$$

for all $S_1 \in \mathscr{B}_1$.

Definition 8.3 A strategy that may have only feasible profiles, that is, $\mathscr{B}_1(S_2) = \mathscr{B}_1$ for the first player and $\mathscr{B}_2(S_1) = \mathscr{B}_2$ for the second one, is called an **essential feasible strategy**.

From Theorem 8.3 follows the next statement.

Corollary 8.1 *If the second player does not have at least one essential feasible strategy in the game Γ^+, then both the games Γ and Γ^+ do not have equilibrium profiles.*

8.3 Solvable Matrix Games on Digraphs

An investigation of digraph matrix games implies solving of three important problems:

1. the problem of determining maximin and minimax profiles;
2. the problem of determining feasible subgraphs in graphs of maximin and minimax profiles;
3. the problem of determining an equilibrium profile.

From relations (8.5) it follows that for an equilibrium profile computing in Γ, Γ^+, Γ_*^+, it is sufficiently to determine and to compare minimax profiles in the games Γ^+ and Γ_2^+.

If $\min\max \Gamma^+ = \min\max \Gamma_2^+$, **then** all three games Γ, Γ^+, Γ_*^+ are solvable and have the same equilibrium profile,
else it is necessary to find $\max\min \Gamma^+$ that is equal to $\max\min \Gamma_2^+$ and to compare it with $\min\max \Gamma^+$ and $\min\max \Gamma_1^+$:

- **if** $\max\min \Gamma^+ = \min\max \Gamma^+$, **then** Γ^+ is solvable;
- **if** $\max\min \Gamma^+ = \min\max \Gamma_1^+$, **then** Γ_*^+ is solvable;
- **if** $\max\min \Gamma^+ \neq \min\max \Gamma^+$ and $\max\min \Gamma^+ \neq \min\max \Gamma_1^+$, **then** all three games Γ, Γ^+, Γ_*^+ are unsolvable.

Consequently, in order to investigate the games Γ^+ and Γ_*^+, the problem of determining maximin and minimax profiles with the corresponding maximin and minimax feasible graphs becomes very important. As the games have limited numbers of profiles, maximin and minimax profiles hypothetically may be found by an exhaustive search, which for large m becomes a hard computational problem. It is obvious, that a game has polynomial complexity if both maximin and minimax profiles may be found in polynomial time on n and m. If the problem of a feasible graph $\overline{M}(S_1, S_2) \neq \varnothing$ construction in G_S for some profile (S_1, S_2) is NP – complete, then the game is at least NP-hard.

The exhaustive search method for solving the game Γ has an exponential complexity, supposing it is necessary to examine $2^{|E|}$ profiles. If the algorithm for construction a feasible subgraph in G has the polynomial complexity $O(n^{k_0} m^{l_0})$ and $|\mathscr{B}_1| = O(n^{k_1} m^{l_1})$, $|\mathscr{B}_2| = O(n^{k_2} m^{l_2})$, where $k_0, k_1, k_2, l_0, l_1, l_2$ are numbers independent of m and n, then the same straightforward method in Γ^+ and Γ_*^+ has the polynomial complexity $O(n^{k_0+k_1+k_2} m^{l_0+l_1+l_2})$.

Thus, depending on properties of G and elements of $\mathscr{D}$, this problems may be essentially simplified in particular games. Further on, we illustrate this for some particular games.

8.3.1 Maximin Directed Tree

Let G be an acyclic digraph and assume that there exist paths in G from every vertex $v \in V$ to vertex v_0. Let $\mathscr{D}$ be the set of all directed trees of G going to v_0; $C : \mathscr{D} \to \mathscr{R}$ be the length of tree (sum of edge lengths). The first player has the aim to maximize the length of tree, the second tries to minimize it.

Take into consideration that feasible graphs are directed trees, we will define admissible strategies so that from every vertex except v_0 exactly one edge is going out. In this case every element belonging to $\mathscr{D}$ at least once is feasible subgraph in Γ^+. Remark, that inadmissible strategies are not advantageous to players because they either not ensure tree construction or they lead to adversary possibility to choice from several alternatives. Therefore, either $\mathscr{B}_1$ and $\mathscr{B}_2$ contain the optimal strategies of the players and the game Γ^+ is right defined, that ensure equality of costs of the games Γ and Γ^+.

Remark further, for all $(S_1, S_2) \in \mathscr{B}_1 \times \mathscr{B}_2$ we have $\overline{M}(S_1, S_2) = (V, S_1 \cup S_2)$. This means, that all profiles of the game Γ^+ are feasible. From theorem 2 follows that Γ, Γ^+, Γ_*^+ have the same equilibrium profile.

To determine the maximin input in v_0 tree we propose the following application of the dynamical programming method [1]:

Algorithm 8.3.1 $M = \{v_0\}, \ T = \varnothing$
while $|M| < n$ do
begin $(u^*, v^*) = \underset{(u,v)\in((V\setminus M)\times M)\cap E}{\arg\max} C(u, v)$
If $(u^* \in V_1)$ or $((u^* \in V_2)$ and $|((V\setminus M) \times M) \cap E| = 1)$
then $M = M \cup u^*, \ T = T \cup (u^*, v^*)$
else $E = E\setminus(u^*, v^*)$.
end
T – maximin tree.

Remark 8.7 Algorithm 8.3.1 determines maximin tree in an arbitrary digraph. But, because maximin may be not equal to minimax in general case, for determining of minimax tree, we must found on every iteration

$$(u^*, v^*) = \underset{(u,v)\in((V-M)\times M)\cap E}{\arg\min} C(u, v).$$

8.3.2 *Maximin Directed Path*

Let G be an acyclic digraph and assume that there exists a path from every vertex $v \in V$ to vertex v_t. Let $\mathscr{D}$ be the set of all directed paths from v_0 to v_t; $C : \mathscr{D} \to \mathscr{R}$ be the length of path (sum of edge lengths). The first player has the aim to maximize the length of path, the second has the aim to minimize it.

We will define admissible strategies so that from every vertex except v_0 exactly one edge is going out. In this case every G_S is an input in v_0 tree, containing a path from v_s to v_t. The set of all feasible profiles of the maximin path game Γ^+ is equivalent to the set of all feasible profiles of the maximin tree game Γ^+. Therefore all three games Γ, Γ^+, Γ^+_* have the same equilibrium profile.

To determine maximin path we may use an adaptation of Dijkstra algorithm. An example of such adaptation is presented in [33].

8.3.3 *Maximin Traveling Salesman Problem with Transportation*

Generally, a Traveling Salesman Problem (TSP) includes diverse mathematical models of distinct real practical problems. Its history may be traced back to the Irish mathematician Sir William Rowan Hamilton and British mathematician Thomas Penyngton Kirkman, who treated it incipiently in 1800s [34, 35]. In the 1930s Karl

Menger studied TSP in general form and Hassler Whitney and Merrill Floodlater promoted promoted TSP later [35].

We consider and investigate an original model of TSP motivated by applications — a synthesis of classical TSP and classical Transportation Problem. Algorithms based on Integer Programming cutting-plane methods and Branch and Bound Techniques are obvious. A maximin traveling salesman problem and the correspondent maximin Hamiltonian cycle problem may be formulated similar to the problem of maximin directed tree. So, in this subsection we expose only the specific features of the traveling salesman problem with transportation. But, why is important this problem in context of considered games? The answer is rather obvious: it suggests an example of games which are simple formulated but are hard to solve.

8.3.3.1 Introduction

The TSP gained notoriety over the past century as the prototype of problem that is easy to state and hard to solve practically. It is simply formulated: *a traveling salesman has to visit exactly once each of n cities and to return to the start city but in a order that minimizes the total cost (it is supposed that the cost c_{ij} of traveling from every city i to every city j is known).* There are other related formulations of this problem [36] and a lot of methods for solving [37–41]. It is well known equivalence of TSP with Hamiltonian circuit problem [42].

The travelling salesman problem is representative for a large class of problems known as NP-complete combinatorial optimization problems [42, 43]. The NP-complete problems have an important property that all of them have or don't have simultaneously polynomial-time algorithms for its solving [42, 43]. Let us recall that the theory of quantum computation and quantum information processing [44], that has progressed very fast during the last decades, will nevertheless have to deal with problems of computational complexity.

To date, no one has found efficient (polynomial-time) algorithms for the TSP. But over the past few years many practical problems of really large size are solved [45]. Thus, at present, the largest solved Norwegian instance of TSP has 24 978 cities (D. Applegate, R. Bixby, V. Chvatal, W. Cook, and K. Helsgaun — 2004). But, the largest solved instance of TSP includes 85 900 cities (points) in an application on chips (2005–2006) [45].

The Transportation Problem is a well known classical problem [46, 47]. There are several efficient methods for its solving [4, 47–49]. Note that there exists also an impressive extension of Transportation Problem in functional spaces [50, 51], that have to highlight once and more a transportation problem significance.

The TSP with Transportation and Fixed Additional Payments (TSPT) generalizes these two problems: TSP [45] and Transportation Problem [47, 52]. The TSPT has some affinities with a Vehicle Routing Problem [53–55].

8.3.3.2 TSPT Formulations

Let a digraph $G = (V, E), |V| = n, |E| = m$ be given. Each node $j \in V$ has its own capacity δ_j (demand, if $\delta_j < 0$, supply, if $\delta_j > 0$), such that $\sum_{j=1}^{n} \delta_j = 0$. A salesman starts his traveling from the node $k \in V$ with $\delta_k > 0$. The unit cost of transportation throw arc $(i, j) \in E$ is equal to c_{ij}. If the arc (i, j) is active, then the additional payment d_{ij} is demanded. If $(i, j) \notin E$, then $c_{ij} = d_{ij} = \infty$. We must find a Hamiltonian circle and a starting node $k \in V$ with a property that the respective salesman travel satisfies all the demands $\delta_j, j = 1, \ldots, n$, and minimizes the total cost.

The TSPT may be formulated as an integer programming problem. Let x_{ij} be a quantity of product which is transported via (i, j). Let $y_{ij} \in \{0; 1\}$ be equal to 1 if $x_{ij} > 0$, and let x_{ij} be equal to 0 if $y_{ij} = 0$. In such notation the TSPT, as stated above, is equivalent to the following problem:

$$\sum_{i=1}^{n} \sum_{j=1}^{n} \left(c_{ij}x_{ij} + d_{ij}y_{ij}\right) \to \min, \tag{8.6}$$

$$\sum_{i=1}^{n} y_{ij} = 1, \quad j = 1, \ldots, n, \tag{8.7}$$

$$\sum_{j=1}^{n} y_{ij} = 1, \quad i = 2, \ldots, n, \tag{8.8}$$

$$\sum_{k=1}^{n} x_{jk} - \sum_{i=1}^{n} x_{ij} = \delta_j, \quad j = 1, \ldots, n, \tag{8.9}$$

$$u_i - u_j + ny_{ij} \leq n - 1, \quad i, j = 2, \ldots, n, \ i \neq j, \tag{8.10}$$

$$x_{ij} \leq My_{ij}, \quad i, j = 1, \ldots, n, \tag{8.11}$$

$$x_{ij} \geq 0, \ y_{ij} \in \{0; 1\}, \ u_j \geq 0, \quad i, j = 1, \ldots, n, \tag{8.12}$$

where $M = \sum_{j=1}^{n} |\delta_j|$.

If $c_{ij} = 0$ for all $i, j = 1, \ldots, n$, then the problem (8.6)–(8.12) becomes a classical TSP. If $d_{ij} = 0$ for all $i, j = 1, \ldots, n$, then the problem (8.6)–(8.12) is simplified to a classical Transportation Problem.

Theorem 8.4 *The TSPT and the problem (8.6)–(8.12) are equivalent.*

Proof The components (8.6)–(8.8), (8.10) and (8.12) of the problem (8.6)–(8.12) define a Hamiltonian circuit [4]. The components (8.6), (8.9) and (8.12) state the transportation problem [46]. The constraint (8.11) realizes a connection between these two *"facets"* of the TSPT. The starting node $k \in V$ may determined by an elementary sequential search. □

A capacity mathematical model of the problem is obtained when any arc $(i, j) \in E$ has an upper bound capacity $u_{ij} > 0$. Constraints

$$x_{ij} \leq u_{ij} y_{ij}, \ (i, j) \in E,$$

substitute (8.11).

The inequalities

$$l_{ij} y_{ij} \leq x_{ij} \leq u_{ij} y_{ij}, \ (i, j) \in E,$$

substitute (8.11) when any arc $(i, j) \in E$ has also a lower bound capacity $l_{ij} > 0$.

Kuhn [4] restrictions (8.10) may be substituted by equivalent restrictions:

$$\sum_{i \in K} \sum_{j \in K} y_{ij} = |K| - 1, \ \forall K \subset V,$$

where K is any proper subset of V.

8.3.3.3 Algorithms for TSPT

It is obvious that the solution of the classical TSP [47] does not solve TSPT.

The branch-and-bound algorithm may be constructed on back-tracking technique for a branch generation and lower bound estimation of the sum of a 1-tree value and T_0, where T_0 is calculated at the first step. T_0 represents the value of a minimal cost flow problem obtained in a relaxed problem without Hamiltonian circuit requirement. For an efficient bounding, T_0 may be substituted, at every step of the algorithm, by the exactly cost of transportation throw the respective fragment of the circuit.

A direct solving of (8.6)–(8.12) with an Gomory type cutting-plane algorithms is rational for a problem with modest size. In recent vogue opinion, the branch-and-cut super-algorithm [48, 49] may be much more recommended for the TSPT.

Finally, note that a dynamic programming approach [56] to solve the TSPT implies some difficulties as the TSPT optimal value depends on first node choosing from which the travel starts. This fact may be simply taken into consideration in previous methods, but not in dynamic programming method.

8.3.3.4 TSP Matrix Games

Finally, let us only remark once again that the TSP and TSPT suggest us an example of matrix games that are simple formulated, but are difficult to solve because of its computation complexity.

8.3.4 Maximin Cost Flow

Let us consider a flow network on a digraph $G = (V, E)$, $|V| = n$, $|E| = m$ with an output (source) vertex $v_s \in V$ and a input vertex (sink) $v_t \in V$, $v_s \neq v_t$, where any edge $(u, v) \in E$ has a capacity $b(u, v) \in Z^+$ and a unit transportation cost $c(u, v) \in Z$. The set V is partitioned into two disjoint sets of player positions:

$$V_1,\ V_2,\ (V_1 \cup V_2 = V,\ V_1 \cap V_2 = \varnothing).$$

Without loss of generality let us assume that $v_s \in V_1$. Thus, the player edge sets are

$$E_1 = \{(u, v) \in E \mid u \in V_1\},\quad E_2 = \{(u, v) \in E \mid u \in V_2\}.$$

Subsets $S_1 \subseteq E_1$, $S_2 \subseteq E_2$ are strategies of the first and second players, correspondingly. Any profile $(S_1, S_2) \in 2^{E_1} \times 2^{E_2}$ generates a net $G_S = (V, S_1 \cup S_2)$. In the net G_S, a flow f of a fixed value φ_0 is defined as the vector $f \in \mathbb{R}^{|S_1 \cup S_2|}$ which satisfies the following properties:

$$1^0. \qquad 0 \le f(u, v) \le b(u, v), \qquad (u, v) \in S_1 \cup S_2,$$

$$2^0. \sum_{(u,v)\in S_1\cup S_2} f(u, v) - \sum_{(v,u)\in S_1\cup S_2} f(v, u) = \begin{cases} 0, & u \notin \{v_s, v_t\}, \\ \varphi_0, & u = v_s, \\ -\varphi_0, & u = v_t. \end{cases}$$

The cost of the flow f is equal to $\sum_{(u,v)\in S_1\cup S_2} c(u, v) f(u, v)$.

Let us suppose that there exists at least one flow with the value φ_0 in the considered net.

For any pair of strategies (S_1, S_2) there is a polyhedron of solutions of system 1^0–2^0, denoted by $F_S = F(G_S)$. Generally, the polyhedron F_S may be an empty set for some pair of strategies if system 1^0–2^0 does not have solutions, but, due to our supposition, there exists at least one pair of strategies for which $F_S \neq \varnothing$. It is known that if the capacity $b(u, v)$ is integer for any $(u, v) \in E$, then all the vertices of the polyhedron F_S have integer components. Thus, as F_S is bounded, the set of all flows, corresponding to (S_1, S_2), is a linear convex combination of a finite number of integer flows. The cardinality of the set F_S may be equal to 0, when system 1^0–2^0 does not have solutions, may be equal to 1, when 1^0–2^0 has one solution, or may be equal to $\aleph$, when 1^0–2^0 has an infinite number of solutions.

Let $\mathscr{D}$ be a set of all subgraphs (subnets) of G that has a flow of a value φ_0 from v_s to v_t. Let

$$M : 2^G \to \mathscr{D},\ M(S_1, S_2) = G_s \cap \mathscr{D}$$

be the choice function,

$$C\colon \mathscr{D} \to R,\ C(\Gamma) = \max_{f \in F(\Gamma)} \sum c(e) f(e)$$

be the choice criterion and

$$\overline{M}(S_1,\ S_2) = \begin{cases} \arg\max\limits_{\Gamma \in M(S_1, S_2)} C(\Gamma), & \text{if } M(S_1,\ S_2) \neq \varnothing, \\ \varnothing, & \text{otherwise,} \end{cases}$$

be a single-valued choice function that chooses the flow of the value φ_0 with a minimal cost in the net G_S. Then, we have the following cost function (payoff matrix):

$$\overline{C}(S_1, S_2) = \begin{cases} C(\overline{M}(S_1, S_2)), & \text{if } \overline{M}(S_1, S_2) \neq \varnothing, \\ -\infty, & \text{if } \overline{M}(S_1, S_2') = \varnothing,\ \forall S_2' \in 2^{E_2}, \\ +\infty, & \text{otherwise.} \end{cases}$$

So, the matrix game Γ is defined. By analogy with the general case, we define games Γ^+ and Γ_*^+ too. A strategy is called admissible if there exists an adversary strategy for which the corresponding profile is feasible (has a φ_0 flow).

Lemma 8.4 *Let the net G has at least one φ_0 flow. Then both the players have at least one feasible strategy in Γ^+.*

Proof Let us order the rows and columns of the matrices of Γ^+ in a non-decreasing order of cardinalities of corresponding strategies. It is obvious that the pair of strategies, which are equal to a union of all admissible strategies of a correspondent player, are feasible. □

The following example shows that for a solvable game Γ there are several non-identical approaches to define games Γ^+, Γ_*^+ so that the game Γ_*^+ may be both solvable and unsolvable.

Example 8.2 Consider the flow games Γ, Γ^+, Γ_*^+, defined on the following graph

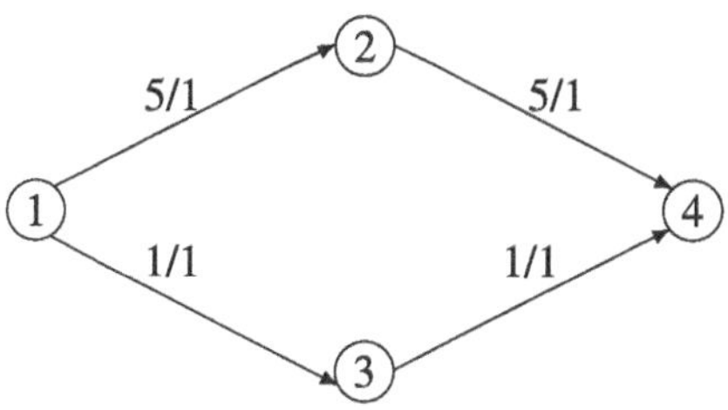

where

$$\varphi_0 = 1;\ v_s = 1;\ v_t = 4;$$
$$V_1 = \{1\};\quad V_2 = \{2;\ 3;\ 4\};$$
$$E_1 = \{(1,\ 2);\ (1,\ 3)\};\quad E_2 = \{(2,\ 4);\ (3,\ 4)\}.$$

The following table contains the payoff matrices of the considered games.

1	2				
	(2, 4)	(3, 4)	(2, 4) (3, 4)	min $\mathscr{B}_2$	min $\mathscr{B}_{2(S_1)}$
(1, 2)	10	$+\infty$	10	10	10
(1, 3)	$+\infty$	2	2	2	2
(1, 2) (1, 3)	10	2	2	2	2
max $\mathscr{B}_1$	$+\infty$	$+\infty$	10	10\10	
max $\mathscr{B}_{1(S_2)}$	10	2	10		2\10

The games $\Gamma,\ \Gamma^+$ have the equilibrium profile. The game Γ_*^+ does not have.

If the sets of admissible strategies are narrowed so that from every vertex, except the fourth, at least one edge is going out, then the payoff matrices are modified

1	2		
	(2, 4) (3, 4)	min $\mathscr{B}_2$	min $\mathscr{B}_{2(S_1)}$
(1, 2)	10	10	10
(1, 3)	2	2	2
(1, 2) (1, 3)	2	2	2
max $\mathscr{B}_1$	10	10\10	
max $\mathscr{B}_{1(S_2)}$	10		10\10

and all the games $\Gamma,\ \Gamma^+,\ \Gamma_*^+$ have the same equilibrium profile.

Thus, for a game Γ there are two pairs of games $\Gamma^+,\ \Gamma_*^+$. For one pair, the game Γ_*^+ does not have an equilibrium profile. For another one, all the games $\Gamma,\ \Gamma^+,\ \Gamma_*^+$ have the same equilibrium profile. □

It is well known that the problem of minimal cost flow may be represented as a linear programming problem [4]. By numbering the vertices and edges of G in a such way that vertices and edges of the first player are the first in the order list, we can define elements of an incidence matrix $A = [a_{ij}]$ of the graph G as

$$a_{ij} = \begin{cases} +1, & \text{if } e_j \text{ exits from vertex } i, \\ -1, & \text{if } e_j \text{ enters in vertex } i, \\ \ \ 0, & \text{otherwise,} \end{cases}$$

$i = 1, \ldots, n;\ j = 1, \ldots, m$. By notation $f = \begin{bmatrix} f_1 \\ f_2 \end{bmatrix} = (f_1, \ldots, f_m)^T \in \mathbb{R}^m, f_j$ — the flow through edge e_j; $b, c \in \mathbb{R}^m$, b_j — the capacity of edge e_j, c_j — the unit cost of edge e_j flow;

$$d \in \mathbb{R}^m, \qquad d_i = \begin{cases} -1, \text{ if } v_i = v_s, \\ +1, \text{ if } v_i = v_t, \\ \ \ 0, \text{ otherwise,} \end{cases}$$

the following minimal cost flow problem in the net G may be formulated

$$c^T f \to \min, \tag{8.13}$$

$$\begin{cases} Af = d\varphi_0, \\ f \leq b, \\ f \geq 0. \end{cases} \tag{8.14}$$

Let us associate with the first n constraints, corresponding to the safety law of the φ_0 flow, dual variables π_i, and with the remains m constraints — dual variables γ_k. Then, the problem (8.13) and (8.14) has the following dual problem

$$\varphi_0(\pi_s - \pi_k) + b^T\gamma \to \max, \tag{8.15}$$

$$\begin{cases} \pi_i - \pi_j + \gamma_k \leq c_k, \text{ for } e_k = (v_i, v_j) \in E, \\ \gamma_k \leq 0, \ k = 1, \ldots, m. \end{cases} \tag{8.16}$$

According to the Strong Duality Theorem of Linear Programming Theory, problems (8.13)–(8.14) and (8.15)–(8.16) have optimal solutions if and only if the system

$$\begin{cases} c^T f = \varphi_0\pi_s - \varphi_0\pi_t + b^T\gamma, \\ Af = d\varphi_0, \\ f \leq b, \\ f \geq 0, \\ \pi_i - \pi_j + \gamma_k \leq c_k, \quad \text{for } e_k = (v_i, v_j) \in E, \\ \gamma_k \leq 0, \quad k = 1, \ldots, m, \end{cases} \tag{8.17}$$

has a solution (remark, the first equality is the binding one).

It is important to observe that for any profile $(S_1, S_2) \in \mathscr{B}_1 \times \mathscr{B}_2$ there is a system of (8.17) type. Let $\Phi(S_1, S_2)$ be the set of all the solutions of the corresponding system (8.17). Then, the cost function may be defined as

$$\overline{C}(S_1, S_2) = \begin{cases} c^T f, \ \text{if } \Phi(S_1, S_2) \neq \varnothing \ \text{ where } (f, \pi, \gamma)^T \in \Phi(S_1, S_2), \\ -\infty, \ \text{if } \Phi(S_1, S_2') = \varnothing \ \text{ for all } S_2' \in \mathscr{B}_2, \\ +\infty, \ \text{otherwise.} \end{cases}$$

By applying linear programming concepts and results, let us show now that problems of finding maximin and minimax profiles in the flow game are equivalent to maximin and minimax linear problems.

Clearly, the set of feasible solutions of problem (8.13) and (8.14) is an polyhedron in $\mathbb{R}^m$, and the minimum is attained on its vertex. Then, the first player purpose is to maximize the flow cost by rational choice of the net G_S structure. This is equivalent to maximisation of the flow cost by optimal choice of the basic columns of the matrix A that corresponds to a choice of edges of E_1. The second player purpose is to minimize the cost of the flow by rational choice of the net G_S structure. This is equivalent to minimisation of the flow cost by optimal choice of the basic columns of the matrix A. Therefore, the first player has to choice a feasible solution for which at least columns, that correspond to edges from E_1, have non-negative dual estimations. The second player has to choice a feasible solution for which at least columns, that correspond to edges from E_2, have non-positive dual estimations. Consequently, a feasible solution is both saddle point and equilibrium profile. So, the problem of an equilibrium profile computing in the flow game is equivalent to a maximin linear programming problem

$$\max_{f_1} \min_{f_2} c^T \begin{bmatrix} f_1 \\ f_2 \end{bmatrix}, \tag{8.18}$$

$$\begin{cases} A \begin{bmatrix} f_1 \\ f_2 \end{bmatrix} = d\varphi_0, \\ 0 \leq \begin{bmatrix} f_1 \\ f_2 \end{bmatrix} \leq b. \end{cases} \tag{8.19}$$

Problem (8.18) and (8.19) is equivalent to

$$\varphi(f_1) = \min_{f_2}(c_1 f_1 + c_2 f_2) = c_1 f_1 + \min_{f_2} c_2 f_2 \to \max. \tag{8.20}$$

$$\begin{cases} A_2 f_2 = d\varphi_0 - A_1 f_1, \\ 0 \leq f_1 \leq b_1, \\ 0 \leq f_2 \leq b_2. \end{cases} \tag{8.21}$$

In (8.21), the function $\varphi(f_1)$ is determined as a solution of a linear parametric program with restrictions (8.21). It is known that solutions of such problems are piecewise linear convex functions [57]. Therefore, $\varphi(f_1)$ is a piecewise-linear convex function on (8.21).

Analogically, the function

$$\psi(f_2) = \max_{f_1}(c_1 f_1 + c_2 f_2) = \max_{f_1} \; c_1 f_1 + c_2 f_2$$

is a piecewise-linear concave function on (8.21).

Theorem 8.5 *The function $\varphi(f_1)$ is a piecewise-linear convex function on (8.21). The function $\psi(f_2)$ is a piecewise-linear concave function on (8.21).*

The problems of maximizing $\varphi(f_1)$ and minimizing $\psi(f_2)$ are the problems of concave programming, which, as it is well known, are *NP* – hard even on a unit hypercube [58]. Consequently, taking into the consideration that (8.20) and (8.21) may be represented as the problem of maximizing a piecewise-linear convex function over a hyper-parallelepiped the following result becomes obvious.

Theorem 8.6 *The maximin (minimax) cost flow problem is an NP-hard problem.*

8.4 Polymatrix Games on Digraphs

Matrix games may be generalized on the case of arbitrary number of players $p \leq n$. The vertex set V is partitioned into disjoint subsets of player positions

$$V_1, V_2, \ldots, V_p \quad \left(\bigcup_{i=1}^{p} V_i = V, \ \ V_i \cap V_j = \emptyset, \ \text{ for } i \neq j \right),$$

which define evidently the corresponding sets of player edges

$$E_i = \{(u, v) \in E \mid u \in V_i\}, \ \ i = 1, \ldots, p.$$

All players independently choose their strategies

$$(S_1, S_2, \ldots, S_p) \in 2^{E_1} \times \cdots \times 2^{E_p}.$$

The value of the choice function $\overline{M}(S)$ (defined analogically as above) is determined, where $S = (S_1, S_2, \ldots, S_p)$. Each player determines his gain

$$c_i(S) = \begin{cases} c_i(\overline{M}(S)), & \text{if } \overline{M}(S) \neq \varnothing, \\ -\infty, & \text{if } \overline{M}(S') = \varnothing, \ \forall S'_k \in 2^{E_k}, \ k \neq i, \\ +\infty, & \text{otherwise}, \end{cases}$$

where $i = 1, \ldots, p$. Thus, the vector payoff function is defined as the mapping

$$c : \ 2^{E_1} \times \cdots \times 2^{E_p} \to \mathbb{R}^p,$$

which sets the correspondence between every profile of player strategies and their gains.

Analogically with the case of matrix games, polymatrix games with feasible strategies can be defined, requiring of course the ordering of player vertices ans edges.

The solution of the polymatrix game may be defined, e.g., as a Nash equilibrium profile [59–61]. If the characteristics of some players have similar tendencies to increase or decrease, coalition games may be considered. Evidently, if $p = 2$ and $c_2(S_1, S_2) = -c(S_1, S_2)$, a zero-sum matrix game is defined.

8.5 Dynamic Games on Digraphs

In this section, we consider games [1] which are closely related to extensive form games [62–66], network and algorithmic games [8, 67–71]. They extend *simultaneous single stage* games considered above to *multi-stage* games.

Consider the above digraph G. Denote by Γ a *digraph polymatrix game* with p players defined on G. It is evident that the digraph matrix game is a particular case of the digraph polymatrix game when $p = 2$ and the gain of one of the players is a loss of his opponent.

The game Γ is a *single stage/single shot* game. Players choose their strategies simultaneously, at the same *single stage/single time moment*. As a result a feasible graph G_S^* is set, the cost of the game is determined and the gain or loss is distributed to players.

Let $\mathbb{G}$ be the set of all possible polymatrix games on a digraph G.

Generally, a *dynamic/multi-stage* game may be seen as a sequence of single stage games. It is denoted by $\Gamma(t)$, and it is defined equivalently both as a mapping

$$\Gamma : \mathbb{N} \to \mathbb{G},$$

and a sequence

$$(\Gamma_t)_{t\in\mathbb{N}} = (\Gamma(t))_{t\in\mathbb{N}} = (\Gamma(1), \Gamma(2), \ldots, \Gamma(t), \ldots).$$

The definition of the dynamic game $\Gamma(t)$ may be completed by a terminus criterion — the criterion which defines conditions for which the game ends/stops/finishes. According to considered types of future horizons, the dynamic games may be divided into two classes:

- the class of *dynamic games with finite time horizon* θ denoted by $\Gamma_\theta(t)$ and defined equivalently both as a mapping

  $$\Gamma : \{1, 2, \ldots, \theta\} \to \mathbb{G},$$

  and a finite sequence

  $$(\Gamma_{\theta 1}, \Gamma_{\theta 2}, \ldots, \Gamma_{\theta\theta}) = (\Gamma(1), \Gamma(2), \ldots, \Gamma(\theta)),$$

- the class of *dynamic games with infinite time horizon* denoted by $\Gamma_\infty(t)$ or simply $\Gamma(t)$, and defined equivalently both as a mapping

$$\Gamma : \mathbb{N} \to \mathbb{G},$$

and an infinite sequence

$$(\Gamma_1, \Gamma_2, \dots) = (\Gamma(1), \Gamma(2), \dots).$$

Remark, the infinite and finite dynamic games $\Gamma_\infty(t)$ and $\Gamma_\theta(t)$ are *repeated games* (supergames or iterated games) [72–78] if the set of all single-stage games $\mathbb{G}$ consists only of one element. A game of a single fixed type is played at every stage t of a repeated game. Evidently, the class of repeated games may be enlarged with dynamic games in which a subsequence of games is repeated.

Games considered in *theory of moves* present an alternative point of view on digraph matrix and polymatrix games [79] based on *dynamics* of player moves.

The strategies of the player $i \in \{1, 2, \dots, p\}$ in the dynamic game $\Gamma(t)$ are defined as sequences of stage game strategies

$$(S_i(t))_{t\in\mathbb{N}} = (S_i(1), S_i(2), \dots, S_i(t), \dots) \in 2^{E_i} \times 2^{E_i} \times \cdots \times 2^{E_i} \times \cdots$$

The game $\Gamma(\tau)$ of any particular stage τ generates a stage profile digraph $G_S^*(\tau)$. Denote by $S(\tau)$ the profile of player strategies at the stage τ, i.e.

$$S(\tau) = (S_1(\tau), \dots, S_p(\tau)).$$

The payoff of the stage game $\Gamma(\tau)$ is a vector denoted equivalently both by $c(\Gamma(\tau))$ and $c(S(\tau))$, with components

$$c_i(S(\tau)) = \begin{cases} c_i(\overline{M}(S(\tau))), & \text{if } \overline{M}(S(\tau)) \neq \varnothing, \\ -\infty, & \text{if } \overline{M}(S'(\tau)) = \varnothing,\ \forall S'_k(\tau) \in 2^{E_k}, k \neq i, \\ +\infty, & \text{otherwise}, \end{cases}$$

$i = 1, \dots, p$.

If the game $\Gamma(t)$ is considered on the discrete time interval $\Theta = \{1, 2, \dots, \theta\}$, then it is obvious that the player strategies are finite sequences of the form

$$(S_i(1), S_i(2), \dots, S_i(\theta)),$$

where $S_i(t),\ t = 1, 2, \dots \theta$, are the strategies of the ith player on the corresponding stages.

The cost/payoff of the dynamic game $\Gamma_\theta(t)$ is determined on the base of the cost/payoff of stage games on the stages $1, 2, \dots, \theta$. In the simplest case, the cost function may be defined by

$$1.\ c(\Gamma_\theta(t)) = \sum_{t=1}^{\theta} c(\Gamma(t));\ 2.\ \overline{c}(\Gamma_\theta(t)) = \frac{1}{\theta}\sum_{t=1}^{\theta} c(\Gamma(t)),$$

interpreted as vector expressions for polymatrix games and scalar expressions for matrix games, or by

$$3.\ c_i(\Gamma_\theta(t)) = \sum_{t=1}^{\theta} c_i(\Gamma(t));\ 4.\ \overline{c}_i(\Gamma_\theta(t)) = \frac{1}{\theta}\sum_{t=1}^{\theta} c_i(\Gamma(t)),$$

$i = 1, \ldots, p$, interpreted as components of the payoff vector function in polymatrix dynamic games.

If we suppose that the type of stage games is fixed, the result of the t-stage game does not depend on results of previous stages and the dynamic game ends at the step θ, then it is obvious that for solvable stage game Γ with optimal strategies (S_1^*, S_2^*) the corresponding dynamic game $\Gamma_\theta(t)$ has optimal strategies

$$(S_1(t))_{t\in\Theta} = \underbrace{(S_1^*, \ldots, S_1^*)}_{\theta}$$

and

$$(S_2(t))_{t\in\Theta} = \underbrace{(S_2^*, \ldots, S_2^*)}_{\theta}$$

with cost $c(\Gamma_\theta(t)) = \theta c(S_1^*, S_2^*)$ or $\overline{c}(\Gamma_\theta(t)) = c(S_1^*, S_2^*)$. Therefore, such dynamic games are generally identical with matrix and polymatrix games.

A dynamic game model acquires more valuable features if the *terminus criterion implies to finish the game* when a feasible subgraph with prescribed structure is constructed in some stage τ graph $G_S(\tau)$. In such case the time horizon may be infinite, but the game nevertheless stops if the terminus criterion is satisfied.

Next, we associate with every stage t a dynamic game state

$$G(t) = (W(t), E(t))$$

as a couple formed by a vertex set and an edge set which depend on t. Evidently, sets of player strategies depend on game states.

If the initial state is

$$G(0) = (W(0), E(0)) \subseteq (V, E)$$

at the initial time moment, the set of all possible strategies of the ith player at the moment t may be defined as:

$$E_i(t) = \{(u, v) \in E | u \in V_i \cap W(t-1)\},\ t = 1, 2, \ldots;\ i = 1, \ldots, p,$$

where $W(t-1)$ is the set of vertices in the graph $G(t-1)$ at the stage $t-1$, V_i — the set of positions of the ith player. Every subset $S_i(t) \subseteq E_i(t)$ is called a strategy of player i at the stage t.

The state of the dynamic game at the stage t is determined after examination of the corresponding game at the previous stage $t-1$ according to formula

$$W(t) = \{v \in V \,|\, \exists (u,v) \in \bigcup_{i=1}^{p} S_i(t)\}, \quad t = 1, \ldots, \tau.$$

As the player strategies at the stage t depend both on their positions and the game state at the moment t, we have to solve games of the same type at consecutive time moments t and $t+1$, but generally with different sets of strategies for every player. After that, as the player strategies are the edges which determine the mapping of $W(t-1)$ in $W(t)$, then players, in antagonistic interests, endeavour to increase (to decrease) the set of their own advantageous positions of the stage $W(t)$, and to decrease (to increase) the set of advantageous (non-advantageous), positions of adversaries on the same state $W(t)$. Therefore, G is the graph of all possible passages (strategies), their number being limited by 2^m. The set V defines the set of all possible states, which cardinality is limited by 2^n.

If the dynamic game is examined on infinite time interval $\{1, 2, \ldots\}$, then it follows from the limited number of states that some states of the game and corresponding strategies of players will be repeated in some sequences. It is obvious, that the payoff function $c(\Gamma(t))$, having the form of the number sequences, will be unlimited on $\tau \to +\infty$. Therefore, the definition of such game must be completed with special ending criterion: or the value cost function is larger then determined limit, or at some moment the graph of determined structure is constructed, etc. In the case of cost function $\overline{c}(\Gamma(t))$ we can examine the limit

$$\overline{c}(\Gamma(t)) = \lim_{\theta \to +\infty} \Big(\frac{1}{\theta} \sum_{t=0}^{\theta} c(\Gamma(t))\Big),$$

for which there exists (as is mentioned above) repeated sequences of game states with limited value of the cost function, such that $\overline{c}(\Gamma(t)$ is equal to fixed number. In this case, the problem to find cycle of the states may be considered.

Next, lengths of edges can be functions depending on t and the cost of a dynamic game is calculated using static game costs only at some stages.

As mentioned, it is clear that the contents and type of dynamic games are depending on:

- static game;
- initial state and restriction of cardinality of the game states;
- cost function;
- edge length function;
- time interval on which the game is examined;

- terminus criterion,
- etc.

In investigation of dynamic games $\Gamma(t)$ it is useful sometimes to use the property that every dynamic game $\Gamma(t)$ may be represented as a matrix game.

8.6 Concluding Remarks

It is necessary to mention that strategies of players may be defined also as subsets of vertices, or the pair of subsets of vertices and edges. The investigation of such games, the determination of solvable games and the elaboration of corresponding algorithms are problems for future work.

References

1. Ungureanu, V. 1998. Games on digraphs and constructing maximin structures. *Computer Science Journal of Moldova* 6 3(18): 313–337.
2. Ungureanu, V. 2016. *Strategic games on digraphs. ROMAI Journal* 12 (1): 133–161.
3. Christofides, N. 1975. *Graph Theory: An Algorithmic Approach*, 415. London: Academic Press.
4. Papadimitriou, C., and K. Steiglitz. 1982. *Combinatorial Optimization: Algorithms and Complexity*, 510. Englewood Cliffs: Prentice-Hall Inc.
5. Berge, C. 1962. *The Theory of Graphs (and its Applications)*. London: Methuen & Co.; New York: Wiley, 272pp.
6. Van den Nouweland, A., P. Borm, W. van Golstein Brouwers, R. Groot Bruinderink, and S. Tijs. 1996. A game theoretic approach to problems in telecommunications. *Management Science* 42 (2): 294–303.
7. Altman, E., T. Boulogne, R. El-Azouzi, T. JimTnez, and L. Wynter. 2006. A survey on networking games in telecommunications. *Computers & Operations Research* 33 (2): 286–311.
8. Tardos, E. 2004. *Network Games*, Proceedings of the Thirty-Sixth Annual ACM Symposium on Theory of Computing (STOC?04), 341–342. Chicago, Illinois, USA, June 13–15 2004.
9. Altman, E., and L. Wynter. 2004. Crossovers between transportation planning and telecommunications, networks and spatial economics. *Editors* 4 (1): 5–124.
10. Jackson, M.O., and Y. Zenou. 2015. Games on networks. In *Handbook of Game Theory with Economic Applications*, vol. 4, ed. H. Peyton Young, and Sh Zamir, 95–163. Amsterdam: North-Holland.
11. Roughgarden, T. 2002. The price of anarchy is independent of the network topology. *Journal of Computer and System Sciences* 67 (2): 341–364.
12. Suri, S., C. Toth, and Y. Zhou. 2005. *Selfish load balancing and atomic congestion games*, Proceedings of the 16th Annual ACM Symposium on Parallel Algorithms and Architectures (SPAA), 188–195.
13. Christodoulou, G., and E. Koutsoupias. 2005. *The price of anarchy of finite congestion games*, Proceedings of the Thirty-Seven Annual ACM Symposium on Theory of Computing (STOC?05), 67–73. New York, USA.
14. Czumaj, A., and B. Vöcking. 2007. Tight bounds for worst-case equilibria. *ACM Transactions on Algorithms* 3 (1): 11. Article 4.
15. Vetta, A. 2002. *Nash equilibria in competitive societies, with applications to facility location, traffic routing and auctions*, Proceedings of the 43rd Annual IEEE Symposium on Foundations of Computer Science (FOCS'02), 416pp.

16. Akella, A., S. Seshan, R. Karp, S. Shenker, and C. Papadimitriou. 2002. *Selfish behavior and stability of the Internet: A game-theoretic analysis of TCP*, ACM SIGCOMM Computer Communication Review - Proceedings of the 2002 SIGCOMM Conference, 32(4): 117–130.
17. Dutta, D., A. Goel, and J. Heidemann. 2002. *Oblivious AQM and nash equilibria*, ACM SIGCOMM Computer Communication Review - Proceedings of the 2002 SIGCOMM conference, 32(3): 20.
18. Fabrikant, A., A. Luthra, E. Maneva, C.H. Papadimitriou and S. Shenker. 2003. *On a network creation game*, Proceedings of the Twenty-Second Annual Symposium on Principles of Distributed Computing (PODC '03), 347–351.
19. Anshelevich, E., A. Dasgupta, E. Tardos, and T. Wexler. 2008. Near-optimal network design with selfish agents. *Theory of Computing* 4: 77–109.
20. Kodialam, M., and T.V. Lakshman. 2003. Detecting network intrusions via sampling: A game theoretic approach. *IEEE INFOCOM* 1–10.
21. Han, Z., D. Niyato, W. Saad, T. Başar, and A. Hjørungnes. 2012. *Game Theory in Wireless and Communication Networks: Theory, Models, and Applications*. Cambridge: Cambridge University Press, XVIII+535pp.
22. Zhang, Y., and M. Guizani (eds.). 2011. *Game Theory for Wireless Communications and Networking*. Boca Raton: CRC Press, XIV+571pp.
23. Kim, S. 2014. *Game Theory Applications in Network Design*. Hershey: IGI Global, XXII+500pp.
24. Mazalov, V. 2014. *Mathematical Game Theory and its Applications*. Tokyo: Wiley, XIV+414pp.
25. Antoniou, J., and A. Pitsillides. 2013. *Game Theory in Communication Networks: Cooperative Resolution of Interactive Networking Scenarios*. Boca Raton: CRC Press, X+137pp.
26. Aubin, J.-P. 2005. Dynamical connectionist network and cooperative games. In *Dynamic Games: Theory and Applications*, ed. A. Haurie, and G. Zaccour, 1–36. US: Springer.
27. El Azouzi, R., E. Altman, and O. Pourtallier. 2005. Braess paradox and properties of wardrop equilibrium in some multiservice networks. In *Dynamic Games: Theory and Applications*, ed. A. Haurie, and G. Zaccour, 57–77. US: Springer.
28. Pavel, L. 2012. *Game Theory for Control of Optical Networks*. New York: Birkhäuser, XIII+261pp.
29. Gurvitch, V., A. Karzanov, and L. Khatchiyan. 1988. Cyclic games: Finding of minimax mean cycles in digraphs. *Journal of Computational Mathematics and Mathematical Physics* 9 (28): 1407–1417. (in Russian).
30. Von Stackelberg, H. 1934. *Marktform und Gleichgewicht (Market Structure and Equilibrium)*. Vienna: Springer, (in German), XIV+134pp.
31. Wolfram, S. 2002. *A New Kind of Science*, 1197. Champaign, IL: Wolfram Media Inc.
32. Wolfram, S. 2016. *An Elementary Introduction to the Wolfram Language*. Champaign, IL: Wolfram Media, Inc., XV+324pp.
33. Boliac, R., and D. Lozovanu. 1996. Finding of minimax paths tree in weighted digraph. *Buletinul Academiei de Ştiinţe a Republicii Moldova* 3 (66): 74–82. (in Russian).
34. Biggs, N.L., E.K. Lloyd, and R.J. Wilson. 1976. *Graph Theory, 1736–1936*, 255. Oxford: Clarendon Press.
35. Schrijver, A. 2005. On the history of combinatorial optimization (Till 1960). *Handbooks in Operations Research and Management Science* 12: 1–68.
36. Gutin, G., and P.P. Abraham (eds.). 2004. *The Traveling Salesman Problem and its Variation*, vol. 2, New-York: : Kluwer Academic Publishers, 831pp.
37. Lawler, E.L., J.K. Lenstra, A.H.G. Rinooy Kan, and D.B. Shmoys (eds.). 1985. *The Traveling Salesman Problem*. Chichester, UK: Wiley.
38. Reinelt, G. 1994. *The Traveling Salesman: Computational Solutions for TSP Applications*, 230. Berlin: Springer.
39. Fleishner, H. 2000. Traversing graphs: The Eulerian and Hamiltonian theme. In *Arc Routing: Theory, Solutions, and Applications*, ed. M. Dror, 19–87. The Netherlands: Kluwer Academic Publishers.

40. Golden, B., S. Raghavan, and E. Wasil (eds.). 2008. *The Vehicle Routing Problem: Latest Advances and New Challenges*. New York: Springer.
41. Ball, M.O., T.L. Magnanti, C.L. Monma, and G.L. Nemhauser (eds.). 1995. *Network Routing*, 779. Elsevier: Amsterdam.
42. Garey, M.R., and D.S. Johnson. 1979. *Computers and Intractability: A Guide to the Theory of NP Completeness*, 351. San Francisco: W.H. Freeman.
43. Sipser, M. 2006. *Introduction to the Theory of Computation*, 2nd ed, Boston, Massachusetts: Thomson Course Technology, XIX+431pp.
44. Nielsen, M.A., and I.L. Chuang. 2010. *Quantum Computation and Quantum Information*, 10th Anniversary ed. Cambridge, UK: Cambridge University Press, XXXII+676pp.
45. Applegate, D.L., R.E. Bixby, V. Chvtal, J. William, and W.J. Cook. 2006. *The Traveling Salesman Problem: A Computational Study*, 606. Princeton: Princeton University Press.
46. Hitchcock, F.L. 1941. The distribution of product from several sources to numerous localities. *Journal of Mathematical Physics* 20 (2): 217–224.
47. Díaz-Parra, O., J.A. Ruiz-Vanoye, B.B. Loranca, A. Fuentes-Penna, and R.A. and Barrera-Cámara. 2014. A survey of transportation problems. *Journal of Applied Mathematics* 2014: 17. Article ID 848129.
48. Hoffman, K.L., and M. Padberg. 1991. LP-based combinatorial problem solving. *Annals of Operations Research* 4: 145–194.
49. Padberg, M., and G. Rinaldi. 1991. A branch-and-cut algorithm for the resolution of large-scale traveling salesman problem. *SIAM Review* 33: 60–100.
50. Villani, C. 2003. *Topics in Optimal Transportation, Graduate Studies in Mathematics*, vol. 58, 382. Providence: American Mathematical Society.
51. Villani, C. 2008. *Optimal Transport, Old and New*, 1000. Berlin: Springer.
52. Ungureanu, V. 2006. Traveling salesman problem with transportation. *Computer Science Journal of Moldova* 14 2(41): 202–206.
53. Caric, T., and H. Gold (eds.). 2008. *Vehicle Routing Problem*, 152. InTech: Croatia.
54. Toth, P., and D. Vigo. 2002. *The Vehicle Routing Problem*, Society for Industrial and Applied Mathematics, 386pp.
55. Labadie, N., and C. Prodhon. 2014. A survey on multi-criteria analysis in logistics: Focus on vehicle routing problems. In *Applications of Multi-Criteria and Game Theory Approaches: Manufacturing and Logistics*, ed. L. Benyoucef, J.-C. Hennet, and M.K. Tiwari, 3–29. New Jersey: Springer.
56. Bellman, R. 1957. *Dynamic Programming*, 365. New Jersey: Princeton University Press.
57. Golshtein, E., and D. Yudin. 1966. *New Directions in Linear Programming*. Moscow: Sovetskoe Radio, 527pp. (in Russian).
58. Ungureanu, V. 1997. Minimizing a concave quadratic function over a hypercube. *Buletinul Academiei de Ştiinţe a Republicii Moldova: Mathematics Series* 2 (24): 69–76. (in Romanian).
59. Yanovskaya, E.B. 1968. Equilibrium points in polymatrix games. *Lithuanian Mathematical Collection (Litovskii Matematicheskii Sbornik)* 8 (2): 381–384. (in Russian).
60. Howson Jr., J.T. 1972. Equilibria of polymatrix games. *Management Science* 18: 312–318.
61. Eaves, B.C. 1973. Polymatrix games with joint constraints. *SIAM Journal of Applied Mathematics* 24: 418–423.
62. Von Neumann, J. 1928. *Zur Theorie der Gesellschaftsspiele. Mathematische Annalen* 100: 295–320. (in German).
63. Kuhn, H.W. 1950. *Extensive games*, Proceedings of the National Academy of Sciences U.S.A., Vol. 36, 570–576.
64. Kuhn, H.W. 1953. Extensive games and the problem of information. *Contributions to the Theory of Games, Vol. II*, vol. 28, 217–243., Annals of Mathematics Study Princeton: Princeton University Press.
65. Kuhn, H.W. 2003. *Lectures on the Theory of Games*, vol. 37, 118. Annals of Mathematics Study Princeton: Princeton University Press.
66. Alós-Ferrer, C., and K. Ritzberger. 2016. *The Theory of Extensive Form Games*. Berlin: Springer, XVI+239pp.

67. Nisan, N., T. Roughgarden, E. Tardos, and V.V. Vazirani (eds.). 2007. *Algorithmic Game Theory*. Cambridge, UK: Cambridge University Press, 775pp.
68. Shoham, Y., and K. Leyton-Brown. 2009. *Multi-Agent Systems: Algorithmic, Game-Theoretic, and Logical Foundations*, 532. Cambridge: Cambridge University Press.
69. Easley, D., and D. Kleinberg. 2010. *Networks, Crowds, and Markets: Reasoning about a Highly Connected World*, 833. Cambridge: Cambridge University Press.
70. Menache, I., and A. Ozdaglar. 2011. *Network Games: Theory, Models, and Dynamics*. Synthesis Lectures on Communication Networks California: Morgan & Claypool Publishers, XIV+143pp.
71. Lozovanu, D. 1993. A strongly polynomial time algorithm for finding minimax paths in network and solving cyclic games. *Cybernetics and System Analysis* 5: 145–151. (in Russian).
72. Mertens, J.F. 1987. *Repeated Games*, Proceedings of the International Congress of Mathematicians, Berkeley, Providence: American Mathematical Society, 1528–1577.
73. Aumann, R.J., M.B. Maschler, and R.E. Stearns. 1995. *Repeated Games with Incomplete Information*, 360. Cambridge, Massachusetts: MIT Press.
74. Fudenberg, D., and J. Tirole. 1991. *Game Theory*, 579. Cambridge: MIT Press.
75. Osborne, M.J., and A. Rubinstein. 1994. *A Course in Game Theory*, 373. Cambridge, Massachusetts: The MIT Press.
76. Mailath, G.J., and L. Samuelson. 2006. *Repeated Games and Reputations: Long-Run Relationships*. New York: Oxford University Press, XVIII+645pp.
77. Sorin, S. 2002. *A First Course on Zero-Sum Repeated Games*. Berlin: Springer, XV+204pp.
78. Osborne, M.J. 2009. *An Introduction to Game Theory*, International ed, 685. Oxford: Oxford University Press.
79. Brams, S.J. 1994. *Theory of Moves*. Cambridge: Cambridge University Press, XII+248pp.

Part II
Mixtures of Simultaneous and Sequential Games

The second part of the monograph is dedicated to Pareto-Nash-Stackelberg games.

Chapter 9
Solution Principles for Mixtures of Simultaneous and Sequential Games

Abstract This chapter unites mathematical models, solution concepts and principles of both simultaneous and sequential games. The names of Pareto, Nash and Stackelberg, are used to identify simply types of game models. Pareto is associated with multi-criteria decision making (Pareto, Manuel d'economie politique, Giard, Paris, 1904, [1]). Nash is associated with simultaneous games (Nash, Ann Math, 54 (2): 280–295, 1951 [2]). Stackelberg is associated with hierarchical games (Von Stackelberg, Marktform und Gleichgewicht (Market Structure and Equilibrium), Springer, Vienna, 1934. [3]). The names have priorities in accordance with their positions in the game title from left to right. For example, the title Pareto-Nash-Stackelberg game (Ungureanu, ROMAI J, 4 (1): 225–242, 2008, [4]) means that players choose their strategies on the basis of multi-criteria Pareto decision making. They play a simultaneous Nash games. Simultaneous Nash games are played at every stage of a hierarchical Stackelberg game. Nash-Stackelberg games may be called also multi-leader-follower games as they are named in Hu's survey (Hu, J Oper Res Soc Jpn, 58: 1–23, 2015, [5]). The chapter has general theoretical meaning both for this second part of the monograph and for the next third part. Investigations are provided by means of the concepts of best response mappings, efficient response mappings, mapping graphs, intersection of graphs, and constrained optimization problems.

9.1 Introduction

Interactive decisions processes, which involve both sequential and simultaneous decision making made by independent and interdependent players with one or more objectives, can be modelled by means of strategic form games: Stackelberg games [3, 6], Nash games [2], Pareto-Nash games [7–12], Nash-Stackelberg games [5, 13, 14], Pareto-Nash-Stackelberg games [4], etc.

At every stage/level of a Nash-Stackelberg game a Nash game is played. Stage profiles (joint decisions) are executed sequentially throughout the hierarchy as a Stackelberg game.

V. Ungureanu, *Pareto-Nash-Stackelberg Game and Control Theory*, Smart Innovation, Systems and Technologies 89,
https://doi.org/10.1007/978-3-319-75151-1_9

At every stage of a multiobjective Pareto-Nash-Stackelberg game a multiobjective Pareto-Nash game is played. Stage profiles are executed sequentially throughout the hierarchy.

Via the notion of a best response mapping graph, we define unsafe and safe Stackelberg equilibria for Stackelberg games, pseudo and multi-stage Nash-Stackelberg equilibria for Nash-Stackelberg games, and Pareto-Nash-Stackelberg equilibria for multiobjective Pareto-Nash-Stackelberg games.

In the following exposition, we continue and extend the research of Nash equilibria as the elements of intersection of best response mapping graphs [15–20].

Consider a noncooperative strategic game

$$\Gamma = \langle N, \{X_p\}_{p \in N}, \{f_p(x)\}_{p \in N} \rangle,$$

where

- $N = \{1, 2, \ldots, n\}$ is a set of players,
- $X_p \subseteq \mathbb{R}^{k_p}$ is a set of strategies of the player $p \in N$,
- $k_p < +\infty, \quad p \in N$,
- and $f_p(x)$ is a pth player cost/payoff function defined on the Cartesian product $X = \underset{p \in N}{\times} X_p$ — the set of profiles. For convenience and without loss of generality **we assume that all the players minimize the values of their cost functions.**

Suppose the players make their moves hierarchically:

the first player chooses his strategy $x_1 \in X_1$ and informs about his choice the second player,
the second player chooses his strategy $x_2 \in X_2$ after observing the moves x_1 of the first player and informs about the choices x_1, x_2 the third player,
and so on
the n**th player** selects his strategy $x_n \in X_n$ after observing the moves $x_1, \ldots, x_{n-1}$ of the preceding players, at last.

On the resulting profile $x = (x_1, \ldots, x_n)$ every player computes the value of his cost function.

When player $p \in N$ **moves, players** $1, 2, \ldots, p-1$ **are leaders or predecessors of player** p **and players** $p+1, \ldots, n$ **are followers or successors of the player** p. **Players have all the information about the predecessors choices and doesn't have information about the choices of the successors, but the** pth **player** $(p < n)$ **has all the information about all strategy sets and cost functions of the players** $p, p+1, \ldots, n$.

By backward induction, every player $(n, n-1, \ldots, 2)$ determines his best move mapping and the first player determines the set of his best moves:

$$B_n(x_1, \ldots, x_{n-1}) = \underset{y_n \in X_n}{\operatorname{Arg\,min}} f_n(x_1, \ldots, x_{n-1}, y_n),$$

$$B_{n-1}(x_1, \ldots, x_{n-2}) = \\ = \underset{y_{n-1},\, y_n :\, (x_1, \ldots, x_{n-2},\, y_{n-1},\, y_n) \in Gr_n}{\operatorname{Arg\,min}} f_{n-1}(x_1, \ldots, x_{n-2}, y_{n-1}, y_n),$$

$$\ldots$$

$$B_2(x_1) = \underset{y_2, \ldots, y_n :\, (x_1, y_2, \ldots, y_n) \in Gr_3}{\operatorname{Arg\,min}} f_2(x_1, y_2, \ldots, y_n),$$

$$\hat{X} = \underset{(y_1, \ldots, y_n) \in Gr_2}{\operatorname{Arg\,min}} f_1(y_1, \ldots, y_n)$$

where

$$Gr_n = \left\{ x \in X : \begin{array}{c} x_1 \in X_1 \\ \ldots \\ x_{n-1} \in X_{n-1} \\ x_n \in B_n(x_1, \ldots, x_{n-1}) \end{array} \right\},$$

$$Gr_{n-1} = \left\{ x \in Gr_n : \begin{array}{c} x_1 \in X_1 \\ \ldots \\ x_{n-2} \in X_{n-2} \\ (x_{n-1}, x_n) \in B_{n-1}(x_1, \ldots, x_{n-2}) \end{array} \right\},$$

$$\ldots$$

$$Gr_2 = \{x \in Gr_3 : x_1 \in X_1, (x_2, \ldots, x_n) \in B_2(x_1)\}.$$

Evidently, the inclusions $Gr_2 \subseteq Gr_3 \subseteq \cdots \subseteq Gr_n$, are truth. The relations mean that all the graphs form a family of nested sets.

Definition 9.1 Any profile $\hat{x} \in \hat{X}$ of the game Γ is called unsafe (optimistic, strong) Stackelberg equilibrium.

Definition 9.1 of the unsafe Stackelberg equilibrium and the correspondent definition from [6] are equivalent.

For $n = 2$, the unsafe Stackelberg equilibrium notion and the original Stackelberg equilibrium notion [3] are equivalent.

9.2 Unsafe Stackelberg Equilibria. Existence and Properties

Properties of unsafe Stackelberg equilibria are more prominent and evident in the case of finite games.

Theorem 9.1 *For any finite hierarchical game the set $\hat{X}$ of unsafe Stackelberg equilibria is non empty.*

Proof It is easy to observe that the graph of best response mapping of the last n^{th} player consists of a finite number of distinct points.

The payoff function of the player $n-1$ achieves its minimal value on a finite set of points and this means that the graph of best response mapping of the player $n-1$ consists of a finite number of points too.

Analogical reasons are valid for players $n-2, n-3, \ldots, 1$.

So, the set of unsafe Stackelberg equilibria is non empty. □

Corollary 9.1 *The unsafe Stackelberg equilibrium notion and the Nash equilibrium notion are not equivalent.*

Proof On the one hand, according to Theorem 9.1 any finite strategic form games has an unsafe Stackelberg equilibrium.

On the other hand, let us recall a very well fact that there are finite strategic form games that do not have pure strategy Nash equilibria.

The above distinction in equilibria existence proves corollary. □

Next Example 9.1 has the aim to illustrate both the unsafe Stackelberg equilibrium concept and the above results.

Example 9.1 Consider a three-player $2 \times 2 \times 2$ game with cost matrices:

$$a_{1**} = \begin{bmatrix} 5 & 2 \\ 1 & 3 \end{bmatrix}, \; a_{2**} = \begin{bmatrix} 1 & 3 \\ 3 & -1 \end{bmatrix},$$

$$b_{1**} = \begin{bmatrix} 5 & 7 \\ 4 & 6 \end{bmatrix}, \; b_{2**} = \begin{bmatrix} 3 & 4 \\ 8 & 5 \end{bmatrix},$$

$$c_{1**} = \begin{bmatrix} 2 & 10 \\ 7 & 3 \end{bmatrix}, \; c_{2**} = \begin{bmatrix} 8 & 6 \\ 4 & 5 \end{bmatrix}.$$

The first player moves the first, the second player moves the second and the third player moves the last.

First of all we must determine the graph Gr_3 of the third player. The graph's elements are emphasized by boxes.

$$c_{1**} = \begin{bmatrix} \boxed{2} & 10 \\ 7 & \boxed{3} \end{bmatrix}, \; c_{2**} = \begin{bmatrix} 8 & \boxed{6} \\ \boxed{4} & 5 \end{bmatrix}.$$

Second player's graph Gr_2 is constructed on the basis of the graph Gr_3. Its elements are emphasized by two frames boxes.

$$b_{1**} = \begin{bmatrix} \boxed{\boxed{5}} & 7 \\ 4 & \boxed{6} \end{bmatrix}, \; b_{2**} = \begin{bmatrix} 3 & \boxed{\boxed{4}} \\ \boxed{8} & 5 \end{bmatrix}.$$

At last, the set of unsafe Stackelberg equilibria is determined on the basis of the graph Gr_2.

$$a_{1**} = \begin{bmatrix} \boxed{\boxed{5}} & 2 \\ 1 & \boxed{3} \end{bmatrix}, \; a_{2**} = \begin{bmatrix} 1 & \boxed{\boxed{\boxed{3}}} \\ \boxed{3} & -1 \end{bmatrix}.$$

The unsafe Stackelberg equilibrium set consists of one equilibrium $(2, 1, 2)$ with players' cost functions values $(3, 4, 6)$. Remark, the profile $(2, 1, 2)$ is not a Nash equilibrium. Moreover, the corresponding three matrix game doesn't have pure strategy Nash equilibria. □

In Example 9.1 player best move mappings are single-valued and the realization of the unique unsafe Stackelberg equilibrium is natural. The situation is more difficult when the mappings are multi-valued. The achieving of the unsafe Stackelberg equilibrium is uncertain. A modification of Example 9.1 illustrates this fact.

Example 9.2 Let us continue Example 9.1 by modifying the elements a_{211}, b_{211} and c_{211} in the cost matrices:

$$a_{1**} = \begin{bmatrix} 5 & 2 \\ 1 & 3 \end{bmatrix}, \; a_{2**} = \begin{bmatrix} +\infty & 3 \\ 3 & -1 \end{bmatrix},$$

$$b_{1**} = \begin{bmatrix} 5 & 7 \\ 4 & 6 \end{bmatrix}, \; b_{2**} = \begin{bmatrix} 4 & 4 \\ 8 & 5 \end{bmatrix},$$

$$c_{1**} = \begin{bmatrix} 2 & 10 \\ 7 & 3 \end{bmatrix}, \; c_{2**} = \begin{bmatrix} 6 & 6 \\ 4 & 5 \end{bmatrix}.$$

The elements of the graph Gr_3 are emphasized by boxes.

$$c_{1**} = \begin{bmatrix} \boxed{2} & 10 \\ 7 & \boxed{3} \end{bmatrix}, \; c_{2**} = \begin{bmatrix} \boxed{6} & \boxed{6} \\ \boxed{4} & 5 \end{bmatrix}.$$

Elements of the graph Gr_2 are emphasized by two frames boxes.

$$b_{1**} = \begin{bmatrix} \boxed{\boxed{5}} & 7 \\ 4 & \boxed{6} \end{bmatrix}, \; b_{2**} = \begin{bmatrix} \boxed{\boxed{4}} & \boxed{\boxed{4}} \\ \boxed{8} & 5 \end{bmatrix},$$

At last, the set of the unsafe Stackelberg equilibria is determined on the graph Gr_2.

$$a_{1**} = \begin{bmatrix} \boxed{\boxed{5}} & 2 \\ 1 & \boxed{3} \end{bmatrix}, \; a_{2**} = \begin{bmatrix} \boxed{\boxed{+\infty}} & \boxed{\boxed{\boxed{3}}} \\ \boxed{3} & -1 \end{bmatrix}.$$

The game consists of the same unique unsafe Stackelberg equilibrium $(2, 1, 2)$ with players' cost functions values $(3, 4, 6)$. Unfortunately, the practical achievement of this equilibrium is uncertain as the first strategy (the profile $(2, 1, 1)$) gives the same value for the cost function as the second strategy (the profile $(2, 1, 2)$) for the third player.

If at the third stage the third player chooses his first strategy, then at the profile $(2, 1, 1)$ the values of the cost functions are $(+\infty, 4, 6)$. It is a disastrous result for the first player. □

Let us consider now continuous games.

Theorem 9.2 *If every strategy set $X_p \subset \mathbb{R}^{k_p}$, $p = 1, \ldots, n$, is compact and every cost function $f_p(x_1, \ldots, x_p, \ldots, x_n)$, $p = 1, \ldots, n$, is continuous in the variables $(x_p, \ldots, x_n)$ on $X_p \times \cdots \times X_n$ for any fixed $x_1 \in X_1, \ldots, x_{p-1} \in X_{p-1}$, then the unsafe Stackelberg equilibria set $\hat{X}$ is non empty.*

The proof follows from the well known Weierstrass theorem.

Theorem 9.3 *If every strategy set $X_p \subseteq \mathbb{R}^{k_p}$, $p = 1, \ldots, n$, is convex and every cost function $f_p(x_1, \ldots, x_p, \ldots, x_n)$, $p = 1, \ldots, n$, is strict convex in the variables $(x_p, \ldots, x_n)$ on $X_p \times \cdots \times X_n$ for any fixed $x_1 \in X_1, \ldots, x_{p-1} \in X_{p-1}$, then the game has a unique unsafe Stackelberg equilibrium with the "guaranteed" achievement/realization property.*

"Guaranteed" achievement/realization means that the unsafe Stackelberg equilibrium is strictly preferred for all players. The proof follows from the properties of the strict convex functions.

Remark that the strict convex requirements in Theorem 9.3 may be substituted by unimodal or other requirements that guaranteed the single-valued characteristics of the involved best-move mappings.

9.3 Safe Stackelberg Equilibria

In order to exclude the case illustrated in Example 9.2 the notion of a safe Stackelberg equilibrium is introduced, which is equivalent to the respective notion from [6].

By backward induction, every player $(n, n-1, \ldots, 2)$ determines his best (minimax) move mapping and the first player determines the set of his best (minimax) moves:

$$B_n(x_1, \ldots, x_{n-1}) = \operatorname*{Arg\,min}_{y_n \in X_n} f_n(x_1, \ldots, x_{n-1}, y_n),$$

$$\begin{aligned}\tilde{B}_{n-1}(x_1, \ldots, x_{n-2}) &= \\ &= \operatorname*{Arg\,min}_{\substack{y_{n-1} \\ (x_1,\ldots,x_{n-2},y_{n-1},y_n)\in Gr_n}} \max_{y_n} f_{n-1}(x_1, \ldots, x_{n-2}, y_{n-1}, y_n),\end{aligned}$$

$$\begin{aligned}\tilde{B}_{n-2}(x_1, \ldots, x_{n-3}) &= \\ &= \operatorname*{Arg\,min}_{\substack{y_{n-2} \\ (x_1,\ldots,x_{n-3},y_{n-2},\ldots,y_n)\in \tilde{G}r_{n-1}}} \max_{y_{n-1},y_n} f_{n-1}(x_1, \ldots, x_{n-3}, y_{n-2}, \ldots, y_n),\end{aligned}$$

$$\ldots$$

$$\tilde{B}_2(x_1) = \operatorname*{Arg\,min}_{\substack{y_2 \\ (x_1,y_2,\ldots,y_n)\in \tilde{G}r_3}} \max_{y_3,\ldots,y_n} f_2(x_1, y_2, \ldots, y_n),$$

$$\tilde{X} = \operatorname*{Arg\,min}_{\substack{y_1 \\ (y_1,\ldots,y_n)\in \tilde{G}r_2}} \max_{y_2,\ldots,y_n} f_1(y_1, \ldots, y_n)$$

where

$$Gr_n = \{x \in X : x_1 \in X_1, \ldots, x_{n-1} \in X_{n-1}, x_n \in B_n(x_1, \ldots, x_{n-1})\},$$

$$\tilde{G}r_{n-1} = \left\{ x \in Gr_n : \begin{array}{c} x_1 \in X_1 \\ \ldots \\ x_{n-2} \in X_{n-2} \\ (x_{n-1}, x_n) \in \tilde{B}_{n-1}(x_1, \ldots, x_{n-2}) \end{array} \right\},$$

$$\ldots$$

$$\tilde{G}r_2 = \left\{ x \in Gr_3 : x_1 \in X_1, (x_2, \ldots, x_n) \in \tilde{B}_2(x_1) \right\}.$$

Evidently, $\tilde{G}r_2 \subseteq \tilde{G}r_3 \subseteq \cdots \subseteq \tilde{G}r_{n-1} \subseteq Gr_n$, too.

Definition 9.2 The profile $\tilde{x} \in \tilde{X}$ of the game is called a safe (pessimistic, weak) Stackelberg equilibrium.

Generally, the unsafe Stackelberg equilibria set is not equivalent to the safe Stackelberg equilibria set, i.e. $\hat{X} \neq \tilde{X}$. Remark that in Example 9.2 the profile $(1, 1, 1)$ (with the costs $(5, 5, 2)$) is a safe Stackelberg equilibrium. The security "*payment*" for the first player is *"supported"* also by the second player because the value of his cost function increases too.

In addition, for $n = 2$ the safe Stackelberg equilibrium is not always and the unsafe Stackelberg equilibrium.

Theorems 9.1–9.3 analogues for the safe Stackelberg equilibrium may be formulated and proved. In conditions of Theorem 9.3, the unsafe and safe Stackelberg equilibria are identical.

9.4 Pseudo-Equilibria. Nash-Stackelberg Equilibria

Consider the strategic game

$$\Gamma = \langle N, \{X_p^l\}_{l \in S, p \in N_l}, \{f_p^l(x)\}_{l \in S, p \in N_l} \rangle,$$

where

- $S = \{1, 2, \ldots, s\}$ is a set of stages,
- $N_l = \{1, 2, \ldots, n_l\}$ is a set of players at stage (level) $l \in S$,
- $X_p^l \subseteq \mathbb{R}^{k_p^l}$ is a set of strategies of player $p \in N_l$ at stage $l \in S$,
- $s < +\infty, \ n_l < +\infty, \ l \in S$,
- and $f_p^l(x)$ is a lth stage pth player cost function defined on the Cartesian product $X = \underset{p \in N_l, l \in S}{\times} X_p^l$.

Elements $x = (x_1^1, x_2^1, \ldots, x_{n_1}^1, x_1^2, x_2^2, \ldots, x_{n_2}^2, \ldots, x_1^s, x_2^s, \ldots, x_{n_s}^s) \in X$ form profiles of the game.

Suppose the players make their moves hierarchically:

- the first stage players $1, 2, \ldots, n_1$ selects their strategies

$$x_1^1 \in X_1^1, \ x_2^1 \in X_2^1, \ldots, \ x_{n_1}^1 \in X_{n_1}^1$$

simultaneously and inform about their selections the players $1, 2, \ldots, n_2$ of the second stage,

- the second stage players $1, 2, \ldots, n_2$ select simultaneously their strategies

$$x_1^2 \in X_1^2, \ x_2^2 \in X_2^2, \ldots, \ x_{n_2}^2 \in X_{n_2}^2$$

after observing the moves $(x_1^1, x_2^1, \ldots, x_{n_1}^1)$ of the first stage players and informs about two stage selection results the third stage players,

- and so on
- the sth stage players $1, 2, \ldots, n_s$ select simultaneously their strategies

$$x_1^s \in X_1^s, \ x_2^s \in X_2^s, \ldots, \ x_{n_s}^s \in X_{n_s}^s$$

in the end after observing the moves

$$(x_1^1, x_2^1, \ldots, x_{n_1}^1, x_1^2, x_2^2, \ldots, x_{n_2}^2, \ldots, x_1^{s-1}, x_2^{s-1}, \ldots, x_{n_{s-1}}^{s-1})$$

of the precedent stages players.

On the resulting profile

$$x = (x_1^1, x_2^1, \ldots, x_{n_1}^1, x_1^2, x_2^2, \ldots, x_{n_2}^2, \ldots, x_1^s, x_2^s, \ldots, x_{n_s}^s)$$

every player computes the value of his cost function.

Suppose the lth stage pth **player has all information about all strategy sets and the cost functions of the players of stages** $l, l+1, \ldots, s$. Without loss of generality **suppose all players minimize the values of their cost functions.**

Definition 9.3 The profile $\hat{x} \in X$ of the game is a pseudo-equilibrium if for any $l \in S$ there exists $y^{l+1} \in X^{l+1}, \ldots, y^n \in X^n$ such that

$$f_p^l(\hat{x}^1, \ldots, \hat{x}^{l-1}, x_p^l \| \hat{x}_{-p}^l, y^{l+1}, \ldots, y^n) \geq f_p^l(\hat{x}^1, \ldots, \hat{x}^l, y^{l+1}, \ldots, y^n),$$

for all $x_p^l \in X_p^l$, $p \in N_l$, where $\hat{x}_{-p}^l = (\hat{x}_1^l, \ldots, \hat{x}_{p-1}^l, \hat{x}_{p+1}^l, \ldots, \hat{x}_{n_l}^l)$.

According to Definition 9.3, the players $1, 2, \ldots, n_l$, at the stages $l = 1, 2, \ldots, s-1, s$, select their pseudo-equilibrium strategies:

$$\begin{aligned}
&B_p^1(\chi_{-p}^1) = \operatorname*{Arg\,min}_{x_p^1 \in X_p^1} f_p^1\left(x_p^1 \| \chi_{-p}^1\right), \, p \in N_1, \\
&(\hat{x}^1, x^2, \ldots, x^s) \in PE^1 = \bigcap_{p \in N_1} Gr_p^1, \\
&B_p^2(\hat{x}^1, \chi_{-p}^2) = \operatorname*{Arg\,min}_{x_p^2 \in X_p^2} f_p^2\left(\hat{x}^1, x_p^2 \| \chi_{-p}^2\right), \, p \in N_2, \\
&(\hat{x}^1, \hat{x}^2, x^3, \ldots, x^s) \in PE^2 = \bigcap_{p \in N_2} Gr_p^2, \\
&\ldots \\
&B_p^s(\hat{x}^1, \hat{x}^2, \ldots, \hat{x}^{s-1}, \chi_{-p}^s) = \operatorname*{Arg\,min}_{x_p^s \in X_p^s} f_p^s\left(\hat{x}^1, \ldots, \hat{x}^{s-1}, x_p^s \| \chi_{-p}^s\right), \\
&\qquad\qquad\qquad\qquad\qquad\qquad\qquad\qquad p \in N_s, \\
&(\hat{x}^1, \hat{x}^2, \ldots, \hat{x}^s) \in PE^s = \bigcap_{p \in N_s} Gr_p^s,
\end{aligned}$$

where

$$
\begin{aligned}
&Gr_p^1 = \left\{(x^1, \ldots, x^s) : x_p^1 \in B_p^1(\chi_{-p}^1)\right\}, p \in N_1,\\
&Gr_p^2 = \left\{(\hat{x}^1, x^2, \ldots, x^s) : x_p^2 \in B_p^2(\hat{x}^1, \chi_{-p}^2)\right\}, p \in N_2,\\
&\ldots\\
&Gr_p^s = \left\{(\hat{x}^1, \ldots, \hat{x}^{s-1}, x^s) : x_p^s \in B_p^s(\hat{x}^1, \ldots, \hat{x}^{s-1}, \chi_{-p}^s)\right\}, p \in N_s,\\
&\chi_{-p}^l = (x_{-p}^l, x^{l+1}, \ldots, x^s) \in X_{-p}^l \times X^{l+1} \times \cdots \times X^s,\\
&X_{-p}^l = X_1^l \times \ldots \times X_{p-1}^l \times X_{p+1}^l \times \ldots \times X_{n_l}^l.
\end{aligned}
$$

Surely, **the set of all pseudo-equilibria is** $PE = PE^s$.

Example 9.3 Consider a four-player $2 \times 2 \times 2 \times 2$ two-stage game with the cost matrices:

$$
a_{**11} = \begin{bmatrix} \mathbf{5} & 8 \\ 6 & \mathbf{4} \end{bmatrix}, a_{**12} = \begin{bmatrix} 6 & \mathbf{4} \\ \mathbf{5} & 9 \end{bmatrix}, a_{**21} = \begin{bmatrix} 3 & 2 \\ \mathbf{1} & \mathbf{1} \end{bmatrix}, a_{**22} = \begin{bmatrix} \mathbf{2} & 3 \\ 3 & \mathbf{2} \end{bmatrix},
$$

$$
b_{**11} = \begin{bmatrix} \mathbf{4} & 7 \\ \mathbf{6} & 8 \end{bmatrix}, b_{**12} = \begin{bmatrix} \mathbf{3} & 7 \\ 6 & \mathbf{4} \end{bmatrix}, b_{**21} = \begin{bmatrix} 7 & \mathbf{3} \\ 5 & \mathbf{3} \end{bmatrix}, b_{**22} = \begin{bmatrix} 8 & \mathbf{3} \\ \mathbf{1} & 9 \end{bmatrix},
$$

$$
c_{**11} = \begin{bmatrix} 5 & 2 \\ 4 & 5 \end{bmatrix}, c_{**12} = \begin{bmatrix} -1 & 3 \\ 3 & 2 \end{bmatrix}, c_{**21} = \begin{bmatrix} 1 & 2 \\ 4 & 3 \end{bmatrix}, c_{**22} = \begin{bmatrix} 3 & 5 \\ 4 & -2 \end{bmatrix},
$$

$$
d_{**11} = \begin{bmatrix} 6 & 2 \\ 1 & 3 \end{bmatrix}, d_{**12} = \begin{bmatrix} -2 & 3 \\ 3 & 1 \end{bmatrix}, d_{**21} = \begin{bmatrix} 7 & 2 \\ 1 & 3 \end{bmatrix}, d_{**22} = \begin{bmatrix} 6 & 3 \\ 3 & -1 \end{bmatrix}.
$$

The first and the second players move at the first stage, the third and the fourth players move at the second stage.

Elements of Gr_1^1 and Gr_2^1 are emphasized in matrices by bold fonts.

Evidently, the set of pseudo-equilibria PE^1 consists of two elements: $(1, 1, 1, 1)$ and $(2, 2, 2, 1)$.

Suppose the first and the second players choose their first strategies.

At the second stage, the third and the fourth players have to play a matrix games with matrices:

$$
c_{11**} = \begin{bmatrix} 5 & \mathbf{-1} \\ \mathbf{1} & 3 \end{bmatrix}, d_{11**} = \begin{bmatrix} 6 & \mathbf{-2} \\ 7 & \mathbf{6} \end{bmatrix}.
$$

The set of pseudo-equilibria PE^2 contains a profile $(1, 1, 1, 2)$. Thus, the profile $(1, 1, 1, 2)$ with costs $(6, 3, -1, -2)$ is a two-stage pseudo-equilibrium.

Suppose the first and the second players choose their second strategies at the first stage.

Then, at the second stage, the third and the fourth players have to play a matrix game with matrices:

$$
c_{22**} = \begin{bmatrix} 5 & 2 \\ \mathbf{3} & \mathbf{-2} \end{bmatrix}, d_{22**} = \begin{bmatrix} 3 & \mathbf{1} \\ 3 & \mathbf{-1} \end{bmatrix}.
$$

The set of pseudo-equilibria PE^2 contains profile $(2, 2, 2, 2)$. Thus, the profile $(2, 2, 2, 2)$ with costs $(2, 9, -2, -1)$ is a two-stage pseudo-equilibrium.

Evidently, both pseudo-equilibria $(1, 1, 1, 2)$ and $(2, 2, 2, 2)$ are not Nash equilibria. □

Pseudo-equilibrium Definition 9.3 does not use the information that at the following stage the stage players will choose the strategies in accordance to the pseudo-equilibrium statement. As the result, the profiles do not obligatory conserve the required statement at all stages.

To exclude this inconvenient, it is reasonable to choose strategies by applying backward induction. As a result, a new equilibrium notion of a Nash-Stackelberg equilibrium may be defined. The new concept has the goals to unite characteristics both of the concept of a Nash equilibrium and a Stackelberg equilibrium. Nevertheless, the new notion has its own individual features.

By stage backward induction, the players $1, 2, \ldots, n_l$, at the stages $l = s, s-1, \ldots, 2, 1$, select their equilibrium strategies:

$$
\begin{aligned}
&B_p^s(x^1, \ldots, x^{s-1}, x_{-p}^s) = \operatorname*{Arg\,min}_{y_p^s \in X_p^s} f_p^s\left(x^1, \ldots, x^{s-1}, y_p^s \| x_{-p}^s\right), \; p \in N_s, \\
&NSE^s = \bigcap_{p \in N_s} Gr_p^s, \\
&B_p^{s-1}(x^1, \ldots, x^{s-2}, x_{-p}^{s-1}) = \\
&\qquad \operatorname*{Arg\,min}_{\substack{y_p^{s-1},\, y^s: \\ (x^1,\ldots,x^{s-2},\, y_p^{s-1} \| x_{-p}^{s-1},\, y^s) \in NSE^s}} f_p^{s-1}\left(x^1, \ldots, x^{s-2}, y_p^{s-1} \| x_{-p}^{s-1}, y^s\right), \; p \in N_{s-1}, \\
&NSE^{s-1} = \bigcap_{p \in N_{s-1}} Gr_p^{s-1}, \\
&B_p^{s-2}(x^1, \ldots, x^{s-2}, x_{-p}^{s-3}) = \\
&\qquad = \operatorname*{Arg\,min}_{\substack{y_p^{s-2},\, y^{s-1},\, y^s: \\ (x^1,\ldots,x^{s-3},\, y_p^{s-2} \| x_{-p}^{s-2},\, y^{s-1},\, y^s) \in NSE^{s-1}}} f_p^{s-2}\left(x^1, \ldots, x^{s-3}, y_p^{s-2} \| x_{-p}^{s-2}, y^{s-1}, y^s\right), \\
&\qquad\qquad p \in N_{s-2}, \\
&NSE^{s-2} = \bigcap_{p \in N_{s-2}} Gr_p^{s-2}, \\
&\ldots \\
&B_p^1(x_{-p}^1) = \operatorname*{Arg\,min}_{y_p^1,\, y^2,\ldots,y^s: (y_p^1 \| x_{-p}^1,\, y^2,\ldots,y^s) \in NSE^2} f_p^1\left(y_p^1 \| x_{-p}^1, y^2, \ldots, y^s\right), \; p \in N_1, \\
&NSE^1 = \bigcap_{p \in N_1} Gr_p^1,
\end{aligned}
$$

where

$$Gr_p^s = \left\{ x \in X : \begin{array}{c} x^l \in X^l, \; l = 1, \ldots, s-1, \\ x_{-p}^s \in X_{-p}^s, \\ x_p^s \in B_p^s(x^1, \ldots, x^{s-1}, x_{-p}^s) \end{array} \right\}, \; p \in N_s,$$

$$Gr_p^{s-1} = \left\{ x \in NSE^s : \begin{array}{c} x^l \in X^l, l = 1, \ldots, s-2, \\ x_{-p}^{s-1} \in X_{-p}^{s-1}, \\ x_p^{s-1} \in B_p^{s-1}(x^1, \ldots, x^{s-2}, x_{-p}^{s-1}) \end{array} \right\}, \; p \in N_{s-1},$$

$$\ldots$$

$$Gr_p^1 = \left\{ x \in NSE^2 : \begin{array}{c} x_{-p}^1 \in X_{-p}^1, \\ x_p^1 \in B_p^1(x_{-p}^1) \end{array} \right\}, \; p \in N_1.$$

Of course, $NSE^1 \subseteq NSE^2 \subseteq \cdots \subseteq NSE^s$.

Definition 9.4 Elements of the set NSE^1 are called Nash-Stackelberg equilibria.

The set of all Nash-Stackelberg equilibria NSE^1 is denoted by NSE too.

If $s = 1$ and $n_1 > 1$, then every Nash-Stackelberg equilibrium is a Nash equilibrium. If $s > 1$ and $n_1 = n_2 = \cdots = n_s = 1$, then every equilibrium is an unsafe Stackelberg equilibrium. Thus, the Nash-Stackelberg equilibrium notion generalizes both Stackelberg and Nash equilibria notions.

Example 9.4 For Nash-Stackelberg equilibrium illustration purposes, let us consider the four-player $2 \times 2 \times 2 \times 2$ two-stage game with the cost matrices from Example 9.3. It is appropriate to represent the cost matrices as follows:

$$a_{11**} = \begin{bmatrix} 5 & 6 \\ 3 & 2 \end{bmatrix}, \; a_{12**} = \begin{bmatrix} 8 & 4 \\ 2 & 3 \end{bmatrix}, \; a_{21**} = \begin{bmatrix} 6 & 5 \\ 1 & 3 \end{bmatrix}, \; a_{22**} = \begin{bmatrix} 4 & 9 \\ 1 & 2 \end{bmatrix},$$

$$b_{11**} = \begin{bmatrix} 4 & 3 \\ 7 & 8 \end{bmatrix}, \; b_{12**} = \begin{bmatrix} 7 & 7 \\ 3 & 3 \end{bmatrix}, \; b_{21**} = \begin{bmatrix} 6 & 6 \\ 5 & 1 \end{bmatrix}, \; b_{22**} = \begin{bmatrix} 8 & 4 \\ 3 & 9 \end{bmatrix},$$

$$c_{11**} = \begin{bmatrix} 5 & \mathbf{-1} \\ \mathbf{1} & 3 \end{bmatrix}, \; c_{12**} = \begin{bmatrix} \mathbf{2} & 3 \\ \mathbf{2} & 5 \end{bmatrix}, \; c_{21**} = \begin{bmatrix} 4 & \mathbf{3} \\ \mathbf{4} & \mathbf{4} \end{bmatrix}, \; c_{22**} = \begin{bmatrix} 5 & \mathbf{2} \\ 3 & \mathbf{-2} \end{bmatrix},$$

$$d_{11**} = \begin{bmatrix} 6 & \mathbf{-2} \\ 7 & \mathbf{6} \end{bmatrix}, \; d_{12**} = \begin{bmatrix} \mathbf{2} & 3 \\ \mathbf{2} & 3 \end{bmatrix}, \; d_{21**} = \begin{bmatrix} \mathbf{1} & 3 \\ \mathbf{1} & 3 \end{bmatrix}, \; d_{22**} = \begin{bmatrix} 3 & \mathbf{1} \\ 3 & \mathbf{-1} \end{bmatrix}.$$

Elements of Gr_3^2, Gr_4^2 are emphasized in matrices by bold faces.

The set of Nash-Stackelberg equilibria NSE^2 consists of profiles: (1, 1, 1, 2), (1, 2, 1, 1), (1, 2, 2, 1), (2, 1, 2, 1), (2, 2, 1, 2), (2, 2, 2, 2).

At the first stage, the first and the second players have to play a specific matrix game with incomplete matrices:

$$a_{11**} = \begin{bmatrix} \cdot & 6 \\ \cdot & \cdot \end{bmatrix}, \; a_{12**} = \begin{bmatrix} 8 & \cdot \\ \mathbf{2} & \cdot \end{bmatrix}, \; a_{21**} = \begin{bmatrix} \cdot & \cdot \\ \mathbf{1} & \cdot \end{bmatrix}, \; a_{22**} = \begin{bmatrix} \cdot & 9 \\ \cdot & \mathbf{2} \end{bmatrix},$$

$$b_{11**} = \begin{bmatrix} \cdot & \mathbf{3} \\ \cdot & \cdot \end{bmatrix}, \; b_{12**} = \begin{bmatrix} 7 & \cdot \\ \mathbf{3} & \cdot \end{bmatrix}, \; b_{21**} = \begin{bmatrix} \cdot & \cdot \\ 5 & \cdot \end{bmatrix}, \; b_{22**} = \begin{bmatrix} \cdot & \mathbf{4} \\ \cdot & 9 \end{bmatrix},$$

Elements of Gr_1^1, Gr_2^1 are emphasized in matrices by bold faces.

The set of Nash-Stackelberg equilibria NSE^1 consists only from one profile $(1, 2, 2, 1)$ with the costs $(2, 3, 2, 2)$.

Evidently, $PE = PE^2 \neq NSE = NSE^1$.

As can be seen, if the first and the second players choose their strategies 1 and 2 at the first stage, then the third and the fourth players can choose also 1 and 2 at the second stage, because the profile $(1, 2, 1, 2)$ has for them the same costs 2 and 2. Unfortunately, for the first and the second players the cost will be 8 and 7, respectively. For the first and second players the profile $(1, 2, 1, 2)$ is worse than $(1, 2, 2, 1)$.

Thus, as in Stackelberg game, it's necessary and reasonable to introduce a notion of a safe Nash-Stackelberg equilibrium. □

By stage backward induction, the players $1, 2, \dots, n_l$, at the stages $l = s, s-1, \dots, 2, 1$, select their equilibrium strategies:

$$B_p^s(x^1, \dots, x^{s-1}, x_{-p}^s) = \operatorname*{Arg\,min}_{y_p^s \in X_p^s} f_p^s\left(x^1, \dots, x^{s-1}, y_p^s \| x_{-p}^s\right), \; p \in N_s,$$

$$SNSE^s = NSE^s = \bigcap_{p \in N_s} Gr_p^s,$$

$$\tilde{B}_p^{s-1}(x^1, \dots, x^{s-2}, x_{-p}^{s-1}) =$$
$$= \operatorname*{Arg\,min}_{y_p^{s-1}} \max_{\substack{y^s \\ (x^1, \dots, x^{s-2}, y_p^{s-1} \| x_{-p}^{s-1}, y^s) \in SNSE^s}} f_p^{s-1}\left(x^1, \dots, x^{s-2}, y_p^{s-1} \| x_{-p}^{s-1}, y^s\right), \; p \in N_{s-1},$$

$$SNSE^{s-1} = \bigcap_{p \in N_{s-1}} \tilde{G}r_p^{s-1},$$

$$\tilde{B}_p^{s-2}(x^1, \dots, x^{s-2}, x_{-p}^{s-3}) =$$
$$= \operatorname*{Arg\,min}_{y_p^{s-2}} \max_{\substack{y^{s-1}, y^s \\ (x^1, \dots, x^{s-3}, y_p^{s-2} \| x_{-p}^{s-2}, y^{s-1}, y^s) \in NSE^{s-1}}} f_p^{s-2}\left(x^1, \dots, x^{s-3}, y_p^{s-2} \| x_{-p}^{s-2}, y^{s-1}, y^s\right),$$

$$p \in N_{s-2}, \; SNSE^{s-2} = \bigcap_{p \in N_{s-2}} \tilde{G}r_p^{s-2},$$

$\dots$

$$\tilde{B}_p^1(x_{-p}^1) = \operatorname*{Arg\,min}_{y_p^1} \max_{\substack{y^2, \dots, y^s \\ (y_p^1 \| x_{-p}^1, y^2, \dots, y^s) \in NSE^2}} f_p^1\left(y_p^1 \| x_{-p}^1, y^2, \dots, y^s\right), \; p \in N_1,$$

$$SNSE^1 = \bigcap_{p \in N_1} \tilde{G}r_p^1,$$

where

$$Gr_p^s = \left\{ x \in X : \begin{array}{l} x^l \in X^l,\ l = 1, \ldots, s-1, \\ x_{-p}^s \in X_{-p}^s, \\ x_p^s \in B_p^s(x^1, \ldots, x^{s-1}, x_{-p}^s) \end{array} \right\},\ p \in N_s,$$

$$\tilde{G}r_p^{s-1} = \left\{ x \in NSE^s : \begin{array}{l} x^l \in X^l, l = 1, \ldots, s-2, \\ x_{-p}^{s-1} \in X_{-p}^{s-1}, \\ x_p^{s-1} \in \tilde{B}_p^{s-1}(x^1, \ldots, x^{s-2}, x_{-p}^{s-1}) \end{array} \right\},\ p \in N_{s-1},$$

...

$$\tilde{G}r_p^1 = \left\{ x \in NSE^2 : \begin{array}{l} x_{-p}^1 \in X_{-p}^1, \\ x_p^1 \in \tilde{B}_p^1(x_{-p}^1) \end{array} \right\},\ p \in N_1.$$

Sure, $SNSE^1 \subseteq SNSE^2 \subseteq \cdots \subseteq SNSE^s$.

Definition 9.5 Elements of the set $SNSE^1$ are called safe Nash-Stackelberg equilibria.

The set $SNSE^1$ of all safe Nash-Stackelberg equilibria is denoted simply by $SNSE$ too.

Example 9.5 Let us continue analysis of Examples 9.3–9.4. We can observe that at the second stage $SNSE^2 = NSE^2$.

At the first stage, the first and the second players have to play a specific matrix game with the same incomplete matrices:

$$a_{11**} = \begin{bmatrix} \cdot & 6 \\ \cdot & \cdot \end{bmatrix},\ a_{12**} = \begin{bmatrix} \mathbf{8} & \cdot \\ 2 & \cdot \end{bmatrix},\ a_{21**} = \begin{bmatrix} \cdot & \cdot \\ \mathbf{1} & \cdot \end{bmatrix},\ a_{22**} = \begin{bmatrix} \cdot & 9 \\ \cdot & 2 \end{bmatrix},$$

$$b_{11**} = \begin{bmatrix} \cdot & \mathbf{3} \\ \cdot & \cdot \end{bmatrix},\ b_{12**} = \begin{bmatrix} 7 & \cdot \\ 3 & \cdot \end{bmatrix},\ b_{21**} = \begin{bmatrix} \cdot & \cdot \\ \mathbf{5} & \cdot \end{bmatrix},\ b_{22**} = \begin{bmatrix} \cdot & 4 \\ \cdot & 9 \end{bmatrix},$$

The elements of $\tilde{G}r_1^1$, $\tilde{G}r_2^1$ are emphasized in matrices by bold faces. (In every 2×2 matrix the maximal element is selected. After that, the first player with the fixed strategy of the second player, chooses the matrix with the minimal among the maximal selected elements. In the same manner, the second player constructs his graph. The graphs intersection consists of a single element.)

The set $SNSE^1$ consists only of one profile $(2, 1, 2, 1)$ with costs $(1, 5, 4, 1)$.

Consequently, for the considered game there are two PE $(1, 1, 1, 2)$ and $(2, 2, 2, 2)$ with costs $(6, 3, -1, -2)$ and $(2, 9, -2, -1)$, respectively, one NSE $(1, 2, 2, 1)$ with costs $(2, 3, 2, 2)$ and one $SNSE$ $(2, 1, 2, 1)$ with costs $(1, 5, 4, 1)$. None of these profiles are better for all players. The problem of the equilibrium principle selection is opened. □

Let us observe that Examples 9.3–9.5 may be considered as an unique uniting example which illustrates both exposed concepts and results.

9.5 Multi-objective Pseudo-Equilibria. Pareto-Nash-Stackelberg Equilibria

Consider a multiobjective strategic game

$$\Gamma = \langle N, \{X_p^l\}_{l\in S, p\in N_l}, \{\mathbf{f}_p^l(x)\}_{l\in S, p\in N_l}\rangle,$$

with vector-cost functions $\{\mathbf{f}_p^l(x)\}_{l\in S, p\in N_l}$.

In the same manner as for Nash-Stackelberg games the equilibrium principles can be introduced. An essential difference in corresponding definitions is the strong requirement that every minimization or maximization operator must be interpreted as Pareto maximization or minimization operator. Evidently, the Pareto optimal response mapping and the graph of the Pareto optimal response mapping are considered for each player. An intersection of the graphs of Pareto optimal response mappings is considered in every definition as the stage profile.

A series of preliminary related results were published earlier with illustrative goals, e.g. the Wolfram language demonstration [21] and paper [22].

9.6 Concluding Remarks

The examined processes of decision making are very often phenomena of real life. Their mathematical modelling as Pareto-Nash-Stackelberg games gives a powerful tool for investigation, analysis and solving hierarchical decision problems. Nevertheless, the problem of equilibrium principle choosing in real situations is a task both for a decision maker and a game theorist. We can mention, e.g. a recent interest in mathematical modelling of airport security as Stackelberg games [23].

Considered models give us recurrent relations, similar with that used traditionally in dynamic programming [24]. It is obvious that every concrete solving process of an n stage dynamic problem may be considered, for example as an n players Stackelberg game such that at every stage the same player moves on the same payoff function basis.

References

1. Pareto, V. 1904. *Manuel d'economie politique*, 504 pp. Giard: Paris. (in French).
2. Nash, J. 1951. Noncooperative games. *Annals of Mathematics* 54 (2): 280–295.

3. Von Stackelberg, H. 1934. *Marktform und Gleichgewicht (Market Structure and Equilibrium)*, XIV+134 pp. Vienna: Springer. (in German).
4. Ungureanu, V. 2008. Solution principles for simultaneous and sequential games mixture. *ROMAI Journal* 4 (1): 225–242.
5. Hu, F. 2015. Multi-leader-follower games: Models, methods and applications. *Journal of the Operations Research Society of Japan* 58: 1–23.
6. Leitmann, G. 1978. On generalized stackelberg strategies. *Journal of Optimization Theory and Applications* 26: 637–648.
7. Blackwell, D. 1956. An analog of the minimax theorem for vector payoffs. *Pacific Journal of Mathematics* 6: 1–8.
8. Shapley, L.S. 1956. Equilibrium points in games with vector payoffs, Rand Corporation Research Memorandum RM-1818, pp. I–III, 1–7.
9. Shapley, L.S. 1959. Equilibrium points in games with vector payoffs. *Naval Research Logistics Quarterly* 6: 57–61.
10. Borm, P., S. Tijs, and J. Van der Aarssen. 1988. Pareto equilibria in multiobjective games. *Methods of Operations Research* 60: 302–312.
11. Podinovskii, V.V., and V.D. Nogin. 1982. *Pareto-Optimal Solutions of the Multi-criteria Problems*, 255 pp. Moscow: Nauka. (in Russian).
12. Wierzbicki, A.P. 1995. Multiple criteria games – theory and applications. *Journal of Systems Engineering and Electronics* 6 (2): 65–81.
13. Sherali, H.D. 1984. A multiple leader Stackelberg model and analysis. *Operations Research* 32: 390–404.
14. Mallozzi, L., and R. Messalli. 2017. Multi-leader multi-follower model with aggregative uncertainty. *Games* 8(25) (3): 1–14.
15. Sagaidac, M., and V. Ungureanu. 2004. *Operational Research*, 296 pp. Chişinău: CEP USM. (in Romanian).
16. Ungureanu, V. 2001. *Mathematical Programming*, 348 pp. Chişinău: USM. (in Romanian).
17. Ungureanu, V., and A. Botnari. 2005. Nash equilibrium sets in mixed extension of $2 \times 2 \times 2$ games. *Computer Science Journal of Moldova* 13 (1(37)): 13–28.
18. Ungureanu, V., and A. Botnari. 2005. Nash equilibrium sets in mixed extended 2×3 games. *Computer Science Journal of Moldova* 13 (2(38)): 136–150.
19. Ungureanu, V. 2006. Nash equilibrium set computing in finite extended games. *Computer Science Journal of Moldova* 14 (3(42)): 345–365.
20. Ungureanu, V. 2007. Nash equilibrium conditions nash equilibrium conditions for strategic form games. *Libertas Mathematica* XXVII: 131–140.
21. Lozan, V., and V. Ungureanu. 2011. Pareto-Nash equilibria in bicriterial dyadic games with mixed strategies, Wolfram Demonstrations Project, Published: 13 October 2011. http://demonstrations.wolfram.com/ParetoNashEquilibriaInBicriterialDyadicGamesWithMixedStrateg/
22. Ungureanu, V., and Lozan, V. Pareto-Nash-Stackelberg equilibrium set in dyadic bimatrix bicriterion two stage mixed-strategy games $2 \times 2 \times 2 \times 2$, Mathematical Modelling, Optimization and Information Technologies, International Conference Proceedings, ATIC, March 25–28, 4th Edition. (2014). Chicsinuau. *Evrica, 2014*, 127–135. (in Romanian).
23. Korzhyk, D., Z. Yin, C. Kiekintveld, V. Conitzer, and M. Tambe. 2011. Stackelberg versus nash in security games: An extended investigation of interchangeability, equivalence, and uniqueness. *Journal of Artificial Intelligence Research* 41: 297–327.
24. Bellman, R. 1957. *Dynamic Programming*, 365. New Jersey: Princeton University Press.

Chapter 10
Computing Pareto–Nash Equilibrium Sets in Finite Multi-Objective Mixed-Strategy Games

Abstract The problem of a Pareto–Nash equilibrium set computing in finite multi-objective mixed-strategy games (Pareto–Nash games) is considered in this chapter. A method for a Pareto–Nash equilibrium set computing is exposed. The method is based on the fact that the set of Pareto–Nash equilibria is identified with the intersection of the graphs of efficient response mappings.

10.1 Introduction

The Pareto–Nash equilibrium set may be determined as the intersection of graphs of efficient response mappings — an approach which is considered above and may be considered a generalization of the earlier published works [1–5] and initiated in [3] for a Nash equilibrium set computing in finite mixed-strategy games. By applying an identical approach in the case of multi-criteria Nash games, the method of the Pareto–Nash equilibrium set computing is constructed for finite mixed-strategy n-player multi-objective games.

Consider a finite multi-criteria strategic game:

$$\Gamma = \langle \mathbf{N},\ \{\mathbf{S_p}\}_{p \in \mathbf{N}},\ \{\mathbf{u_p(s)}\}_{p \in \mathbf{N}} \rangle,$$

where

- $\mathbf{N} = \{1, 2, \ldots, n\}$ is a set of players;
- $\mathbf{S_p} = \{1, 2, \ldots, m_p\}$ is a set of strategies of player $p \in \mathbf{N}$;
- $\mathbf{u_p(s)} : \mathbf{S} \mapsto \mathbb{R}^{\mathbf{k_p}}, \mathbf{u_p(s)} = \left(u_p^1(\mathbf{s}), u_p^2(\mathbf{s}), \ldots, u_p^{k_p}(\mathbf{s})\right)$ is the utility vector-function of the player $p \in \mathbf{N}$;
- $\mathbf{s} = (s_1, s_2, \ldots, s_n) \in \mathbf{S} = \underset{\mathbf{p} \in \mathbf{N}}{\times} \mathbf{S_p}$, where $\mathbf{S}$ is the set of profiles;
- $k_p, m_p < +\infty$, $p \in \mathbf{N}$.

Let us associate with the utility vector-function $\mathbf{u_p(s)}$, $p \in \mathbf{N}$, its matrix representation

$$\mathbf{u_p(s)} = \mathbf{A_s^p} = \left[a^{pi}_{s_1 s_2 \ldots s_n}\right]_{\mathbf{s} \in \mathbf{S}}^{i=1,\ldots,k_p} \in \mathbb{R}^{\mathbf{k_p} \times \mathbf{m_1} \times \mathbf{m_2} \times \cdots \times \mathbf{m_n}}.$$

V. Ungureanu, *Pareto-Nash-Stackelberg Game and Control Theory*, Smart Innovation, Systems and Technologies 89,
https://doi.org/10.1007/978-3-319-75151-1_10

The pure-strategy multi-criteria game Γ defines in an evident manner a mixed-strategy multi-criteria game:

$$\Gamma' = \langle \mathbf{N}, \{\mathbf{X_p}\}_{p\in\mathbf{N}}, \{\mathbf{f_{p(x)}}\}_{p\in\mathbf{N}} \rangle,$$

where

- $\mathbf{X_p} = \{\mathbf{x^p} \in \mathbb{R}^{\mathbf{m_p}}_{\geq} : x_1^p + x_2^p + \cdots + x_{m_p}^p = 1\}$ is a set of mixed strategies of player $p \in \mathbf{N}$;
- $\mathbf{f_p(x)} : \mathbf{X} \mapsto \mathbb{R}^{\mathbf{k_p}}$, $\mathbf{f_p(x)} = \left(f_p^1(\mathbf{x}), f_p^2(\mathbf{x}), \ldots, f_p^{k_p}(\mathbf{x})\right)$ is the utility vector-function of the player $p \in \mathbf{N}$ defined on the Cartesian product $\mathbf{X} = \underset{\mathbf{p}\in\mathbf{N}}{\times} \mathbf{X_p}$ and

$$f_p^i(\mathbf{x}) = \sum_{s_1=1}^{m_1} \sum_{s_2=1}^{m_2} \cdots \sum_{s_n=1}^{m_n} a^{pi}_{s_1 s_2 \ldots s_n} x^1_{s_1} x^2_{s_2} \ldots x^n_{s_n}.$$

Remark that each player has to solve solely the multi-criteria parametric optimization problem, where the parameters are strategic choices of the others players.

Definition 10.1 The strategy $\mathbf{x^p} \in \mathbf{X_{-p}}$ is "*better*" than the strategy $\mathbf{y^p} \in \mathbf{X_{-p}}$ if

$$\mathbf{f_p(x^p, x^{-p})} \geq \mathbf{f_p(y^p, x^{-p})},$$

for all $x^{-p} \in X_{-p}$ and there exists an index $i \in \{1, \ldots, k_p\}$ and a joint strategy $\mathbf{x^{-p}} \in \mathbf{X_{-p}}$ for which

$$f_p^i(\mathbf{x^p, x^{-p}}) > f_p^i(\mathbf{y^p, x^{-p}}).$$

The defined relationship is denoted $\mathbf{x^p} \succeq \mathbf{y^p}$.

Player problem. The player p selects from his strategy set $\mathbf{X_p}$ a strategy $\mathbf{\hat{x}^p} \in \mathbf{X_p}$, $p \in \mathbf{N}$, for which every component of the utility vector-function $\mathbf{f_p(x^p, \hat{x}^{-p})}$ achieves its maximum possible value.

10.2 Pareto Optimality

Let us recall the Pareto optimality definition and the main results [6–9] concerning the Pareto optimality notion used in this chapter.

Definition 10.2 The strategy $\mathbf{\hat{x}^p}$ is called efficient or optimal in the sense of Pareto if there is no a strategy $\mathbf{x^p} \in \mathbf{X_p}$ such that $\mathbf{x^p} \succeq \mathbf{\hat{x}^p}$ [6].

The set of efficient strategies of the player p is denoted by **ef** $\mathbf{X_p}$. Any two efficient strategies are or equivalent or incomparable.

Theorem 10.1 *If the sets* $\mathbf{X_p} \subseteq \mathbb{R}^{\mathbf{k_p}}$, $p = 1, \ldots, n$, *are compact and the cost functions are continuous, i.e.* $f_p^i(\mathbf{x}) \in C(\mathbf{X_p})$, $i = 1, \ldots, m_p$, $p = 1, \ldots, n$, *then the sets* $\mathbf{ef}\,\mathbf{X_p}$, $p = 1, \ldots, n$, *are non empty.*

Proof The proof follows from the known results [10]. A linear synthesis function is continuous for the continuous component functions and according to the well known Weierstrass theorem it achieves its global optimal values on a compact set. So, there exists at least one efficient point. □

Theorem 10.2 *Every element* $\hat{\mathbf{x}} = (\hat{x}^1, \hat{x}^2, \ldots, \hat{x}^n) \in \mathbf{ef}\,\mathbf{X} = \underset{\mathbf{p} \in \mathbf{N}}{\times}\ \mathbf{ef}\,\mathbf{X_p}$ *is efficient.*

Proof In Theorem 10.1 efficiency is considered in relations with criteria of every player. In this theorem, efficiency is examined in relations with all the players and their criteria. In this context, the proof follows from the definition of an efficient joint strategy. □

10.3 Pareto–Nash Equilibria

Let us unify the concepts of Pareto optimality and the Nash equilibrium by the means of a Pareto–Nash equilibrium notion and let us highlight its main characteristics.

Definition 10.3 The outcome $\hat{\mathbf{x}} \in \mathbf{X}$ of the game Γ is a Pareto–Nash equilibrium [11–13] if

$$\mathbf{f_p}(\mathbf{x^p}, \hat{\mathbf{x}}^{-\mathbf{p}}) \leq \mathbf{f_p}(\hat{\mathbf{x}}^{\mathbf{p}}, \hat{\mathbf{x}}^{-\mathbf{p}}),$$

for all $\mathbf{x^p} \in \mathbf{X_p}$ and for all $\mathbf{p} \in \mathbf{N}$, where

$$\hat{\mathbf{x}}^{-\mathbf{p}} = (\hat{x}^1, \hat{x}^2, \ldots, \hat{x}^{p-1}, \hat{x}^{p+1}, \ldots, \hat{x}^n),$$

$$\hat{\mathbf{x}}^{-\mathbf{p}} \in \mathbf{X_{-p}} = \mathbf{X_1} \times \mathbf{X_2} \times \ldots \times \mathbf{X_{p-1}} \times \mathbf{X_{p+1}} \times \ldots \times \mathbf{X_n},$$

$$\hat{\mathbf{x}} = \hat{\mathbf{x}}^{\mathbf{p}} \parallel \hat{\mathbf{x}}^{-\mathbf{p}} = (\hat{x}^{\mathbf{p}}, \hat{\mathbf{x}}^{-\mathbf{p}}) = (\hat{x}^1, \hat{x}^2, \ldots, \hat{x}^{p-1}, \hat{x}^p, \hat{x}^{p+1}, \ldots, \hat{x}^n) \in \mathbf{X}.$$

It is well known that there are games in pure strategies that do not have Pareto–Nash equilibria, but all the extended games Γ' have at least one Pareto–Nash equilibrium. The proof of this sentence, based on scalarization technique, is presented below in the next section. The same scalarization technique may served as a bases for diverse alternative formulations of a Pareto–Nash equilibrium, as well as for a Nash equilibrium: as a fixed point of the efficient response correspondence, as a fixed point of a synthesis sum of functions, as a solution of a nonlinear complementarity problem, as a solution of a stationary point problem, as a maximum of a synthesis sum of functions on a polytope, as a semi-algebraic set. The set of Pareto–Nash equilibria may be considered as well as an intersection of graphs of efficient response multi-valued mappings [4, 5]:

$$\operatorname*{Arg\,max}_{\mathbf{x^p} \in \mathbf{X_p}} \mathbf{f_p}(\mathbf{x^p}, \mathbf{x^{-p}}) : \mathbf{X_{-p}} \multimap \mathbf{X_p},\ p = \overline{1, n} :$$

$$\mathbf{PNES(\Gamma')} = \bigcap_{\mathbf{p} \in \mathbf{N}} \mathbf{Gr_p},$$

$$\mathbf{Gr_p} = \left\{ (\mathbf{x^p}, \mathbf{x^{-p}}) \in \mathbf{X} : \begin{array}{c} \mathbf{x^{-p}} \in \mathbf{X_{-p}}, \\ \mathbf{x^p} \in \operatorname*{Arg\,max}_{\mathbf{x^p} \in \mathbf{X_p}} \mathbf{f_p}(\mathbf{x^p}, \mathbf{x^{-p}}) \end{array} \right\}$$

The problem of Pareto–Nash equilibrium sets computing in mixed-strategy two-person games was studied in [5]. In this chapter a method for Pareto–Nash equilibrium sets computing in bimatrix mixed-strategy games and polymatrix mixed-strategy games is proposed.

The computational complexity of the considered problem may be established on the bases of the problem of a Nash equilibrium computing. Let us recall Remark 3.4 hich stands that finding an equilibrium of a bimatrix game with maximum payoff sum is NP-hard [14]. Consequently, the problem of Pareto–Nash equilibrium set computing has at least the same complexity as the problem of a Nash equilibrium computing. Let us recall the fact that the problem of a Nash equilibrium computing in two-player game is PPAD-complete [15] (PPAD is an abbreviation for a Polynomial Parity Argument for Directed graphs [16]). **These facts enforce conclusion that the problem of a Pareto–Nash equilibrium set computing is computationally very hard (unless P = NP).**

As it is easy to see, the algorithm for a Pareto–Nash equilibrium set computing in polymatrix mixed-strategy games solves, particularly, the problem of a Pareto–Nash equilibrium set computing in $m \times n$ mixed-strategy games. But, bimatrix games have peculiar features that permit to give a more expedient algorithm. Presented examples have to give the reader the opportunity to easy and prompt grasp of the method.

10.4 Scalarization Technique

The solution of the multi-criteria problem may be found by applying the scalarization technique (weighted sum method), which may interpret the weighted sum of the player utility functions as the unique utility (synthesis) function of the player p $(p = 1, \ldots, n)$:

$$F_p(\mathbf{x}, \lambda^{\mathbf{p}}) = \lambda_1^p \sum_{s_1=1}^{m_1} \sum_{s_2=1}^{m_2} \cdots \sum_{s_n=1}^{m_n} a^{p1}_{s_1 s_2 \ldots s_n} x^1_{s_1} x^2_{s_2} \ldots x^n_{s_n} + \ldots +$$
$$+ \lambda_{k_p}^p \sum_{s_1=1}^{m_1} \sum_{s_2=1}^{m_2} \cdots \sum_{s_n=1}^{m_n} a^{pk_p}_{s_1 s_2 \ldots s_n} x^1_{s_1} x^2_{s_2} \ldots x^n_{s_n},$$

$$\begin{aligned}
&\mathbf{x^p} \in \mathbf{X_p},\\
&\boldsymbol{\lambda}^{\mathbf{p}} = (\lambda_1^p, \lambda_2^p, \ldots, \lambda_{k_p}^p) \in \mathbf{\Lambda_p},\\
&\mathbf{\Lambda_p} = \left\{ \boldsymbol{\lambda}^{\mathbf{p}} \in \mathbb{R}^{\mathbf{k_p}} : \begin{array}{l} \lambda_1^p + \lambda_2^p + \cdots + \lambda_{k_p}^p = 1, \\ \lambda_i^p \geq 0, i = 1, \ldots, k_p, \end{array} \right\},\\
&p = 1, \ldots, n.
\end{aligned}$$

Theorem 10.3 *Assume that* $\mathbf{x^{-p}} \in \mathbf{X_{-p}}$.

1. *If* $\hat{\mathbf{x}}^{\mathbf{p}}$ *is the solution of the single-objective problem* $\max\limits_{\mathbf{x^p} \in \mathbf{X_p}} F_p(\mathbf{x}, \boldsymbol{\lambda}^{\mathbf{p}})$, *for some* $\boldsymbol{\lambda}^{\mathbf{p}} \in \Lambda_{\mathbf{p}}, \lambda^{\mathbf{p}} > 0$, *then* $\hat{\mathbf{x}}^{\mathbf{p}}$ *is the efficient point for player* $p \in \mathbf{N}$ *for the fixed* $\mathbf{x^{-p}}$.
2. *The solution* $\hat{\mathbf{x}}^{\mathbf{p}}$ *of the single-criterion problem* $\max\limits_{\mathbf{x^p} \in \mathbf{X_p}} F_p(\mathbf{x}, \boldsymbol{\lambda}^{\mathbf{p}})$, *with* $\boldsymbol{\lambda}^{\mathbf{p}} \geq 0$, $p \in \mathbf{N}$, *is an efficient point for the player* $p \in \mathbf{N}$, *if it is unique.*

Proof Theorem's truth follows from the sufficient Pareto condition with linear synthesis function [10]. □

Let us define the single-objective game

$$\Gamma''(\boldsymbol{\lambda}^1, \boldsymbol{\lambda}^2, \ldots, \boldsymbol{\lambda}^n) = \langle \mathbf{N}, \{\mathbf{X_p}\}_{p \in \mathbf{N}}, \{F_p(\mathbf{x}, \boldsymbol{\lambda}^{\mathbf{p}})\}_{p \in \mathbf{N}} \rangle,$$

where

- $\boldsymbol{\lambda}^{\mathbf{p}} \in \Lambda_{\mathbf{p}}, p \in \mathbf{N}$,
- $\mathbf{X_p} = \{\mathbf{x^p} \in \mathbb{R}_{\geq}^{\mathbf{m_p}} : x_1^p + x_2^p + \cdots + x_{m_p}^p = 1\}$ is a set of mixed strategies of the player $p \in \mathbf{N}$;
- $F_p(\mathbf{x}, \boldsymbol{\lambda}^{\mathbf{p}}) : \mathbf{X} \mapsto \mathbb{R}^{\mathbf{k_p}}$, is the utility synthesis function of the player p $\in \mathbf{N}$; the function is defined on $\mathbf{X}$ and $\mathbf{\Lambda_p}$.

For the simplicity, let us introduce the notation:

$$\Gamma''(\boldsymbol{\lambda}) = \Gamma''(\boldsymbol{\lambda}^1, \boldsymbol{\lambda}^2, \ldots, \boldsymbol{\lambda}^n),$$

$$\boldsymbol{\lambda} = (\lambda^1, \lambda^2, \ldots, \lambda^n) \in \mathbf{\Lambda} = \Lambda_1 \times \Lambda_2 \times \ldots \times \Lambda_n.$$

Evidently, the game $\Gamma''(\boldsymbol{\lambda})$ represents a polymatrix mixed-strategy single-objective game for a fixed $\boldsymbol{\lambda} \in \Lambda$. It is well known that such a game has at least one Nash equilibrium [17]. Consequently, from this well known result and from the precedent theorem next ones follow.

Theorem 10.4 *The outcome* $\hat{\mathbf{x}} \in \mathbf{X}$ *is a Pareto–Nash equilibrium in the game* Γ' *if and only if there exists such a* $\boldsymbol{\lambda} \in \Lambda, \lambda > 0$, $p \in \mathbf{N}$, *for which the point* $\hat{\mathbf{x}} \in \mathbf{X}$ *is a Nash equilibrium in the game* $\Gamma''(\boldsymbol{\lambda})$.

Proof As stated above, the truth of the theorem follows from the Nash theorem and the relations between scalar optimization problem and multi-criteria optimization problem. □

Theorem 10.5 *The set of Pareto–Nash equilibria in the game* Γ' *is non empty and*

$$\mathbf{PNES}(\boldsymbol{\Gamma}') = \bigcup_{\boldsymbol{\lambda} \in \boldsymbol{\Lambda}, \boldsymbol{\lambda} > 0} \mathbf{NES}(\Gamma''(\boldsymbol{\lambda})).$$

Proof Based on the Nash theorem and relations between the games Γ' and Γ'', there is at least one Pareto–Nash equilibrium.

Based on the Pareto–Nash equilibrium definition and the relations between the games Γ' and Γ'', the above expression follows. □

Let us denote the graphs of best response mappings

$$\operatorname{Arg}\max_{\mathbf{x^p} \in \mathbf{X_p}} F_p(\mathbf{x^p}, \mathbf{x^{-p}}, \lambda^{\mathbf{p}}) : \mathbf{X_{-p}} \multimap \mathbf{X_p},\ p = 1, \ldots, n,$$

by

$$\mathbf{Gr_p}(\boldsymbol{\lambda}^{\mathbf{p}}) = \left\{ \boldsymbol{\Gamma}(\mathbf{x^p}, \mathbf{x^{-p}}) \in \mathbf{X} : \begin{array}{l} \mathbf{x^{-p}} \in \mathbf{X_{-p}}, \\ \mathbf{x^p} \in \operatorname{Arg}\max\limits_{\mathbf{x^p} \in \mathbf{X_p}} F_p(\mathbf{x^p}, \mathbf{x^{-p}}, \lambda^{\mathbf{p}}) \end{array} \right\},$$

$$\mathbf{Gr_p} = \bigcup_{\boldsymbol{\lambda}^{\mathbf{p}} \in \boldsymbol{\Lambda}_{\mathbf{p}}, \boldsymbol{\lambda}^{\mathbf{p}} > \mathbf{0}} \mathbf{Gr_p}(\boldsymbol{\lambda}^{\mathbf{p}}).$$

Remark 10.1 Let us observe that we use the same notation for the graphs Gr_p in all the games, but from the context it must be understood that for different games the single-objective and multi-objective problems are solved to define the respective graphs.

From the above, we are able to establish the truth of Theorem 10.6, which permit to compute practically the set of Pareto–Nash equilibria in the game Γ'. The next sections are essentially based on Theorem 10.6, as well as on the rest obtained above results.

Theorem 10.6 *The set of Pareto–Nash equilibria is non empty in the game* Γ' *and it may computed as*

$$\mathbf{PNES}(\boldsymbol{\Gamma}') = \bigcap_{p=1}^{n} \mathbf{Gr_p}.$$

Proof The theorem's truth follows from the Pareto–Nash equilibrium definition, the definition of the graph of an efficient response mapping, and Theorems 10.4 and 10.5. □

10.5 Pareto–Nash Equilibrium Set in Two-Player Mixed-Strategy Games

Consider a two-player $m \times n$ multi-criteria game Γ with the matrices:

$$A^q = (a^q_{ij}),\;\; B^r = (b^r_{ij}),$$

where $i = 1, \dots, m$, $j = 1, \dots, n$, $q = 1, \dots, k_1$, $r = 1, \dots, k_2$.

Let

$$A^{iq},\; i = 1, \dots, m,\; q = 1, \dots, k_1,$$

be the lines of the matrices A^q, $q = 1, \dots, k_1$,

$$b^{jr},\; j = 1, \dots, n,\; r = 1, \dots, k_2,$$

be the columns of the matrices B^r, $r = 1, \dots, k_2$, and

$$\mathbf{X} = \{\mathbf{x} \in \mathbb{R}^{\mathbf{m}}_{\geq} : x_1 + x_2 + \dots + x_m = 1\},$$

$$\mathbf{Y} = \{\mathbf{y} \in \mathbb{R}^{\mathbf{n}}_{\geq} : y_1 + y_2 + \dots + y_n = 1\},$$

be the players' sets of mixed strategies.

As in the precedent subsection, let us consider the mixed-strategy games Γ' and $\Gamma''(\boldsymbol{\lambda}^1, \boldsymbol{\lambda}^2)$ with the following synthesis functions of the players:

$$\begin{aligned} F_1(\mathbf{x}, \mathbf{y}, \boldsymbol{\lambda}^1) &= \lambda^1_1 \sum_{i=1}^{m} \sum_{j=1}^{n} a^1_{ij} x_i y_j + \dots + \lambda^1_{k_1} \sum_{i=1}^{m} \sum_{j=1}^{n} a^{k_1}_{ij} x_i y_j = \\ &= \left[\left(\lambda^1_1 A^{11} + \lambda^1_2 A^{12} + \dots + \lambda^1_{k_1} A^{1k_1}\right) \mathbf{y}\right] x_1 + \dots + \\ &\quad + \left[\left(\lambda^1_1 A^{m1} + \lambda^1_2 A^{m2} + \dots + \lambda^1_{k_1} A^{mk_1}\right) \mathbf{y}\right] x_m, \end{aligned}$$

$$\begin{aligned} F_2(\mathbf{x}, \mathbf{y}, \boldsymbol{\lambda}^2) &= \lambda^2_1 \sum_{i=1}^{m} \sum_{j=1}^{n} b^1_{ij} x_i y_j + \dots + \lambda^2_{k_2} \sum_{i=1}^{m} \sum_{j=1}^{n} b^{k_2}_{ij} x_i y_j = \\ &= \left[\mathbf{x}^T \left(\lambda^2_1 b^{11} + \lambda^2_2 b^{12} + \dots + \lambda^2_{k_2} b^{1k_2}\right)\right] y_1 + \dots + \\ &\qquad + \left[\mathbf{x}^T \left(\lambda^2_1 b^{n1} + \lambda^2_2 b^{n2} + \dots + \lambda^2_{k_2} b^{nk_2}\right)\right] y_n, \end{aligned}$$

and

$$\lambda^1_1 + \lambda^1_2 + \dots + \lambda^1_{k_1} = 1,\; \lambda^1_q \geq 0,\; q = 1, \dots, k_1,$$

$$\lambda^2_1 + \lambda^2_2 + \dots + \lambda^2_{k_2} = 1,\; \lambda^2_r \geq 0,\; r = 1, \dots, k_2.$$

The game $\Gamma'' = \langle \mathbf{X}, \mathbf{Y}; F_1, F_2 \rangle$ is a scalarization of the mixed-strategy multi-criteria two-player game Γ'.

If the strategy of the second player is fixed, then the first player has to solve a linear programming parametric problem:

$$F_1(\mathbf{x}, \mathbf{y}, \boldsymbol{\lambda}^1) \to max, \ \mathbf{x} \in \mathbf{X}, \tag{10.1}$$

where $\boldsymbol{\lambda}^1 \in \boldsymbol{\Lambda}_1$ and $\mathbf{y} \in \mathbf{Y}$.

Analogically, the second player has to solve the linear programming parametric problem:

$$F_2(\mathbf{x}, \mathbf{y}, \boldsymbol{\lambda}^2) \to max, \ \mathbf{y} \in \mathbf{Y}, \tag{10.2}$$

with the parameter-vector $\boldsymbol{\lambda}^2 \in \boldsymbol{\Lambda}_2$ and $\mathbf{x} \in \mathbf{X}$.

Theorem 10.7 *The set of Pareto–Nash equilibria in the game* Γ'' *is non empty and is equal to*

$$\mathbf{PNES}(\Gamma'') = \mathbf{Gr}_1 \bigcap \mathbf{Gr}_2 =$$

$$= \bigcup_{\substack{\boldsymbol{\lambda}^1 \in \boldsymbol{\Lambda}_1,\ \lambda^1 > 0 \\ \boldsymbol{\lambda}^2 \in \boldsymbol{\Lambda}_2,\ \lambda^2 > 0}} \ \bigcup_{\substack{i \in U,\ I \in 2^{U \setminus \{i\}} \\ j \in V,\ J \in 2^{V \setminus \{j\}}}} X_{iI}^{jJ}(\boldsymbol{\lambda}^2) \times Y_{jJ}^{iI}(\boldsymbol{\lambda}^1).$$

Proof Consider the following notation

$$\mathbf{ex}^T = (1, \ldots, 1) \in \mathbb{R}^{\mathbf{m}}, \ \mathbf{ey}^T = (1, \ldots, 1) \in \mathbb{R}^{\mathbf{n}}.$$

Let us recall that the optimal solution of a linear programming problem is realized on the vertices of the polyhedral set of feasible solutions. In problems (10.1) and (10.2), the sets $\mathbf{X}$ and $\mathbf{Y}$ have m and, respectively, n vertices — the axis unit vectors $\mathbf{e}^{x_i} \in \mathbb{R}^{\mathbf{m}}$, $i = 1, \ldots, m$, and $\mathbf{e}^{y_j} \in \mathbb{R}^{\mathbf{n}}$, $j = 1, \ldots, n$. Thus, in accordance with the simplex method and its optimality criterion, in parametric problem (10.1) the parameter set $\mathbf{Y}$ is partitioned into the such m subsets

$$Y^i(\boldsymbol{\lambda}^1) = \left\{ \mathbf{y} \in \mathbb{R}^{\mathbf{n}} : \begin{array}{l} \left(\sum\limits_{q=1}^{k_1} \lambda_q^1 (A^{kq} - A^{iq}) \right) \mathbf{y} \le 0, \ k = 1, \ldots, m, \\ \lambda_1^1 + \lambda_2^1 + \cdots + \lambda_{k_1}^1 = 1, \ \boldsymbol{\lambda}^1 > 0, \\ \mathbf{ey}^T \mathbf{y} = 1, \\ \mathbf{y} \ge \mathbf{0}. \end{array} \right\},$$

$i = 1, \ldots, m$, for which one of the optimal solutions of linear programming problem (10.1) is $\mathbf{e}^{x_i}$ — the unit vector corresponding to the x_i-axis.

Assume that $U = \left\{ i \in \{1, 2, \ldots, m\} : Y^i(\boldsymbol{\lambda}^1) \neq \emptyset \right\}$.

In conformity with the optimality criterion of the simplex method, for all $i \in U$, and for all $I \in 2^{U\setminus\{i\}}$, all the points of

$$\mathbf{Conv}\{\mathbf{e}^{x_k}, k \in I \cup \{i\}\} = \left\{ \mathbf{x} \in \mathbb{R}^{\mathbf{m}} : \begin{array}{l} \mathbf{ex}^T\mathbf{x} = 1, \\ \mathbf{x} \geq 0, \\ x_k = 0, k \notin I \cup \{i\} \end{array} \right\}$$

are optimal for the the parameters

$$\mathbf{y} \in Y^{iI}(\boldsymbol{\lambda}^1) = \left\{ \mathbf{y} \in \mathbb{R}^{\mathbf{n}} : \begin{array}{l} \left(\sum_{q=1}^{k_1} \lambda_q^1 (A^{kq} - A^{iq}) \right) \mathbf{y} = 0, k \in I, \\ \left(\sum_{q=1}^{k_1} \lambda_q^1 (A^{kq} - A^{iq}) \right) \mathbf{y} \leq 0, k \notin I \cup \{i\}, \\ \lambda_1^1 + \lambda_2^1 + \cdots + \lambda_{k_1}^1 = 1, \boldsymbol{\lambda}^1 > 0, \\ \mathbf{ey}^T\mathbf{y} = 1, \\ \mathbf{y} \geq \mathbf{0}. \end{array} \right\}.$$

Evidently, $Y^{i\emptyset}(\boldsymbol{\lambda}^1) = Y^i(\boldsymbol{\lambda}^1)$. Hence,

$$\mathbf{Gr}_1(\boldsymbol{\lambda}^1) = \bigcup_{i \in U,\, I \in 2^{U\setminus\{i\}}} \mathbf{Conv}\{\mathbf{e}^{x_k}, k \in I \cup \{i\}\} \times Y^{iI}(\boldsymbol{\lambda}^1).$$

$$\mathbf{Gr}_1 = \bigcup_{\boldsymbol{\lambda}^1 \in \boldsymbol{\Lambda}_1,\, \boldsymbol{\lambda}^1 > 0} \mathbf{Gr}_1(\boldsymbol{\lambda}^1).$$

In parametric problem (10.2), the parameter set $\mathbf{X}$ is partitioned into the such n subsets

$$X^j(\boldsymbol{\lambda}^2) = \left\{ \mathbf{x} \in \mathbb{R}^{\mathbf{m}} : \begin{array}{l} \left(\sum_{r=1}^{k_2} \lambda_r^2 (b^{kr} - b^{jr}) \right) \mathbf{x} \leq 0, \; k = 1, \ldots, n, \\ \lambda_1^2 + \lambda_2^2 + \cdots + \lambda_{k_2}^2 = 1, \boldsymbol{\lambda}^2 > 0, \\ \mathbf{ex}^T\mathbf{x} = 1, \\ \mathbf{x} \geq \mathbf{0}. \end{array} \right\},$$

$j = 1, \ldots, n$, for which one of the optimal solution of linear programming problem (10.2) is $\mathbf{e}^{y_j}$ - the unit vector corresponding to the y_j - axis.

Assume that $V = \left\{ j \in \{1, 2, \ldots, n\} : X^j(\boldsymbol{\lambda}^2) \neq \emptyset \right\}$.

In conformity with the optimality criterion of the simplex method, for all $j \in V$ and for all $J \in 2^{V\setminus\{j\}}$, all the points of

$$\mathbf{Conv}\{\mathbf{e}^{y_k}, k \in J \cup \{j\}\} = \left\{ \mathbf{y} \in \mathbb{R}^{\mathbf{n}} : \begin{array}{l} \mathbf{e}\mathbf{y}^T\mathbf{y} = 1, \\ \mathbf{y} \geq 0, \\ y_k = 0, k \notin J \cup \{j\} \end{array} \right\}$$

are optimal for parameters

$$\mathbf{x} \in X^{jJ}(\boldsymbol{\lambda}^2) = \left\{ \mathbf{x} \in \mathbb{R}^{\mathbf{m}} : \begin{array}{l} \left(\sum_{r=1}^{k_2} \lambda_r^2 (b^{kr} - b^{jr}) \right) \mathbf{x} = 0, k \in J, \\ \left(\sum_{r=1}^{k_2} \lambda_r^2 (b^{kr} - b^{jr}) \right) \mathbf{x} \leq 0, k \notin J \cup \{j\}, \\ \lambda_1^2 + \lambda_2^2 + \cdots + \lambda_{k_2}^2 = 1, \ \boldsymbol{\lambda}^2 > 0, \\ \mathbf{e}\mathbf{x}^T\mathbf{x} = 1, \\ \mathbf{x} \geq \mathbf{0}. \end{array} \right\}.$$

Evidently $X^{j\emptyset}(\boldsymbol{\lambda}^2) = X^j(\boldsymbol{\lambda}^2)$. Hence,

$$\mathbf{Gr_2}(\boldsymbol{\lambda^2}) = \bigcup_{j \in V, J \in 2^{V \setminus \{j\}}} X^{jJ}(\boldsymbol{\lambda}^2) \times \mathbf{Conv}\{\mathbf{e}^{y_k}, k \in J \cup \{j\}\}.$$

$$\mathbf{Gr_2} = \bigcup_{\boldsymbol{\lambda}^2 \in \boldsymbol{\Lambda}_2, \ \lambda^2 > 0} \mathbf{Gr_2}(\boldsymbol{\lambda^2}).$$

Finally,

$$\mathbf{PNES}(\boldsymbol{\Gamma}'') = \mathbf{Gr_1} \bigcap \mathbf{Gr_2} =$$

$$= \bigcup_{\substack{\boldsymbol{\lambda}^1 \in \Lambda_1, \ \lambda^1 > 0 \\ \boldsymbol{\lambda}^2 \in \Lambda_2, \ \lambda^2 > 0}} \bigcup_{\substack{i \in U, I \in 2^{U \setminus \{i\}} \\ j \in V, J \in 2^{V \setminus \{j\}}}} X_{iI}^{jJ}(\boldsymbol{\lambda}^2) \times Y_{jJ}^{iI}(\boldsymbol{\lambda}^1),$$

where $X_{iI}^{jJ}(\boldsymbol{\lambda}^2) \times Y_{jJ}^{iI}(\boldsymbol{\lambda}^1)$ is a convex component of the set of Pareto–Nash equilibria,

$$X_{iI}^{jJ}(\boldsymbol{\lambda}^2) = \mathbf{Conv}\{\mathbf{e}^{x_k}, k \in I \cup \{i\}\} \cap X^{jJ}(\boldsymbol{\lambda}^2),$$

$$Y_{jJ}^{iI}(\boldsymbol{\lambda}^1) = \mathbf{Conv}\{\mathbf{e}^{y_k}, k \in J \cup \{j\}\} \cap Y^{iI}(\boldsymbol{\lambda}^1),$$

$$X_{iI}^{jJ}(\boldsymbol{\lambda}^2) = \left\{ \mathbf{x} \in \mathbb{R}^{\mathbf{m}} : \begin{array}{l} \left(\sum_{r=1}^{k_2} \lambda_r^2 (b^{kr} - b^{jr}) \right) \mathbf{x} = 0, k \in J, \\ \left(\sum_{r=1}^{k_2} \lambda_r^2 (b^{kr} - b^{jr}) \right) \mathbf{x} \leq 0, k \notin J \cup \{j\}, \\ \lambda_1^2 + \lambda_2^2 + \cdots + \lambda_{k_2}^2 = 1, \, \boldsymbol{\lambda}^2 > 0, \\ \mathbf{e}\mathbf{x}^T \mathbf{x} = 1, \mathbf{x} \geq \mathbf{0}, \\ x_k = 0, k \notin I \cup \{i\}. \end{array} \right\}$$

is a set of strategies $\mathbf{x} \in \mathbf{X}$ with the support from $\{i\} \cup I$ and for which points of $\mathbf{Conv}\{\mathbf{e}^{y_k}, \; k \in J \cup \{j\}\}$ are optimal,

$$Y_{jJ}^{iI}(\boldsymbol{\lambda}^1) = \left\{ \mathbf{y} \in \mathbb{R}^{\mathbf{n}} : \begin{array}{l} \left(\sum_{q=1}^{k_1} \lambda_q^1 (A^{kq} - A^{iq}) \right) \mathbf{y} = 0, k \in I, \\ \left(\sum_{q=1}^{k_1} \lambda_q^1 (A^{kq} - A^{iq}) \right) \mathbf{y} \leq 0, k \notin I \cup \{i\}, \\ \lambda_1^1 + \lambda_2^1 + \cdots + \lambda_{k_1}^1 = 1, \, \boldsymbol{\lambda}^1 > 0, \\ \mathbf{e}\mathbf{y}^T \mathbf{y} = 1, \mathbf{y} \geq \mathbf{0}, \\ y_k = 0, k \notin J \cup \{j\}. \end{array} \right\}.$$

is a set of strategies $\mathbf{y} \in \mathbf{Y}$ with the support from $\{j\} \cup J$ and for which points of $\mathbf{Conv}\{\mathbf{e}^{x_k}, k \in I \cup \{i\}\}$ are optimal. □

The following two theorems are essentially based on the proof of Theorem 10.7 and highlight the properties of the components of the set of Pareto–Nash equilibria.

Theorem 10.8 *If* $X_{iI}^{j\emptyset}(\boldsymbol{\lambda}^2) = \emptyset$, *then* $X_{iI}^{jJ}(\boldsymbol{\lambda}^2) = \emptyset$ *for all* $J \in 2^V$.

Proof For the proof it is sufficient to maintain that $X_{iI}^{jJ}(\boldsymbol{\lambda}^2) \subseteq X_{iI}^{j\emptyset}(\boldsymbol{\lambda}^2)$ for $J \neq \emptyset$. □

Theorem 10.9 *If* $Y_{jJ}^{i\emptyset}(\boldsymbol{\lambda}^1) = \emptyset$, *then* $Y_{jJ}^{iI}(\boldsymbol{\lambda}^1) = \emptyset$ *for all* $I \in 2^U$.

Proof It is sufficient to observe that Theorem 10.9 establishes the same properties as Theorem 10.8, but for the second player graph. □

From the above, the algorithm for a Pareto–Nash equilibrium set computing follows:

Algorithm 10.5.1

```
PNES = ∅;  U = {i ∈ {1, 2, ..., m} : Y^i(λ^1) ≠ ∅};  UX = U;
V = {j ∈ {1, 2, ..., n} : X^j(λ^2) ≠ ∅};
for i ∈ U do
  {
     UX = UX \ {i};
     for I ∈ 2^UX do
       {
          VY = V;
          for j ∈ V do
             {
                if (X_{iI}^{j∅}(λ^2) = ∅) break;
                VY = VY \ {j};
                for J ∈ 2^VY do
                    if (Y_{jJ}^{iI}(λ^1) ≠ ∅)
                        PNES = PNES ∪ (X_{iI}^{jJ}(λ^2) × Y_{jJ}^{iI}(λ^1));
             }
       }
  }
```

Theorem 10.10 *Algorithm 10.5.1 examines no more than*

$$(2^m - 1)(2^n - 1)$$

polytopes of the $X_{iI}^{jJ}(\boldsymbol{\lambda}^2) \times Y_{jJ}^{iI}(\boldsymbol{\lambda}^1)$ *type.*

Proof Let us observe that Algorithm 10.5.1 executes the interior **if** no more than

$$\begin{gathered}2^{m-1}(2^{n-1} + 2^{n-2} + \cdots + 2^1 + 2^0)+\\ +2^{m-2}(2^{n-1} + 2^{n-2} + \cdots + 2^1 + 2^0)+\\ \cdots\\ +2^1(2^{n-1} + 2^{n-2} + \cdots + 2^1 + 2^0)+\\ +2^0(2^{n-1} + 2^{n-2} + \cdots + 2^1 + 2^0) =\\ = (2^m - 1)(2^n - 1)\end{gathered}$$

times. So, the truth of Theorem 10.10 is established. □

If all the players' strategies are equivalent, then the set of Pareto–Nash equilibria is a union of $(2^m - 1)(2^n - 1)$ polytopes.

Evidently, for practical reasons Algorithm 10.5.1 may be improved by identifying equivalent, dominant and dominated strategies in pure game [1–3, 10] with the following pure and extended game simplification, but the difficulty is connected with multi-criteria nature of the initial game. "In a non-degenerate game, both players use the same number of pure strategies in equilibrium, so only supports of equal cardinality need to be examined" [18]. This property may be used to minimize essentially the number of components $X_{iI}^{jJ}(\boldsymbol{\lambda}^2) \times Y_{jJ}^{iI}(\boldsymbol{\lambda}^1)$ examined in nondegenerate game.

Example 10.1 Matrices of the two-person two-criterion game are

$$A = \begin{bmatrix} 1,0 & 0,2 & 4,1 \\ 0,2 & 2,1 & 3,3 \end{bmatrix}, \quad B = \begin{bmatrix} 0,1 & 2,3 & 3,3 \\ 6,4 & 5,1 & 3,0 \end{bmatrix}$$

The exterior cycle in the above algorithm is executed for the value $i = 1$.

As

$$X_{1\emptyset}^{1\emptyset}(\boldsymbol{\lambda}^2) = \left\{ \mathbf{x} \in \mathbb{R}^2 : \begin{array}{l} (2\lambda_1^2 + 2\lambda_2^2)x_1 + (-\lambda_1^2 - 3\lambda_2^2)x_2 \le 0, \\ (3\lambda_1^2 + 2\lambda_2^2)x_1 + (-3\lambda_1^2 - 4\lambda_2^2)x_2 \le 0 \\ \lambda_1^2 + \lambda_2^2 = 1,\ \boldsymbol{\lambda}^2 > 0, \\ x_1 + x_2 = 1, \\ x_1 \ge 0, x_2 = 0. \end{array} \right\} = \emptyset,$$

then the cycle for $j = 1$ is omitted.

Since

$$X_{1\emptyset}^{2\emptyset}(\boldsymbol{\lambda}^2) = \left\{ \mathbf{x} \in \mathbb{R}^2 : \begin{array}{l} (-2\lambda_1^2 - 2\lambda_2^2)x_1 + (\lambda_1^2 + 3\lambda_2^2)x_2 \le 0, \\ \lambda_1^2 x_1 + (-2\lambda_1^2 - \lambda_2^2)x_2 \le 0, \\ \lambda_1^2 + \lambda_1^2 = 1,\ \boldsymbol{\lambda}^2 > 0, \\ x_1 + x_2 = 1, \\ x_1 \ge 0, x_2 = 0. \end{array} \right\} \ne \emptyset,$$

$$Y_{2\emptyset}^{1\emptyset}(\boldsymbol{\lambda}^1) = \left\{ \mathbf{y} \in \mathbb{R}^3 : \begin{array}{l} (-\lambda_1^1 + 2\lambda_2^1)y_1 + (2\lambda_1^1 - \lambda_2^1)y_2 + \\ \quad +(-\lambda_1^1 + 2\lambda_2^1)y_3 \le 0, \\ \lambda_1^1 + \lambda_2^1 = 1,\ \boldsymbol{\lambda}^1 > 0, \\ y_1 + y_2 + y_3 = 1, \\ y_1 = 0,\ y_2 \ge 0,\ y_3 = 0. \end{array} \right\} \ne \emptyset,$$

the point $(1, 0) \times (0, 1, 0)$ is a Pareto–Nash equilibrium for which the payoff values are $(0, 2)$ and $(2, 3)$.

$$X_{1\emptyset}^{2\{3\}}(\boldsymbol{\lambda}^2) = \left\{ \mathbf{x} \in \mathbb{R}^2 : \begin{array}{l} (-2\lambda_1^2 - 2\lambda_2^2)x_1 + (\lambda_1^2 + 3\lambda_2^2)x_2 \le 0, \\ \lambda_1^2 x_1 + (-2\lambda_1^2 - \lambda_2^2)x_2 = 0, \\ \lambda_1^2 + \lambda_2^2 = 1,\ \boldsymbol{\lambda}^2 > 0, \\ x_1 + x_2 = 1, \\ x_1 \ge 0, x_2 = 0. \end{array} \right\} \ne \emptyset,$$

$$Y^{1\emptyset}_{2\{3\}}(\boldsymbol{\lambda}^1) = \left\{ \mathbf{y} \in \mathbb{R}^3 : \begin{array}{l} (-\lambda^1_1 + 2\lambda^1_2)y_1 + (2\lambda^1_1 - \lambda^1_2)y_2 + \\ \quad +(-\lambda^1_1 + 2\lambda^1_2)y_3 \le 0, \\ \lambda^1_1 + \lambda^1_2 = 1,\ \boldsymbol{\lambda}^1 > 0, \\ y_1 + y_2 + y_3 = 1, \\ y_1 = 0,\ y_2 \ge 0,\ y_3 \ge 0. \end{array} \right\} \ne \emptyset,$$

the set $\left\{ \begin{pmatrix} 1 \\ 0 \end{pmatrix} \times \begin{pmatrix} 0 \\ 0 \le y_2 \le \frac{1}{3} \\ \frac{2}{3} \le y_3 \le 1 \end{pmatrix}, \begin{pmatrix} 1 \\ 0 \end{pmatrix} \times \begin{pmatrix} 0 \\ \frac{2}{3} \le y_2 \le 1 \\ 0 \le y_3 \le \frac{1}{3} \end{pmatrix} \right\}$
is a component of the set of Pareto–Nash equilibria.

Since

$$X^{3\emptyset}_{1\emptyset}(\boldsymbol{\lambda}^2) = \left\{ \mathbf{x} \in \mathbb{R}^2 : \begin{array}{l} (-3\lambda^2_1 - 2\lambda^2_2)x_1 + (3\lambda^2_1 + 4\lambda^2_2)x_2 \le 0, \\ -\lambda^2_1 x_1 + (2\lambda^2_1 + \lambda^2_2)x_2 \le 0, \\ \lambda^2_1 + \lambda^2_2 = 1,\ \boldsymbol{\lambda}^2 > 0, \\ x_1 + x_2 = 1, \\ x_1 \ge 0, x_2 = 0. \end{array} \right\} \ne \emptyset,$$

$$Y^{1\emptyset}_{3\emptyset}(\boldsymbol{\lambda}^1) = \left\{ \mathbf{y} \in \mathbb{R}^3 : \begin{array}{l} (-\lambda^1_1 + 2\lambda^1_2)y_1 + (2\lambda^1_1 - \lambda^1_2)y_2 + \\ \quad +(-\lambda^1_1 + 2\lambda^1_2)y_3 \le 0, \\ \lambda^1_1 + \lambda^1_2 = 1,\ \boldsymbol{\lambda}^1 > 0, \\ y_1 + y_2 + y_3 = 1, \\ y_1 = 0,\ y_2 = 0,\ y_3 \ge 0. \end{array} \right\} \ne \emptyset,$$

the point $(1, 0) \times (0, 0, 1)$ is a Pareto–Nash equilibrium for which the payoff values are $(4, 1)$ and $(3, 3)$.

Since

$$X^{1\emptyset}_{1\{2\}}(\boldsymbol{\lambda}^2) = \left\{ \mathbf{x} \in \mathbb{R}^2 : \begin{array}{l} (2\lambda^2_1 + 2\lambda^2_2)x_1 + (-\lambda^2_1 - 3\lambda^2_2)x_2 \le 0, \\ (3\lambda^2_1 + 2\lambda^2_2)x_1 + (-3\lambda^2_1 - 4\lambda^2_2)x_2 \le 0, \\ \lambda^2_1 + \lambda^2_2 = 1,\ \boldsymbol{\lambda}^2 > 0, \\ x_1 + x_2 = 1, \\ x_1 \ge 0, x_2 \ge 0. \end{array} \right\} \ne \emptyset,$$

$$Y^{1\{2\}}_{1\emptyset}(\boldsymbol{\lambda}^1) = \left\{ \mathbf{y} \in \mathbb{R}^3 : \begin{array}{l} (-\lambda^1_1 + 2\lambda^1_2)y_1 + (2\lambda^1_1 - \lambda^1_2)y_2 + \\ \quad +(-\lambda^1_1 + 2\lambda^1_2)y_3 = 0, \\ \lambda^1_1 + \lambda^1_2 = 1,\ \boldsymbol{\lambda}^1 > 0, \\ y_1 + y_2 + y_3 = 1, \\ y_1 \ge 0,\ y_2 = 0,\ y_3 = 0. \end{array} \right\} \ne \emptyset,$$

the set $\left\{ \left(0 \le x_1 \le \frac{1}{3}, \frac{2}{3} \le x_2 \le 1\right) \times (1, 0, 0) \right\}$ is a component of the Pareto–Nash equilibrium set.

Since

$$X^{1\{2\}}_{1\{2\}}(\boldsymbol{\lambda}^2) = \left\{ \mathbf{x} \in \mathbb{R}^2 : \begin{array}{l} (2\lambda^2_1 + 2\lambda^2_2)x_1 + (-\lambda^2_1 - 3\lambda^2_2)x_2 = 0, \\ (3\lambda^2_1 + 2\lambda^2_2)x_1 + (-3\lambda^2_1 - 4\lambda^2_2)x_2 \leq 0, \\ \lambda^2_1 + \lambda^2_2 = 1,\ \boldsymbol{\lambda}^2 > 0, \\ x_1 + x_2 = 1, \\ x_1 \geq 0, x_2 \geq 0. \end{array} \right\} \neq \emptyset,$$

$$Y^{1\{2\}}_{1\{2\}}(\boldsymbol{\lambda}^1) = \left\{ \mathbf{y} \in \mathbb{R}^3 : \begin{array}{l} (-\lambda^1_1 + 2\lambda^1_2)y_1 + (2\lambda^1_1 - \lambda^1_2)y_2 + \\ \quad +(-\lambda^1_1 + 2\lambda^1_2)y_3 = 0, \\ \lambda^1_1 + \lambda^1_2 = 1,\ \boldsymbol{\lambda}^1 > 0, \\ y_1 + y_2 + y_3 = 1 \\ y_1 \geq 0, y_2 \geq 0, y_3 = 0. \end{array} \right\} \neq \emptyset,$$

the set $\left\{\left(\frac{1}{3} \leq x_1 \leq \frac{3}{5}, \frac{2}{5} \leq x_2 \leq \frac{2}{3}\right) \times \left(0 \leq y_1 \leq \frac{1}{3}, \frac{2}{3} \leq y_2 \leq 1, 0\right)\right\} \bigcup$ $\left\{\left(\frac{1}{3} \leq x_1 \leq \frac{3}{5}, \frac{2}{5} \leq x_2 \leq \frac{2}{3}\right) \times \left(\frac{2}{3} \leq y_1 \leq 1, 0 \leq y_2 \leq \frac{1}{3}, 0\right)\right\}$ is a component of the Pareto–Nash equilibrium set.

$X^{1\{3\}}_{1\{2\}}(\boldsymbol{\lambda}^2) = \emptyset$, $X^{1\{2,3\}}_{1\{2\}}(\boldsymbol{\lambda}^2) = \emptyset$.

Since

$$X^{2\emptyset}_{1\{2\}}(\boldsymbol{\lambda}^2) = \left\{ \mathbf{x} \in \mathbb{R}^2 : \begin{array}{l} (-2\lambda^2_1 - 2\lambda^2_2)x_1 + (\lambda^2_1 + 3\lambda^2_2)x_2 \leq 0, \\ \lambda^2_1 x_1 + (-2\lambda^2_1 - \lambda^2_2)x_2 \leq 0, \\ \lambda^2_1 + \lambda^2_2 = 1,\ \boldsymbol{\lambda}^2 > 0, \\ x_1 + x_2 = 1, \\ x_1 \geq 0, x_2 \geq 0. \end{array} \right\} \neq \emptyset,$$

$$Y^{1\{2\}}_{2\emptyset}(\boldsymbol{\lambda}^1) = \left\{ \mathbf{y} \in \mathbb{R}^3 : \begin{array}{l} (-\lambda^1_1 + 2\lambda^1_2)y_1 + (2\lambda^1_1 - \lambda^1_2)y_2 + \\ \quad +(-\lambda^1_1 + 2\lambda^1_2)y_3 = 0, \\ \lambda^1_1 + \lambda^1_2 = 1,\ \boldsymbol{\lambda}^1 > 0, \\ y_1 + y_2 + y_3 = 1, \\ y_1 = 0, y_2 \geq 0, y_3 = 0. \end{array} \right\} \neq \emptyset,$$

the set $\left\{\left(\frac{3}{5} \leq x_1 \leq \frac{2}{3}, \frac{1}{3} \leq x_2 \leq \frac{2}{5}\right) \times (0, 1, 0)\right\}$ is a component of the Pareto–Nash equilibrium set.

Since

$$X^{2\{3\}}_{1\{2\}}(\boldsymbol{\lambda}^2) = \left\{ \mathbf{x} \in \mathbb{R}^2 : \begin{array}{l} (-2\lambda^2_1 - 2\lambda^2_2)x_1 + (\lambda^2_1 + 3\lambda^2_2)x_2 \leq 0, \\ \lambda^2_1 x_1 + (-2\lambda^2_1 - \lambda^2_2)x_2 = 0, \\ \lambda^2_1 + \lambda^2_2 = 1,\ \boldsymbol{\lambda}^2 > 0, \\ x_1 + x_2 = 1, \\ x_1 \geq 0, x_2 \geq 0. \end{array} \right\} \neq \emptyset,$$

$$Y_{2\{3\}}^{1\{2\}}(\boldsymbol{\lambda}^1) = \left\{ \mathbf{y} \in \mathbb{R}^3 : \begin{array}{l} (-\lambda_1^1 + 2\lambda_2^1)y_1 + (2\lambda_1^1 - \lambda_2^1)y_2 + \\ \quad +(-\lambda_1^1 + 2\lambda_2^1)y_3 = 0, \\ \lambda_1^1 + \lambda_2^1 = 1,\ \boldsymbol{\lambda}^1 > 0, \\ y_1 + y_2 + y_3 = 1, \\ y_1 = 0,\ y_2 \geq 0,\ y_3 \geq 0. \end{array} \right\} \neq \emptyset,$$

the set $\left\{\left(\frac{2}{3} \leq x_1 \leq 1, 0 \leq x_2 \leq \frac{1}{3}\right) \times \left(0, 0 \leq y_2 \leq \frac{1}{3}, \frac{2}{3} \leq y_3 \leq 1\right)\right\} \bigcup$ $\left\{\left(\frac{2}{3} \leq x_1 \leq 1, 0 \leq x_2 \leq \frac{1}{3}\right) \times \left(0, \frac{2}{3} \leq y_2 \leq 1, 0 \leq y_3 \leq \frac{1}{3}\right)\right\}$ is a component of the Pareto–Nash equilibrium set.

$$X_{1\{2\}}^{3\emptyset}(\boldsymbol{\lambda}^2) = \left\{ \mathbf{x} \in \mathbb{R}^2 : \begin{array}{l} (-3\lambda_1^2 - 2\lambda_2^2)x_1 + (3\lambda_1^2 + 4\lambda_2^2)x_2 \leq 0, \\ -\lambda_1^2 x_1 + (2\lambda_1^2 + \lambda_2^2)x_2 \leq 0, \\ \lambda_1^2 + \lambda_2^2 = 1,\ \boldsymbol{\lambda}^2 > 0, \\ x_1 + x_2 = 1, \\ x_1 \geq 0, x_2 \geq 0. \end{array} \right\} \neq \emptyset,$$

$$Y_{3\emptyset}^{1\{2\}}(\boldsymbol{\lambda}^1) = \left\{ \mathbf{y} \in \mathbb{R}^3 : \begin{array}{l} (-\lambda_1^1 + 2\lambda_2^1)y_1 + (2\lambda_1^1 - \lambda_2^1)y_2 + \\ \quad +(-\lambda_1^1 + 2\lambda_2^1)y_3 = 0, \\ \lambda_1^1 + \lambda_2^1 = 1,\ \boldsymbol{\lambda}^1 > 0, \\ y_1 + y_2 + y_3 = 1, \\ y_1 = 0,\ y_2 = 0,\ y_3 \geq 0. \end{array} \right\} \neq \emptyset,$$

the set $\left\{\left(\frac{2}{3} \leq x_1 \leq 1, 0 \leq x_2 \leq \frac{1}{3}\right) \times (0, 0, 1)\right\}$ is a component of the Pareto–Nash equilibrium set.

The exterior cycle is executed for the value $i = 2$.

$$X_{2\emptyset}^{1\emptyset}(\boldsymbol{\lambda}^2) = \left\{ \mathbf{x} \in \mathbb{R}^2 : \begin{array}{l} (2\lambda_1^2 + 2\lambda_2^2)x_1 + (-\lambda_1^2 - 3\lambda_2^2)x_2 \leq 0, \\ (3\lambda_1^2 + 2\lambda_2^2)x_1 + (-3\lambda_1^2 - 4\lambda_2^2)x_2 \leq 0, \\ \lambda_1^2 + \lambda_2^2 = 1,\ \boldsymbol{\lambda}^2 > 0, \\ x_1 + x_2 = 1, \\ x_1 = 0, x_2 \geq 0. \end{array} \right\} \neq \emptyset$$

$$Y_{1\emptyset}^{2\emptyset}(\boldsymbol{\lambda}^1) = \left\{ \mathbf{y} \in \mathbb{R}^3 : \begin{array}{l} (\lambda_1^1 - 2\lambda_2^1)y_1 + (-2\lambda_1^1 + \lambda_2^1)y_2 + \\ \quad +(\lambda_1^1 - 2\lambda_2^1)y_3 \leq 0, \\ \lambda_1^1 + \lambda_2^1 = 1,\ \boldsymbol{\lambda}^1 > 0, \\ y_1 + y_2 + y_3 = 1, \\ y_1 \geq 0,\ y_2 = 0,\ y_3 = 0. \end{array} \right\} \neq \emptyset$$

the point $(0, 1) \times (1, 0, 0)$ is a Pareto–Nash equilibrium for which payoff values are $(0, 2)$ and $(6, 4)$.

$X_{2\emptyset}^{1\{2\}}(\boldsymbol{\lambda}^2) = \emptyset$, $X_{2\emptyset}^{1\{3\}}(\boldsymbol{\lambda}^2) = \emptyset$, $X_{2\emptyset}^{1\{2,3\}}(\boldsymbol{\lambda}^2) = \emptyset$.

Because

$$X_{2\emptyset}^{2\emptyset}(\boldsymbol{\lambda}^2) = \left\{ \mathbf{x} \in \mathbb{R}^2 : \begin{array}{l} (-2\lambda_1^2 - 2\lambda_2^2)x_1 + (\lambda_1^2 + 3\lambda_2^2)x_2 \le 0, \\ \lambda_1^2 x_1 + (-2\lambda_1^2 - \lambda_2^2)x_2 \le 0, \\ \lambda_1^2 + \lambda_2^2 = 1,\ \boldsymbol{\lambda}^2 > 0, \\ x_1 + x_2 = 1, \\ x_1 = 0, x_2 \ge 0. \end{array} \right\} = \emptyset,$$

the cycle for $j = 2$ is omitted.

$$X_{2\emptyset}^{3\emptyset}(\boldsymbol{\lambda}^2) = \left\{ \mathbf{x} \in \mathbb{R}^2 : \begin{array}{l} (-3\lambda_1^2 - 2\lambda_2^2)x_1 + (3\lambda_1^2 + 4\lambda_2^2)x_2 \le 0, \\ -\lambda_1^2 x_1 + (2\lambda_1^2 + \lambda_2^2)x_2 \le 0, \\ \lambda_1^2 + \lambda_2^2 = 1,\ \boldsymbol{\lambda}^2 > 0, \\ x_1 + x_2 = 1, \\ x_1 = 0, x_2 \ge 0. \end{array} \right\} = \emptyset.$$

Thus, the Pareto–Nash equilibrium set consists of nine elements. Three of them are pure Pareto–Nash equilibria.

Let us add one more utility function in Example 10.1 for each player.

Example 10.2 Matrices of the two person game are

$$A = \begin{bmatrix} 1,0,2 & 0,2,1 & 4,1,3 \\ 0,2,1 & 2,1,0 & 3,3,1 \end{bmatrix},\quad B = \begin{bmatrix} 0,1,0 & 2,3,1 & 3,3,2 \\ 6,4,5 & 5,1,3 & 3,0,1 \end{bmatrix}.$$

Algorithm 10.5.1 examines $(2^2 - 1)(2^3 - 1) = 21$ cases for this game.

The set of Pareto–Nash equilibria consists of eleven components. It is enlarged, comparatively with Example 10.1, and it coincides with the graph of best response mapping of the second player.

Corollary 10.1 *The number of criteria increases the total number of arithmetic operations, but the number of investigated cases remains intact.*

Example 10.3 Let us examine the game with matrices:

$$A = \begin{bmatrix} 2,0 & 1,2 & 6,-1 \\ 3,5 & 2,0 & -1,2 \\ -1,3 & 2,3 & 1,1 \end{bmatrix},\ B = \begin{bmatrix} 1,2 & 0,1 & 3,2 \\ -1,3 & 1,-1 & -2,0 \\ 2,0 & -1,3 & 2,1 \end{bmatrix}.$$

Algorithm 10.5.1 examines $(2^3 - 1)(2^3 - 1) = 49$ polyhedra $X_{iI}^{jJ}(\boldsymbol{\lambda}^2) \times Y_{jJ}^{iI}(\boldsymbol{\lambda}^1)$.

In this game thirty-seven components $X_{iI}^{jJ}(\boldsymbol{\lambda}^2)$ and eighteen components $Y_{jJ}^{iI}(\boldsymbol{\lambda}^1)$ are non empty. The Pareto–Nash equilibrium set consists of twenty-three elements.

Examples 10.1–10.3 illustrate not only Algorithm 10.5.1, but particular features of the Pareto–Nash games too. Examples are a good starting point to begin investigation of the games examined in the following section.

10.6 Pareto–Nash Equilibrium Sets in Multi-Criterion Polymatrix Mixed-Strategy Games

Consider an n-player $m_1 \times m_2 \times \cdots \times m_n$ mixed-strategy game

$$\Gamma''(\boldsymbol{\lambda}) = \langle \mathbf{N}, \{\mathbf{X_p}\}_{p \in \mathbf{N}}, \{F_p(\mathbf{x}, \boldsymbol{\lambda}^{\mathbf{p}})\}_{p \in \mathbf{N}} \rangle,$$

formulated in Sect. 10.4. The utility synthesis function of the player p is linear if the strategies of the remaining players are fixed:

$$\begin{aligned} F_p(\mathbf{x}, \boldsymbol{\lambda}^{\mathbf{p}}) = & \left(\lambda_1^p \sum_{s_{-p} \in S_{-p}} a_{1\|s_{-p}}^{p1} \prod_{q=1,\dots,n, q \neq p} x_{s_q}^q + \cdots + \right. \\ & \left. \lambda_{k_p}^p \sum_{s_{-p} \in S_{-p}} a_{1\|s_{-p}}^{pk_p} \prod_{q=1,\dots,n, q \neq p} x_{s_q}^q \right) x_1^p \\ & + \cdots + \\ & \left(\lambda_1^p \sum_{s_{-p} \in S_{-p}} a_{m_p\|s_{-p}}^{p1} \prod_{q=1,\dots,n, q \neq p} x_{s_q}^q + \cdots + \right. \\ & \left. \lambda_{k_p}^p \sum_{s_{-p} \in S_{-p}} a_{m_p\|s_{-p}}^{pk_p} \prod_{q=1,\dots,n, q \neq p} x_{s_q}^q \right) x_{m_p}^p, \end{aligned}$$

$$\lambda_1^p + \lambda_2^p + \cdots + \lambda_{k_p}^p = 1, \ \lambda_i^p \geq 0, i = 1, \dots, k_p,$$

It means that the player p has to solve a linear parametric problem with the parameter vector $\mathbf{x}^{-\mathbf{p}} \in \mathbf{X_{-p}}$ and $\boldsymbol{\lambda}^p \in \boldsymbol{\Lambda}_p$:

$$F_p(\mathbf{x}^{\mathbf{p}}, \mathbf{x}^{-\mathbf{p}}, \boldsymbol{\lambda}^{\mathbf{p}}) \to max, \ \mathbf{x}^{\mathbf{p}} \in \mathbf{X_p}, \ \boldsymbol{\lambda}^{\mathbf{p}} \in \boldsymbol{\Lambda}_{\mathbf{p}}, \ p = 1, \dots, n. \tag{10.3}$$

Theorem 10.11 *The set of Pareto–Nash equilibria in the game Γ'' is non empty and identical to*

$$\mathbf{PNES}(\Gamma''(\boldsymbol{\lambda})) = \bigcap_{p=1}^{n} \mathbf{Gr_p} =$$

$$= \bigcup_{\boldsymbol{\lambda}\in\boldsymbol{\Lambda},\,\boldsymbol{\lambda}>0} \quad \bigcup_{\substack{i_1 \in U_1,\ I_1 \in 2^{U_1\setminus\{i_1\}} \\ \dots \\ i_n \in U_n,\ I_n \in 2^{U_n\setminus\{i_n\}}}} X(i_1 I_1 \dots i_n I_n)(\boldsymbol{\lambda}).$$

Proof The optimal solutions of parametric problem (10.3) are realized on vertices of the polytope $\mathbf{X_p}$ which has m_p vertices — the unit vectors $\mathbf{e}^{x_i^p} \in \mathbb{R}^{\mathbf{m_i}}$, $i = 1, \dots, m_p$, of the x_i^p-axes, $i = 1, \dots, m_p$. In accordance with the simplex method and its optimality criterion, the parameter set $\mathbf{X_{-p}}$ is partitioned into the such m_p subsets $X_{-p}(i_p)(\boldsymbol{\lambda}^p)$:

$$\begin{cases} \sum\limits_{s_{-p}\in S_{-p}} \left(\sum\limits_{i=1,\dots,k_p} \lambda_i^p (a^{pi}_{k\|s_{-p}} - a^{pi}_{i_p\|s_{-p}}) \right) \prod\limits_{q=1,\dots,n, q\neq p} x^q_{s_q} \le 0, \\ k = 1, \dots, m_p, \\ \lambda_1^p + \lambda_2^p + \dots + \lambda_{k_p}^p = 1,\ \boldsymbol{\lambda}^p > 0, \\ x_1^q + x_2^q + \dots + x_{m_q}^q = 1, q = 1, \dots, n, q \neq p, \\ \mathbf{x^{-p}} \ge \mathbf{0}, \end{cases},$$

corresponding to $\mathbf{x^{-p}} \in \mathbb{R}^{\mathbf{m-m_p}}$, $i_p = 1, \dots, m_p$, for which one of the optimal solution of linear programming problem (10.3) is $\mathbf{e}^{x_i^p}$.

Assume that

$$U_p = \{i_p \in \{1, 2, \dots, m_p\} : X_{-p}(i_p)(\boldsymbol{\lambda}^p) \neq \emptyset\}, \mathbf{ep}^T = (1, \dots, 1) \in \mathbb{R}^{\mathbf{m_p}}.$$

In conformity with the optimality criterion of the simplex method, for all $i_p \in U_p$ and $I_p \in 2^{U_p\setminus\{i_p\}}$, all the points of

$$\mathbf{Conv}\{\mathbf{e}^{x_k^p}, k \in I_p \cup \{i_p\}\} = \left\{ \mathbf{x} \in \mathbb{R}^{\mathbf{m_p}} : \begin{array}{l} \mathbf{ep}^T \mathbf{x}^p = 1, \\ \mathbf{x}^p \ge 0, \\ x_k^p = 0, k \notin I_p \cup \{i_p\} \end{array} \right\}$$

are optimal for the parameters $\mathbf{x^{-p}} \in X_{-p}(i_p I_p)(\boldsymbol{\lambda}^p) \subset \mathbb{R}^{\mathbf{m-m_p}}$, where $X_{-p}(i_p I_p)(\boldsymbol{\lambda}^p)$ is the set of solutions of the system:

$$\begin{cases} \sum\limits_{s_{-p}\in S_{-p}} \left(\sum\limits_{i=\overline{1,k_p}} \lambda_i^p (a^{pi}_{k\|s_{-p}} - a^{pi}_{i_p\|s_{-p}}) \right) \prod\limits_{q=1,\dots,n, q\neq p} x^q_{s_q} = 0, k \in I_p, \\ \sum\limits_{s_{-p}\in S_{-p}} \left(\sum\limits_{i=1,\dots,k_p} \lambda_i^p (a^{pi}_{k\|s_{-p}} - a^{pi}_{i_p\|s_{-p}}) \right) \prod\limits_{q=1,\dots,n, q\neq p} x^q_{s_q} \leq 0, \\ k \notin I_p \cup \{i_p\}, \\ \lambda_1^p + \lambda_2^p + \dots + \lambda_{k_p}^p = 1, \ \boldsymbol{\lambda}^p > 0, \\ \mathbf{e}\mathbf{r}^T \mathbf{x}^r = 1, r = 1, \dots, n, r \neq p, \\ \mathbf{x^r} \geq \mathbf{0}, r = 1, \dots, n, r \neq p. \end{cases}$$

Evidently, $X_{-p}(i_p\emptyset)(\boldsymbol{\lambda}^p) = X_{-p}(i_p)(\boldsymbol{\lambda}^p)$. Hence,

$$\mathbf{Gr_p}(\boldsymbol{\lambda}^p) = \bigcup_{i_p \in U_p, I_p \in 2^{U_p \setminus \{i_p\}}} \mathbf{Conv}\{\mathbf{e}^{x_k^p}, k \in I_p \cup \{i_p\}\} \times X_{-p}(i_p I_p)(\boldsymbol{\lambda}^p),$$

$$\mathbf{Gr_p} = \bigcup_{\boldsymbol{\lambda}^p \in \boldsymbol{\Lambda}_p, \, \boldsymbol{\lambda}^p > 0} \mathbf{Gr_p}(\boldsymbol{\lambda}^{\mathbf{p}}).$$

Finally,

$$\mathbf{PNES}(\Gamma''(\boldsymbol{\lambda})) = \bigcap_{p=1}^{n} \mathbf{Gr_p},$$

$$\bigcap_{p=1}^{n} \mathbf{Gr_p} = \bigcup_{\boldsymbol{\lambda} \in \boldsymbol{\Lambda}, \, \boldsymbol{\lambda} > 0} \ \bigcup_{\substack{i_1 \in U_1, \, I_1 \in 2^{U_1 \setminus \{i_1\}} \\ \dots \\ i_n \in U_n, \, I_n \in 2^{U_n \setminus \{i_n\}}}} X(i_1 I_1 \dots i_n I_n)(\boldsymbol{\lambda}),$$

where $X(i_1 I_1 \dots i_n I_n)(\boldsymbol{\lambda}) = \mathbf{PNES}(i_1 I_1 \dots i_n I_n)(\boldsymbol{\lambda})$ is the set of solutions of the system:

$$\begin{cases} \sum\limits_{s_{-r}\in S_{-r}} \left(\sum\limits_{i=1,\dots,k_r} \lambda_i^r (a^{ri}_{k\|s_{-r}} - a^{ri}_{i_r\|s_{-r}}) \right) \prod\limits_{q=1,\dots,n, q\neq r} x^q_{s_q} = 0, k \in I_r, \\ \sum\limits_{s_{-r}\in S_{-r}} \left(\sum\limits_{i=1,\dots,k_r} \lambda_i^r (a^{ri}_{k\|s_{-r}} - a^{ri}_{i_r\|s_{-r}}) \right) \prod\limits_{q=1,\dots,n, q\neq r} x^q_{s_q} \leq 0, \\ k \notin I_r \cup \{i_r\}, \\ \lambda_1^r + \lambda_2^r + \dots + \lambda_{k_r}^r = 1, \ \boldsymbol{\lambda} > 0, \ r = 1, \dots, n, \\ \mathbf{e}\mathbf{r}^T \mathbf{x}^r = 1, \mathbf{x^r} \geq \mathbf{0}, r = 1, \dots, n, \\ x_k^r = 0, k \notin I_r \cup \{i_r\}, r = 1, \dots, n. \end{cases}$$

So, the theorem is proved. □

Let us observe that Theorem 10.11 is an extension for multi-player games of Theorem 10.7.

An exponential algorithm for Pareto–Nash equilibrium set computing in an n-player multi-criteria game follows from the expression established by Theorem 10.11.

Remark 10.2 The exposed bellow algorithm is mainly conceptual and abstract because of parameters set Λ, which is necessary in algorithm exposition and is essential in computing the entire set of Pareto–Nash equilibria, and because of pseudo programming language and code application for the algorithm exposition.

Algorithm 10.6.1

$PNES = \emptyset;$
for $\lambda \in \mathbf{\Lambda}$
{
  for $p \in N$ **do** $U_p = \{i_p \in \{1, 2, \ldots, m_p\} : X_{-p}(i_p)(\boldsymbol{\lambda}^p) \neq \emptyset\};$

  $V_1 = U_1;$
  for $i_1 \in U_1$ **do**
    {
      $V_1 = V_1 \setminus \{i_1\};$
      for $I_1 \in 2^{V_1}$ **do**
        {
          ⋮
            $V_n = U_n;$
            for $i_n \in U_n$ **do**
              {
                $V_n = V_n \setminus \{i_n\};$
                for $I_n \in 2^{V_n}$ **do**
                    if $(X(i_1 I_1 \ldots i_n I_n)(\boldsymbol{\lambda})) \neq \emptyset)$
                      $PNES = PNES \cup X(i_1 I_1 \ldots i_n I_n)(\boldsymbol{\lambda});$
              }
        }
    }
}

Remark 10.3 Algorithm 10.6.1 may be improved by including loop breaks when $X(i_1 I_1 \ldots i_n I_n)(\boldsymbol{\lambda})) = \emptyset$. We intentionally do not include breaks to emphasize the structure and logics of the algorithm.

The next theorem highlights the complexity of Algorithm 10.6.1. It might be seen and related as a corollary of Theorem 10.11 and a generalization of Theorem 10.10.

Theorem 10.12 **PNES**$(\Gamma''(\boldsymbol{\lambda}))$ *consists of no more than*

$$(2^{m_1}-1)(2^{m_2}-1)\dots(2^{m_n}-1)$$

components of the type $X(i_1 I_1 \dots i_n I_n)(\boldsymbol{\lambda})$.

Proof Let us observe that Algorithm 10.5.1 follows from Algorithm 10.6.1 when $n = 2$. For $n = 2$ the complexity of the algorithm is established by Theorem 10.10 and is equal to

$$(2_1^m - 1)(2_2^m - 1).$$

For $n = 3$, the complexity of Algorithm 10.6.1 is increased by $(2_3^m - 1)$ and is equal to

$$(2_1^m - 1)(2_2^m - 1)(2_3^m - 1).$$

Evidently, each additional player increases the complexity of Algorithm 10.6.1 by a similar multiplier as above. So, we can conclude that if the number of player is n, then the complexity of the algorithm is equal to

$$(2_1^m - 1)(2_2^m - 1)\dots(2_n^m - 1).$$

The truth of Theorem 10.12 is proved. □

Remark 10.4 In a game for which all the players have the strategies equivalent, the set of Pareto–Nash equilibria is partitioned into the maximal possible number $(2^{m_1}-1)(2^{m_2}-1)\dots(2^{m_n}-1)$ of components.

Remark 10.5 Generally, the components $X(i_1 I_1 \dots i_n I_n)(\boldsymbol{\lambda})$ are non-convex in an n-player game if $n \geq 3$.

Algorithm 10.6.1 requires to solve $(2^{m_1}-1)(2^{m_2}-1)\dots(2^{m_n}-1)$ finite systems of multi-linear ($n-1$-linear) and linear equations and inequalities in m variables. The last problem is itself a difficult one.

Example 10.4 Consider a three-player mixed-strategy two-criterion $2\times 2\times 2$ game with matrices:

$$a_{1**} = \begin{bmatrix} 9,6 & 0,0 \\ 0,0 & 3,2 \end{bmatrix},\ a_{2**} = \begin{bmatrix} 0,0 & 3,4 \\ 9,3 & 0,0 \end{bmatrix},$$

$$b_{*1*} = \begin{bmatrix} 8,3 & 0,0 \\ 0,0 & 4,6 \end{bmatrix},\ b_{*2*} = \begin{bmatrix} 0,0 & 4,3 \\ 8,6 & 0,0 \end{bmatrix},$$

$$c_{**1} = \begin{bmatrix} 12,6 & 0,0 \\ 0,0 & 2,4 \end{bmatrix},\ c_{**2} = \begin{bmatrix} 0,0 & 6,6 \\ 4,2 & 0,0 \end{bmatrix}.$$

The payoff functions in mixed-strategy games are:

$$F_1(\mathbf{x}, \mathbf{y}, \mathbf{z}, \lambda^1) = ((9\lambda_1^1 + 6\lambda_2^1)y_1z_1 + (3\lambda_1^1 + 2\lambda_2^1)y_2z_2)x_1 + \\ +((9\lambda_1^1 + 3\lambda_2^1)y_2z_1 + (3\lambda_1^1 + 4\lambda_2^1)y_1z_2)x_2,$$

$$F_2(\mathbf{x}, \mathbf{y}, \mathbf{z}, \lambda^2) = ((8\lambda_1^2 + 3\lambda_2^2)x_1z_1 + (4\lambda_1^2 + 6\lambda_2^2)x_2z_2)y_1 + \\ +((8\lambda_1^2 + 6\lambda_2^2)x_2z_1 + (4\lambda_1^2 + 3\lambda_2^2)x_1z_2)y_2,$$

$$F_3(\mathbf{x}, \mathbf{y}, \mathbf{z}, \lambda^3) = ((12\lambda_1^3 + 6\lambda_2^3)x_1y_1 + (2\lambda_1^3 + 4\lambda_2^3)x_2y_2)z_1 + \\ +((4\lambda_1^3 + 2\lambda_2^3)x_2y_1 + (6\lambda_1^3 + 6\lambda_2^3)x_1y_2)z_2.$$

By applying substitutions:

$\lambda_1^1 = \lambda_1 > 0$ and $\lambda_2^1 = 1 - \lambda_1 > 0$,
$\lambda_1^2 = \lambda_2 > 0$ and $\lambda_2^2 = 1 - \lambda_2 > 0$,
$\lambda_1^3 = \lambda_3 > 0$ and $\lambda_2^3 = 1 - \lambda_3 > 0$,

we obtain a simplified game with the following payoff functions:

$$F_1(\mathbf{x}, \mathbf{y}, \mathbf{z}, \lambda_1) = ((6 + 3\lambda_1)y_1z_1 + (2 + \lambda_1)y_2z_2)x_1 + \\ +((3 + 6\lambda_1)y_2z_1 + (4 - \lambda_1)y_1z_2)x_2,$$

$$F_2(\mathbf{x}, \mathbf{y}, \mathbf{z}, \lambda_2) = ((3 + 5\lambda_2)x_1z_1 + (6 - 2\lambda_2)x_2z_2)y_1 + \\ +((6 + 2\lambda_2)x_2z_1 + (3 + \lambda_2)x_1z_2)y_2,$$

$$F_3(\mathbf{x}, \mathbf{y}, \mathbf{z}, \lambda_3) = ((6 + 6\lambda_3)x_1y_1 + (4 - 2\lambda_3)x_2y_2)z_1 + \\ +((2 + 2\lambda_3)x_2y_1 + 6x_1y_2)z_2.$$

Totally, we have to consider $(2^2 - 1)(2^2 - 1)(2^2 - 1) = 27$ components. For simplicity, we will enumerate only non-empty components.

Thus, the component

$$\mathbf{PNES}(1\emptyset1\emptyset1\emptyset)(\boldsymbol{\lambda}) = (1, 0) \times (1, 0) \times (1, 0)$$

with the payoffs (9, 6), (8, 3), (12, 6), is the solution of the system:

$$\begin{cases} (3 + 6\lambda_1)y_2z_1 + (4 - \lambda_1)y_1z_2 - (6 + 3\lambda_1)y_1z_1 - (2 + \lambda_1)y_2z_2 \le 0, \\ (6 + 2\lambda_2)x_2z_1 + (3 + \lambda_2)x_1z_2 - (3 + 5\lambda_2)x_1z_1 - (6 - 2\lambda_2)x_2z_2 \le 0, \\ (2 + 2\lambda_3)x_2y_1 + 6x_1y_2 - (6 + 6\lambda_3)x_1y_1 - (4 - 2\lambda_3)x_2y_2 \le 0, \\ \lambda_1,\ \lambda_2,\ \lambda_3 \in (0,\ 1), \\ x_1 + x_2 = 1, x_1 \ge 0, x_2 = 0, \\ y_1 + y_2 = 1, y_1 \ge 0, y_2 = 0, \\ z_1 + z_2 = 1, z_1 \ge 0, z_2 = 0. \end{cases}$$

The component

$$\mathbf{PNES(1\emptyset1\{2\}1\{2\})}(\lambda) = (1,0) \times \begin{pmatrix} \frac{1}{3} \le y_1 \le 1 \\ 1 - y_1 \end{pmatrix} \times \begin{pmatrix} \frac{1}{3} \le z_1 \le \frac{2}{5} \\ 1 - z_1 \end{pmatrix}$$

is the solution of the system:

$$\begin{cases} (3+6\lambda_1)y_2z_1 + (4-\lambda_1)y_1z_2 - (6+3\lambda_1)y_1z_1 - (2+\lambda_1)y_2z_2 \le 0, \\ (6+2\lambda_2)x_2z_1 + (3+\lambda_2)x_1z_2 - (3+5\lambda_2)x_1z_1 - (6-2\lambda_2)x_2z_2 = 0, \\ (2+2\lambda_3)x_2y_1 + 6x_1y_2 - (6+6\lambda_3)x_1y_1 - (4-2\lambda_3)x_2y_2 = 0, \\ \lambda_1,\ \lambda_2,\ \lambda_3 \in (0,\ 1), \\ x_1 + x_2 = 1, x_1 \ge 0, x_2 = 0, \\ y_1 + y_2 = 1, y_1 \ge 0, y_2 \ge 0, \\ z_1 + z_2 = 1, z_1 \ge 0, z_2 \ge 0. \end{cases}$$

The component $\mathbf{PNES(1\emptyset2\emptyset2\emptyset)}(\boldsymbol{\lambda}) = (1,0) \times (0,1) \times (0,1)$, with the payoffs $(3,2)$, $(4,3)$, $(6,6)$, is the solution of the system:

$$\begin{cases} (3+6\lambda_1)y_2z_1 + (4-\lambda_1)y_1z_2 - (6+3\lambda_1)y_1z_1 - (2+\lambda_1)y_2z_2 \le 0, \\ -(6+2\lambda_2)x_2z_1 - (3+\lambda_2)x_1z_2 + (3+5\lambda_2)x_1z_1 + (6-2\lambda_2)x_2z_2 \le 0, \\ -(2+2\lambda_3)x_2y_1 - 6x_1y_2 + (6+6\lambda_3)x_1y_1 + (4-2\lambda_3)x_2y_2 \le 0, \\ \lambda_1,\ \lambda_2,\ \lambda_3 \in (0,\ 1), \\ x_1 + x_2 = 1, x_1 \ge 0, x_2 = 0, \\ y_1 + y_2 = 1, y_1 = 0, y_2 \ge 0, \\ z_1 + z_2 = 1, z_1 = 0, z_2 \ge 0. \end{cases}$$

The component

$$\mathbf{PNES(1\{2\}1\emptyset1\{2\})}(\boldsymbol{\lambda}) = \begin{pmatrix} \frac{1}{4} \le x_1 \le 1 \\ 1 - x_1 \end{pmatrix} \times \begin{pmatrix} 1 \\ 0 \end{pmatrix} \times \begin{pmatrix} \frac{1}{3} \le z_1 \le \frac{2}{5} \\ 1 - z_1 \end{pmatrix}$$

is the solution of the system:

$$\begin{cases} (3+6\lambda_1)y_2z_1 + (4-\lambda_1)y_1z_2 - (6+3\lambda_1)y_1z_1 - (2+\lambda_1)y_2z_2 = 0, \\ (6+2\lambda_2)x_2z_1 + (3+\lambda_2)x_1z_2 - (3+5\lambda_2)x_1z_1 - (6-2\lambda_2)x_2z_2 \le 0, \\ (2+2\lambda_3)x_2y_1 + 6x_1y_2 - (6+6\lambda_3)x_1y_1 - (4-2\lambda_3)x_2y_2 = 0, \\ \lambda_1,\ \lambda_2,\ \lambda_3 \in (0,\ 1), \\ x_1 + x_2 = 1, x_1 \ge 0, x_2 \ge 0, \\ y_1 + y_2 = 1, y_1 \ge 0, y_2 = 0, \\ z_1 + z_2 = 1, z_1 \ge 0, z_2 \ge 0. \end{cases}$$

The component

$$\mathbf{PNES(1\{2\}1\{2\}1\emptyset)(\lambda)} = \begin{pmatrix} \frac{1}{2} \le x_1 \le \frac{2}{3} \\ 1 - x_1 \end{pmatrix} \times \begin{pmatrix} \frac{1}{3} \le y_1 \le \frac{1}{2} \\ 1 - y_1 \end{pmatrix} \times (1, 0)$$

is the solution of the system:

$$\begin{cases} (3 + 6\lambda_1)y_2z_1 + (4 - \lambda_1)y_1z_2 - (6 + 3\lambda_1)y_1z_1 - (2 + \lambda_1)y_2z_2 = 0, \\ (6 + 2\lambda_2)x_2z_1 + (3 + \lambda_2)x_1z_2 - (3 + 5\lambda_2)x_1z_1 - (6 - 2\lambda_2)x_2z_2 = 0, \\ (2 + 2\lambda_3)x_2y_1 + 6x_1y_2 - (6 + 6\lambda_3)x_1y_1 - (4 - 2\lambda_3)x_2y_2 \le 0, \\ \lambda_1,\ \lambda_2,\ \lambda_3 \in (0,\ 1), \\ x_1 + x_2 = 1, x_1 \ge 0, x_2 \ge 0, \\ y_1 + y_2 = 1, y_1 \ge 0, y_2 \ge 0, \\ z_1 + z_2 = 1, z_1 \ge 0, z_2 = 0. \end{cases}$$

The component

$$\mathbf{PNES(1\{2\}1\{2\}1\{2\})(\lambda)} =$$

$$= \left\{ \begin{pmatrix} \frac{1}{2} \le x_1 \le 1 \\ 1 - x_1 \end{pmatrix} \times \begin{pmatrix} \frac{1}{3} \le y_1 \le \frac{1}{2} \\ 1 - y_1 \end{pmatrix} \times \begin{pmatrix} \frac{1}{3} \le z_1 \le \frac{2}{5} \\ 1 - z_1 \end{pmatrix} \right\} \bigcup$$

$$\bigcup \left\{ \begin{pmatrix} \frac{1}{4} \le x_1 \le \frac{2}{5} \\ 1 - x_1 \end{pmatrix} \times \begin{pmatrix} 0 \le y_1 \le 1 \\ 1 - y_1 \end{pmatrix} \times \begin{pmatrix} \frac{1}{3} \le z_1 \le \frac{2}{5} \\ 1 - z_1 \end{pmatrix} \right\} \bigcup$$

$$\bigcup \left\{ \begin{pmatrix} 0 \le x_1 \le \frac{1}{4} \\ 1 - x_1 \end{pmatrix} \times \begin{pmatrix} 0 \le y_1 \le \frac{5x_1 - 2}{9x_1 - 3} \\ 1 - y_1 \end{pmatrix} \times \begin{pmatrix} \frac{1}{3} \le z_1 \le \frac{2}{5} \\ 1 - z_1 \end{pmatrix} \right\} \bigcup$$

$$\bigcup \left\{ \begin{pmatrix} \frac{2}{5} \le x_1 \le \frac{1}{2} \\ 1 - x_1 \end{pmatrix} \times \begin{pmatrix} \frac{5x_1 - 2}{9x_1 - 3} \le y_1 \le 1 \\ 1 - y_1 \end{pmatrix} \times \begin{pmatrix} \frac{1}{3} \le z_1 \le \frac{2}{5} \\ 1 - z_1 \end{pmatrix} \right\}$$

is the solution of the system:

$$\begin{cases} (3 + 6\lambda_1)y_2z_1 + (4 - \lambda_1)y_1z_2 - (6 + 3\lambda_1)y_1z_1 - (2 + \lambda_1)y_2z_2 = 0, \\ (6 + 2\lambda_2)x_2z_1 + (3 + \lambda_2)x_1z_2 - (3 + 5\lambda_2)x_1z_1 - (6 - 2\lambda_2)x_2z_2 = 0, \\ (2 + 2\lambda_3)x_2y_1 + 6x_1y_2 - (6 + 6\lambda_3)x_1y_1 - (4 - 2\lambda_3)x_2y_2 = 0, \\ \lambda_1,\ \lambda_2,\ \lambda_3 \in (0,\ 1), \\ x_1 + x_2 = 1, x_1 \ge 0, x_2 \ge 0, \\ y_1 + y_2 = 1, y_1 \ge 0, y_2 \ge 0, \\ z_1 + z_2 = 1, z_1 \ge 0, z_2 \ge 0. \end{cases}$$

The component

$$\mathbf{PNES(1\{2\}1\{2\}2\emptyset)(\lambda)} = \begin{pmatrix} \frac{1}{2} \le x_1 \le \frac{2}{3} \\ 1 - x_1 \end{pmatrix} \times \begin{pmatrix} \frac{1}{3} \le y_1 \le \frac{1}{2} \\ 1 - y_1 \end{pmatrix} \times \begin{pmatrix} 0 \\ 1 \end{pmatrix}$$

is the solution of the system:

$$\begin{cases} (3+6\lambda_1)y_2z_1 + (4-\lambda_1)y_1z_2 - (6+3\lambda_1)y_1z_1 - (2+\lambda_1)y_2z_2 = 0, \\ (6+2\lambda_2)x_2z_1 + (3+\lambda_2)x_1z_2 - (3+5\lambda_2)x_1z_1 - (6-2\lambda_2)x_2z_2 = 0, \\ -(2+2\lambda_3)x_2y_1 - 6x_1y_2 + (6+6\lambda_3)x_1y_1 + (4-2\lambda_3)x_2y_2 \le 0, \\ \lambda_1, \lambda_2, \lambda_3 \in (0, 1), \\ x_1 + x_2 = 1, x_1 \ge 0, x_2 \ge 0, \\ y_1 + y_2 = 1, y_1 \ge 0, y_2 \ge 0, \\ z_1 + z_2 = 1, z_1 = 0, z_2 \ge 0. \end{cases}$$

The component

$$\mathbf{PNES(1\{2\}2\emptyset 1\{2\})(\boldsymbol{\lambda})} = \begin{pmatrix} 0 \le x_1 \le \frac{2}{5} \\ 1 - x_1 \end{pmatrix} \times \begin{pmatrix} 0 \\ 1 \end{pmatrix} \times \begin{pmatrix} \frac{1}{3} \le z_1 \le \frac{2}{5} \\ 1 - z_1 \end{pmatrix}$$

is the solution of the system:

$$\begin{cases} (3+6\lambda_1)y_2z_1 + (4-\lambda_1)y_1z_2 - (6+3\lambda_1)y_1z_1 - (2+\lambda_1)y_2z_2 = 0, \\ -(6+2\lambda_2)x_2z_1 - (3+\lambda_2)x_1z_2 + (3+5\lambda_2)x_1z_1 + (6-2\lambda_2)x_2z_2 \le 0, \\ (2+2\lambda_3)x_2y_1 + 6x_1y_2 - (6+6\lambda_3)x_1y_1 - (4-2\lambda_3)x_2y_2 = 0, \\ \lambda_1, \lambda_2, \lambda_3 \in (0, 1), \\ x_1 + x_2 = 1, x_1 \ge 0, x_2 \ge 0, \\ y_1 + y_2 = 1, y_1 = 0, y_2 \ge 0, \\ z_1 + z_2 = 1, z_1 \ge 0, z_2 \ge 0. \end{cases}$$

The component

$$\mathbf{PNE(2\emptyset 1\emptyset 2\emptyset)(\boldsymbol{\lambda})} = (0, 1) \times (1, 0) \times (0, 1)$$

with the payoffs (3, 4), (4, 6), (4, 2), is the solution of the system:

$$\begin{cases} -(3+6\lambda_1)y_2z_1 - (4-\lambda_1)y_1z_2 + (6+3\lambda_1)y_1z_1 + (2+\lambda_1)y_2z_2 \le 0, \\ (6+2\lambda_2)x_2z_1 + (3+\lambda_2)x_1z_2 - (3+5\lambda_2)x_1z_1 - (6-2\lambda_2)x_2z_2 \le 0, \\ -(2+2\lambda_3)x_2y_1 - 6x_1y_2 + (6+6\lambda_3)x_1y_1 + (4-2\lambda_3)x_2y_2 \le 0, \\ \lambda_1, \lambda_2, \lambda_3 \in (0, 1), \\ x_1 + x_2 = 1, x_1 = 0, x_2 \ge 0, \\ y_1 + y_2 = 1, y_1 \ge 0, y_2 = 0, \\ z_1 + z_2 = 1, z_1 = 0, z_2 \ge 0. \end{cases}$$

The component

$$\mathbf{PNES(2\emptyset 1\{2\}1\{2\})(\boldsymbol{\lambda})} = \begin{pmatrix} 0 \\ 1 \end{pmatrix} \times \begin{pmatrix} 0 \le y_1 \le \frac{2}{3} \\ 1 - y_1 \end{pmatrix} \times \begin{pmatrix} \frac{1}{3} \le z_1 \le \frac{2}{5} \\ 1 - z_1 \end{pmatrix}$$

is the solution of the system:

$$\begin{cases} -(3+6\lambda_1)y_2z_1 - (4-\lambda_1)y_1z_2 + (6+3\lambda_1)y_1z_1 + (2+\lambda_1)y_2z_2 \le 0, \\ (6+2\lambda_2)x_2z_1 + (3+\lambda_2)x_1z_2 - (3+5\lambda_2)x_1z_1 - (6-2\lambda_2)x_2z_2 = 0, \\ (2+2\lambda_3)x_2y_1 + 6x_1y_2 - (6+6\lambda_3)x_1y_1 - (4-2\lambda_3)x_2y_2 = 0, \\ \lambda_1,\ \lambda_2,\ \lambda_3 \in (0,\ 1), \\ x_1 + x_2 = 1, x_1 = 0, x_2 \ge 0, \\ y_1 + y_2 = 1, y_1 \ge 0, y_2 \ge 0, \\ z_1 + z_2 = 1, z_1 \ge 0, z_2 \ge 0. \end{cases}$$

The component

$$\mathbf{PNES}(\mathbf{2}\emptyset\mathbf{2}\emptyset\mathbf{1}\emptyset)(\boldsymbol{\lambda}) = (0, 1) \times (0, 1) \times (1, 0)$$

with the payoffs (9, 3), (8, 6), (2, 4), is the solution of the system:

$$\begin{cases} -(3+6\lambda_1)y_2z_1 - (4-\lambda_1)y_1z_2 + (6+3\lambda_1)y_1z_1 + (2+\lambda_1)y_2z_2 \le 0, \\ -(6+2\lambda_2)x_2z_1 - (3+\lambda_2)x_1z_2 + (3+5\lambda_2)x_1z_1 + (6-2\lambda_2)x_2z_2 \le 0, \\ (2+2\lambda_3)x_2y_1 + 6x_1y_2 - (6+6\lambda_3)x_1y_1 - (4-2\lambda_3)x_2y_2 \le 0, \\ \lambda_1,\ \lambda_2,\ \lambda_3 \in (0,\ 1), \\ x_1 + x_2 = 1, x_1 = 0, x_2 \ge 0, \\ y_1 + y_2 = 1, y_1 = 0, y_2 \ge 0, \\ z_1 + z_2 = 1, z_1 \ge 0, z_2 = 0. \end{cases}$$

As a consequence, the set of Pareto–Nash equilibria consists of eleven components enumerated above.

10.7 Conclusions

The idea to consider the set of Pareto–Nash equilibria as an intersection of the graphs of efficient response mappings yields to a method of its computing, an extension of the method proposed initially in [3] for a Nash equilibrium set computing (see Chap. 3). Taking into account the computational complexity of the problem, proposed exponential algorithms are pertinent.

The set of Pareto–Nash equilibria in bimatrix mixed-strategy games may be partitioned into a finite number of polytopes, no more then $(2^m - 1)(2^n - 1)$. The proposed algorithm examines, generally, a much more small number of sets of the type $X_{iI}^{jJ}(\boldsymbol{\lambda}^2) \times Y_{jJ}^{iI}(\boldsymbol{\lambda}^1)$.

The set of Pareto–Nash equilibria in polymatrix mixed-strategy games may be partitioned into finite number of components, no more than $(2^{m_1} - 1) \ldots (2^{m_n} - 1)$, but they, generally, are non-convex and moreover non-polytopes. The algorithmic realization of the method is closely related with the problem of solving the systems of multi-linear ($n - 1$-linear and simply linear) equations and inequalities, that itself represents a serious obstacle to efficient computing of the set of Pareto–Nash equilibria.

References

1. Ungureanu, V., and A. Botnari. 2005. Nash equilibrium sets in mixed extension of $2 \times 2 \times 2$ games. *Computer Science Journal of Moldova* 13 (1(37)): 13–28.
2. Ungureanu, V., and A. Botnari. 2005. Nash equilibrium sets in mixed extended 2×3 games. *Computer Science Journal of Moldova* 13 (2(38)): 136–150.
3. Ungureanu, V. 2006. Nash equilibrium set computing in finite extended games. *Computer Science Journal of Moldova* 14 (3(42)): 345–365.
4. Ungureanu, V. 2008. Solution principles for simultaneous and sequential games mixture. *ROMAI Journal* 4 (1): 225–242.
5. Lozan, V., and V. Ungureanu. 2012. The Set of Pareto-nash Equilibria in Multi-criteria Strategic Games. *Computer Science Journal of Moldova* 20 (1(58)): 3–15.
6. Pareto, V. 1904. *Manuel d'economie politique*, 504 pp. Paris: Giard. (in French).
7. Zeleny, M. 1974. *Linear Multiobjective Programming*, X+222 pp. Berlin: Springer.
8. Cohon, J.L. 1978. *Multiobjective Programming and Planning*, XIV+333 pp. New York: Academic Press.
9. Jahn, J. 2004. *Vector Optimization: Theory, Applications and Extensions*, XV+481 pp., Operation Research Berlin: Springer.
10. Ehrgott, M. 2005. *Multicriteria optimization*, 328 pp. Berlin: Springer.
11. Blackwell, D. 1956. An analog of the minimax theorem for vector payoffs. *Pacific Journal of Mathematics* 6: 1–8.
12. Borm, P., S. Tijs, and J. Van der Aarssen. 1988. Pareto equilibria in multiobjective games. *Methods of Operations Research* 60: 302–312.
13. Shapley, L.S. 1959. Equilibrium Points in Games with Vector Payoffs. *Naval Research Logistics Quarterly* 6: 57–61.
14. Hermelin, D., C.C. Huang, S. Kratsch, and M. Wahlström. 2013. Parameterized two-player nash equilibrium. *Algorithmica* 65 (4): 802–816.
15. Daskalakis, C., P.W. Goldberg, and C.H. Papadimitriou. 2009. The complexity of computing a Nash equilibrium. *Communications of the ACM* 52 (2): 89–97.
16. Papadimitriou, C.H. 1994. On the Complexity of the Parity Argument and Other Inefficient Proofs of Existence. *Journal of Computer and System Sciences* 48 (3): 498–532.
17. Nash, J. 1951. Noncooperative games. *Annals of Mathematics* 54 (2): 280–295.
18. Von Stengel, B. 2002. Computing equilibria for two-person games. In *Handbook of Game Theory with Economic Applications*, ed. R.J. Aumann, and S. Hart, 1723–1759. Elsevier Science B.V: North-Holland.

Chapter 11
Sets of Pareto–Nash Equilibria in Dyadic Two-Criterion Mixed-Strategy Games

Abstract In this chapter, the notion of Pareto–Nash equilibrium is investigated as a continuation of the precedent chapter as well as a continuation of prior works (Sagaidac and Ungureanu, Operational research, CEP USM, Chişinău, 296 pp, 2004 (in Romanian), [1]; Ungureanu, Comp Sci J Moldova, 14(3(42)):345–365, 2006, [2]; Ungureanu, ROMAI J, 4(1):225–242, 2008, [3]). First, problems and needed basic theoretical results are exposed. The method of intersection of graphs of best response mappings presented above and initiated in Ungureanu (Comp Sci J Moldova, 14(3(42)):345–365, 2006, [2]) is applied to solve dyadic two-criterion mixed-strategy games. To avoid misunderstanding, some previous results, which are applied in this chapter, are briefly exposed, too.

11.1 Introduction

Consider a noncooperative strategic form game:

$$\Gamma = \langle \mathbf{N},\ \{\mathbf{X_p}\}_{p\in\mathbf{N}},\ \{f_p^i(\mathbf{x})\}_{i=1}^{m_p},\ p \in \mathbf{N}\rangle,$$

where

- $\mathbf{N} = \{1, 2, \ldots, n\}$ is a set of players;
- $\mathbf{X_p} \in \mathbf{R}^{k_p}$ is a set of strategies of the player $p \in \mathbf{N}$;
- $k_p < +\infty$, $p \in \mathbf{N}$;
- and $\{f_p^i(\mathbf{x})\}_{i=1}^{m_p}$ are the pth player cost/payoff functions defined on the Cartesian product $\mathbf{X} = \underset{p\in\mathbf{N}}{\times} \mathbf{X_p}$.

Remark that each player has to solve solitary the multi-criteria parametric optimization problem in which the parameters are strategic choices of the others players.

Definition 11.1 The strategy x'_p is "better" than the strategy x''_p, if

$$\{f_p^i(x'_p, x_{-p})\}_{i=1}^{m_p} \geq \{f_p^i(x''_p, x_{-p})\}_{i=1}^{m_p},$$

V. Ungureanu, *Pareto-Nash-Stackelberg Game and Control Theory*,
Smart Innovation, Systems and Technologies 89,
https://doi.org/10.1007/978-3-319-75151-1_11

for all $x_{-p} \in \mathbf{X}_{-p}$ and exist at least one index $j \in \{1, \ldots, m_p\}$ and a joint strategy $x_{-p} \in \mathbf{X}_{-p}$ for which

$$f_p^j(x'_p, x_{-p}) > f_p^j(x''_p, x_{-p});$$

the last relationship is denoted $x'_p \succeq x''_p$.

The problem of the player p. The player p selects from his set of strategies the strategy $x_p^* \in \mathbf{X}_p$, $p \in \mathbf{N}$, for which all of his cost functions $\{f_p^i(x_p, x_{-p}^*)\}_{i=1}^{m_p}$ achieve their maximum values.

11.2 Pareto Optimality

Definition 11.2 The strategy x_p^* is called efficient or optimal in the sense of Pareto if there is no another strategy $x_p \in \mathbf{X}_p$ such that $x_p \succeq x_p^*$.

Let us denote the set of efficient strategies of the player p by $\mathbf{ef}\,\mathbf{X_p}$. Any two efficient strategies are equivalent or incomparable.

A refinement of equilibria for multi-criteria games based on the perfectness concept of Selten [4] may be found in [5]. Other models of multi-criteria games may be considered too [6].

Theorem 11.1 *If the sets* $\mathbf{X_p} \in \mathbf{R}^{k_p}$, $p = 1, \ldots, n$, *are compact and the cost functions are continuous*

$$f_p^i(\mathbf{x}) \in C(\mathbf{X_p}),\ i = 1, \ldots, m_p,\ p = 1, \ldots, n,$$

then the sets $\mathbf{ef}\,\mathbf{X_p}$, $p = 1, \ldots, n$, *are non empty, i.e.* $\mathbf{ef}\,\mathbf{X} \neq \emptyset$.

The proof follows from the known results [1, 3, 7, 8].

Definition 11.3 Every element

$$\mathbf{x}^* = (x_1^*, x_2^*, \ldots, x_n^*) \in \mathbf{ef}\,\mathbf{X} = \underset{p \in \mathbf{N}}{\times}\ \mathbf{ef}\,\mathbf{X_p}$$

is called efficient or Pareto outcome/situation.

11.3 Synthesis Function

The solution of a multi-criteria problem may be found by means of the synthesis function which may be interpreted as a unique payoff function of the player p ($p = 1, \ldots, n$):

$$F_p(x) = \sum_{i=1,\dots,m_p} \lambda_i f_p^i(x_p, x_{-p}) \longrightarrow \max,$$

$$x_p \in \mathbf{X}_p,$$

$$\sum_{i=1,\dots,m_p} \lambda_i = 1, \lambda_i \geq 0, i = 1, \dots, m_p.$$

Theorem 11.2 *If x_p^* is the solution of single-objective problem*

$$F_p(x) = \sum_{i=\overline{1,m_p}} \lambda_i f_p^i(x_p, x_{-p}) \longrightarrow \max, x_p \in \mathbf{X}_p,$$

with

$$\lambda_i > 0,\ i = 1, \dots, m_p,\ \sum_{i=1,\dots,m_p} \lambda_i = 1,$$

then x_p^* is an efficient point for the given $x_{-p} \in \mathbf{X}_{-p}$.

Theorem's proof follows from the sufficient Pareto condition with linear synthesis function [1, 7, 8].

11.4 Pareto–Nash Equilibrium

Consider the convex game Γ for which sets of strategies are convex and payoff functions are concave in respect to respective player strategies when the strategies of the other players are fixed.

Definition 11.4 The point $\mathbf{x}^* = (x_1^*, x_2^*, \dots, x_n^*) \in \mathbf{X}$ is a Pareto–Nash equilibrium, if and only if for every player p the relations

$$F_p(x_p, x_{-p}^*) \leq F_p(x_p^*, x_{-p}^*) \equiv F_p(x^*),$$

are verified for all $x_p \in \mathbf{X}_p$.

As a corollary of the precedent two theorems follows

Theorem 11.3 *If the sets $\mathbf{X_p}$, $p = 1, \dots, n$, of the convex game Γ are compact and the functions $\{f_p^i(\mathbf{x})\}_{i=1}^{m_p}$ are continuous on $\underset{p \in \mathbf{N}}{\times} \mathbf{X_p}$, then the convex game Γ has a Pareto–Nash equilibrium.*

The proof follows from the known result [2].

The above definition may be formulated in other equivalent form:

Definition 11.5 The point $\mathbf{x}^* = (x_1^*, x_2^*, \ldots, x_n^*) \in \mathbf{X}$ is a Pareto–Nash equilibrium, if and only if

$$F(x^*) = \left(\max_{x_1 \in \mathbf{X}_1} F_1(x_1, x_{-1}^*), \ldots, \max_{x_n \in \mathbf{X}_n} F_n(x_n, x_{-n}^*) \right),$$

where $(x_p, x_{-p}^*) \equiv (x_1^*, x_2^*, \ldots, x_{p-1}^*, x_p, x_{p+1}^*, \ldots, x_n^*)$, $p = 1, \ldots, n$.

So, the Pareto–Nash equilibrium requires from each player to choose his own strategy as the Pareto best response to the strategies chosen by other players.

Let us denote the graph of the mapping

$$\operatorname*{Arg\,max}_{x_p \in \mathbf{X_p}} F_p(x_p, x_{-p}) : \mathbf{X_{-p}} \multimap \mathbf{X_p}$$

by

$$\mathbf{Gr_p} = \{(x_p, x_{-p}) \in \mathbf{X} : x_{-p} \in \mathbf{X_{-p}}, x_p = \operatorname*{arg\,max}_{y_p \in \mathbf{X_p}} F_p(y_p, x_{-p}).$$

In such notation, by [3], the set of Pareto–Nash equilibria is equal to

$$\mathbf{PNES} = \bigcap_{p=1,\ldots,n} \mathbf{Gr_p},$$

where

$$\mathbf{X_{-p}} = \mathop{\times}_{i \in \mathrm{N} \setminus \{p\}} \mathbf{X}_i, \; x_{-p} = (x_1, x_2, \ldots, x_{p-1}, x_{p+1}, \ldots, x_n).$$

Let us consider the following example as an illustration of the previous concepts and the method of Pareto–Nash equilibrium set computing.

Example 11.1 Consider the discrete game of two players. Each player has two strategies and two payoff functions. The players have to maximize the values of the both payoff functions.

The values of the payoff functions are associated with the elements of the following matrices:

$$A = \begin{pmatrix} 4,3 & 7,7 \\ 6,6 & 8,4 \end{pmatrix},$$

$$B = \begin{pmatrix} 5,-1 & 2,4 \\ 4,3 & 6,2 \end{pmatrix}.$$

First, we determine the sets of efficient strategies **ef X** and **ef Y**. Elements of **ef X** and **ef Y** are highlighted boxes.

$$A = \begin{pmatrix} 4,3 & \boxed{7,7} \\ 6,6 & \boxed{8,4} \end{pmatrix},$$

$$B = \begin{pmatrix} 5,-1 & \boxed{2,4} \\ \boxed{4,3} & \boxed{6,2} \end{pmatrix}.$$

The set of Pareto–Nash equilibria is

$$\mathbf{PNES} = \mathbf{ef\, X} \bigcap \mathbf{ef\, Y} = \{(1,2),(2,2)\}$$

with the payoffs $\{((7,7),(2,4)),((8,4),(6,2))\}$.

Theorem 11.4 *If in the convex game Γ the sets $\mathbf{X_p}$, $p = 1,\ldots,n$, are compact and the functions $F_p(x)$ are continuous on $\mathbf{X} = \underset{p\in\mathbf{N}}{\times} \mathbf{X_p}$, then the convex game Γ has a Pareto–Nash equilibrium.*

The proof follows from [1, 3, 9] and the previous theorems too.

11.5 Dyadic Two-Criterion Mixed-Strategy Games

Consider a dyadic two-criterion mixed-strategy game. The sets of strategies are

$$\mathbf{X} = \{(x_1,x_2) : x_1 + x_2 = 1, x_1 \geq 0, x_2 \geq 0\},$$

$$\mathbf{Y} = \{(y_1,y_2) : y_1 + y_2 = 1, y_1 \geq 0, y_2 \geq 0\}.$$

The payoff functions are bilinear (for a fixed opponent's strategy are linear):

$$f_1^1(\mathbf{x},\mathbf{y}) = \mathbf{x}^T A\mathbf{y},$$

$$f_1^2(\mathbf{x},\mathbf{y}) = \mathbf{x}^T B\mathbf{y},$$

$$f_2^1(\mathbf{x},\mathbf{y}) = \mathbf{x}^T C\mathbf{y},$$

$$f_2^2(\mathbf{x},\mathbf{y}) = \mathbf{x}^T D\mathbf{y},$$

where $\mathbf{x},\mathbf{y} \in \mathbf{R^2}$, $A, B, C, D \in \mathbf{R^{2\times 2}}$.

For each player consider his synthesis function:

$$F_1(x) = \lambda_1 f_1^1(x) + \lambda_2 f_1^2(x) \longrightarrow \max,$$

$$F_2(x) = \mu_1 f_2^1(x) + \mu_2 f_2^2(x) \longrightarrow \max.$$

By applying substitutions: $\lambda_1 = \lambda > 0$, and $\lambda_2 = 1-\lambda > 0$, $\mu_1 = \mu > 0$ and $\mu_2 = 1-\mu > 0$, we obtain the first equivalent game:

$$F_1(\mathbf{x}, \mathbf{y}) = \lambda f_1^1(\mathbf{x}, \mathbf{y}) + (1-\lambda) f_1^2(\mathbf{x}, \mathbf{y}) = \lambda \mathbf{x}^T A \mathbf{y} + (1-\lambda)\mathbf{x}^T B \mathbf{y},$$

$$F_2(\mathbf{x}, \mathbf{y}) = \mu f_2^1(\mathbf{x}, \mathbf{y}) + (1-\mu) f_2^2(\mathbf{x}, \mathbf{y}) = \mu \mathbf{x}^T C \mathbf{y} + (1-\mu)\mathbf{x}^T D \mathbf{y}.$$

By applying obvious transformations:

$$x_1 = x, x_2 = 1 - x, 1 \geq x \geq 0,$$

$$y_1 = y, y_2 = 1 - y, 1 \geq y \geq 0,$$

the second equivalent game is obtained:

$$F_1(x, y) = (\alpha(\lambda)y + \beta(\lambda))x + \alpha_0(\lambda)y + \beta_0(\lambda),$$

$$F_2(x, y) = (\gamma(\mu)x + \delta(\mu))y + \gamma_0(\mu)x + \delta_0(\mu),$$

$$x, y \in [0, 1], \lambda, \mu \in [0, 1],$$

where:

$$\alpha(\lambda) = (a_{11} - a_{12} - a_{21} + a_{22} - b_{11} + b_{12} + b_{21} - b_{22})\lambda + b_{11} - b_{12} - b_{21} + b_{22},$$

$$\beta(\lambda) = (a_{12} - a_{22} - b_{12} + b_{22})\lambda + b_{12} - b_{22},$$

$$\alpha_0(\lambda) = (a_{21} - a_{22} - b_{21} + b_{22})\lambda + b_{21} - b_{22},$$

$$\beta_0(\lambda) = (a_{22} - b_{22})\lambda + b_{22},$$

$$\gamma(\mu) = (c_{11} - c_{12} - c_{21} + c_{22} - d_{11} + d_{12} + d_{21} - d_{22})\mu + d_{11} - d_{12} - d_{21} + d_{22},$$

$$\delta(\mu) = (c_{21} - c_{22} - d_{21} + d_{22})\mu + d_{21} - d_{22},$$

$$\gamma_0(\mu) = (c_{12} - c_{22} - d_{12} + d_{22})\mu + d_{12} - d_{22},$$

$$\delta_0(\mu) = (c_{22} - d_{22})\mu + d_{22}.$$

The graphs of Pareto best response mappings are:

$$\mathbf{Gr_1} = \begin{cases} (1, y), & \text{if } \alpha(\lambda)y + \beta(\lambda) > 0, \\ (0, y) & \text{if } \alpha(\lambda)y + \beta(\lambda) < 0, \\ [0, 1] \times y, & \text{if } \alpha(\lambda)y + \beta(\lambda) = 0, \end{cases}$$

$$\mathbf{Gr_2} = \begin{cases} (x, 1), & \text{if } \gamma(\mu)x + \delta(\mu) > 0, \\ (x, 0), & \text{if } \gamma(\mu)x + \delta(\mu) < 0, \\ x \times [0, 1], & \text{if } \gamma(\mu)x + \delta(\mu) = 0. \end{cases}$$

The solutions of equations $\alpha(\lambda)y + \beta(\lambda) = 0$ and $\gamma(\mu)x + \delta(\mu) = 0$ are $y(\lambda) = -\frac{\beta(\lambda)}{\alpha(\lambda)}$ and $x(\mu) = -\frac{\delta(\mu)}{\gamma(\mu)}$. Vertical asymptotes of the respective hyperboles are determined from the relations $\alpha(\lambda) = 0$ and $\gamma(\mu) = 0$ and they are denoted by λ_α and μ_γ, respectively.

If the solution λ_α does not belong to the interval (0, 1), then y belongs to the interval with extremities $[y(0), y(1)]$. If the extremity value is negative, it is replaced by 0, if it is greater than 1, it is replaced by 1.

If λ_α belong to the interval (0, 1), then the graph $\mathbf{Gr_1}$ will be represented by two rectangles and an edge ([(0, 0), (0, 1)] or [(1, 0), (1, 1)]) of the square $[0, 1] \times [0, 1]$ or two edges and a rectangle of the square $[0, 1] \times [0, 1]$. Other alternatives graphs are possible in the case when λ_α does not belong to the interval (0, 1) and they are described bellow.

Similar reasoning is applied for μ_γ and the graph $\mathbf{Gr_2}$.

For the first player and his graph $\mathbf{Gr_1}$ the following cases are possible too:

1. If $\alpha(\lambda) > 0, \beta(\lambda) < 0, \alpha(\lambda) > -\beta(\lambda)$, then $\mathbf{Gr_1} = [(0, 0), (0, y(\lambda))] \bigcup [(0, y(\lambda)), (1, y(\lambda))] \bigcup [(1, y(\lambda)), (1, 1)]$;
2. If $\alpha(\lambda) < 0, \beta(\lambda) > 0, -\alpha(\lambda) > \beta(\lambda)$, then $\mathbf{Gr_1} = [(0, 1), (0, y(\lambda))] \bigcup [(0, y(\lambda)), (1, y(\lambda))] \bigcup [(1, y(\lambda)), (1, 0)]$;
3. If $\alpha(\lambda) > 0, \beta(\lambda) < 0, \alpha(\lambda) = -\beta(\lambda)$, then $\mathbf{Gr_1} = [(0, 0), (0, 1)] \bigcup [(0, 1), (1, 1)]$;
4. If $\alpha(\lambda) < 0, \beta(\lambda) > 0, -\alpha(\lambda) = \beta(\lambda)$, then $\mathbf{Gr_1} = [(0, 1), (1, 1)] \bigcup [(1, 1), (1, 0)]$;
5. If $\alpha(\lambda) > 0, \beta(\lambda) = 0$, then $\mathbf{Gr_1} = [(0, 0), (1, 0)] \bigcup [(1, 0), (1, 1)]$;
6. If $\alpha(\lambda) < 0, \beta(\lambda) = 0$, then $\mathbf{Gr_1} = [(0, 1), (0, 0)] \bigcup [(0, 0), (1, 0)]$;
7. If $\alpha(\lambda) > 0, \beta(\lambda) < 0, \alpha(\lambda) < -\beta(\lambda)$ or $\alpha(\lambda) < 0, \beta(\lambda) < 0$ or $\alpha(\lambda) = 0$, $\beta(\lambda) < 0$, then $\mathbf{Gr_1} = [(0, 0), (0, 1)]$;
8. If $\alpha(\lambda) < 0, \beta(\lambda) > 0, -\alpha(\lambda) < \beta(\lambda)$ or $\alpha(\lambda) > 0, \beta(\lambda) > 0$ or $\alpha(\lambda) = 0$, $\beta(\lambda) > 0$, then $\mathbf{Gr_1} = [(1, 0), (1, 1)]$;
9. If $\alpha(\lambda) = 0, \beta(\lambda) = 0$, then $\mathbf{Gr_1} = [0, 1] \times [0, 1]$.

For the second player the following cases are possible:

1. If $\gamma(\mu) > 0, \delta(\mu) < 0, \gamma(\mu) > -\delta(\mu)$, then $\mathbf{Gr_2} = [(0, 0), (x(\mu), 0)] \bigcup [(x(\mu), 0), (x(\mu), 1)] \bigcup [(x(\mu), 1), (1, 1)]$;
2. If $\gamma(\mu) < 0, \delta(\mu) > 0, -\gamma(\mu) > \delta(\mu)$, then $\mathbf{Gr_2} = [(0, 1), (x(\mu), 1)] \bigcup [(x(\mu), 1), (x(\mu), 0)] \bigcup [(x(\mu), 0), (1, 0)]$;
3. If $\gamma(\mu) > 0, \delta(\mu) < 0, \gamma(\mu) = -\delta(\mu)$, then $\mathbf{Gr_2} = [(0, 0), (1, 0)] \bigcup [(1, 0), (1, 1)]$;
4. If $\gamma(\mu) < 0, \delta(\mu) > 0, -\gamma(\mu) = \delta(\mu)$, then $\mathbf{Gr_2} = [(0, 1), (1, 1)] \bigcup [(1, 1), (1, 0)]$;

5. If $\gamma(\mu) > 0, \delta(\mu) = 0$, then $\mathbf{Gr_2} = [(0,0),(0,1)] \bigcup [(0,1),(1,1)]$;
6. If $\gamma(\mu) < 0, \delta(\mu) = 0$, then $\mathbf{Gr_2} = [(0,1),(0,0)] \bigcup [(0,0),(1,0)]$;
7. If $\gamma(\mu) > 0$, $\delta(\mu) < 0$, $\gamma(\mu) < -\delta(\mu)$ or $\gamma(\mu) < 0$, $\delta(\mu) < 0$ or $\gamma(\mu) = 0$, $\delta(\mu) < 0$, then $\mathbf{Gr_2} = [(0,0),(1,0)]$;
8. If $\gamma(\mu) < 0$, $\delta(\mu) > 0$, $-\gamma(\mu) < \delta(\mu)$ or $\gamma(\mu) > 0$, $\delta(\mu) > 0$ or $\gamma(\mu) = 0$, $\delta(\mu) > 0$, then $\mathbf{Gr_2} = [(0,1),(1,1)]$;
9. If $\gamma(\mu) = 0, \delta(\mu) = 0$, then $\mathbf{Gr_2} = [0,1] \times [0,1]$.

Remark 11.1 For the graph computing, the expressions $\alpha(0)$, $\beta(0)$, $y(0)$ and $\alpha(1)$, $\beta(1)$, $y(1)$ are calculated for the first player and $\gamma(0), \delta(0), x(0)$ and $\gamma(1), \delta(1), x(1)$ for the second player. When the player graph depends on only one of the matrix, it's constructed exactly as in the case of Nash equilibrium [1]. If the expressions $y(0)$ and $y(1)$ do not depend on parameter λ and $y(0) = y(1)$, the graph of first player will be all the square $[0,1] \times [0,1]$. Similar argument is true for the second player.

Based on the above, $\mathbf{Gr_1}$ and $\mathbf{Gr_2}$ may be constructed.

The set of Pareto–Nash equilibria (**PNE**) is obtained as the intersection of the player graphs, that is $\mathbf{PNES} = \mathbf{Gr_1} \bigcap \mathbf{Gr_2}$.

Example 11.2 Consider the following matrices:

$$(A, B) = \begin{pmatrix} 4,3 & 7,7 \\ 6,6 & 8,4 \end{pmatrix}; (C, D) = \begin{pmatrix} 5,-1 & 2,4 \\ 4,3 & 6,2 \end{pmatrix}.$$

After the simplifications, the synthesis functions of the players are:

$$F_1(x, y) = [(5\lambda - 6)y - 4\lambda + 3]x + (4\lambda + 10)y + 4\lambda + 4,$$
$$F_2(x, y) = [(11\mu - 6)x - 3\mu + 1]y + (2\mu + 6)x + 4\mu + 2.$$

In conformity with the described method, the following 4 steps are provided:

1. $\alpha(\lambda) = 5\lambda - 6, \beta(\lambda) = -4\lambda + 3, y(\lambda) = \frac{4\lambda-3}{5\lambda-6}; \gamma(\mu) = 11\mu - 6, \delta(\mu) = -3\mu + 1$, $x(\mu) = \frac{3\mu-1}{11\mu-6}$.
2. The values λ_α and μ_γ are the solutions of the equations $\alpha(\lambda) = 0$ and $\gamma(\mu) = 0$, respectively. $\lambda_\alpha = \frac{6}{5} \notin (0, 1)$ and $\mu_\gamma = \frac{6}{11} \in (0, 1)$.
3. The values on the interval extremities are calculated:

   a. $y(0) = \frac{1}{2}$, $\alpha(0) = -6 < 0$, $\beta(0) = 3 > 0$ and $-\alpha(0) > \beta(0)$ - the case 2; $y(1) = -1$, $\alpha(1) = -1 < 0$ and $\beta(1) = -1 < 0$ - the case 7. The lines are drawn and the interval between them is highlighted. The following result is obtained

$$\mathbf{Gr_1} = \mathbf{Rectangle}[(0,0), (0, y(0)), (1, y(0)), (1,0)] \bigcup$$

$$[(0,0), (0,1)]$$

where $y(0) = \frac{1}{2}$.

b. $x(0) = \frac{1}{6} \in (0,1)$, $\gamma(0) = -6 < 0$, $\delta(0) = 1 > 0$ and $-\gamma(0) > \delta(0)$ - the case 2; $x(1) = \frac{2}{5} \in (0,1)$, $\gamma(1) = 5 > 0$, $\delta(0) = -2 < 0$ and $\gamma(0) > -\delta(0)$ - the case 1. The respective lines are drawn and the interval between them is highlighted; the respective edges of the square $[0,1] \times [0,1]$ are highlighted, too. The following result is obtained

$$\mathbf{Gr_2} = \mathbf{Rectangle}[(0,0), (0,1), (x(0), 1), (x(0), 0)] \bigcup$$

$$\mathbf{Rectangle}[(1,0), (1,1), (x(0), 1), (x(0), 0)] \bigcup$$

$$[(0,0), (1,0)]$$

where $x(0) = \frac{1}{6}$ and $x(1) = \frac{2}{5}$.

4. By determining the intersection of the graphs obtained above, the following set of Pareto–Nash equilibria in mixed strategies is obtained:

$$\mathbf{PNES} = [(0,1), (0,0)] \bigcup$$

$$[(0,0), (1,0)] \bigcup$$

$$\mathbf{Rectangle}\left[(0,0), \left(0, \frac{1}{2}\right), \left(\frac{1}{6}, \frac{1}{2}\right), \left(\frac{1}{6}, 0\right)\right] \bigcup$$

$$\mathbf{Rectangle}\left[\left(\frac{2}{5}, 0\right), \left(\frac{2}{5}, \frac{1}{2}\right), \left(1, \frac{1}{2}\right), (1,0)\right].$$

11.6 The Wolfram Language Program

The method described above is realized as an Wolfram Language Program for the dyadic two-criterion mixed-strategy games. The program is published on the Wolfram Demonstration Project [10]. It may be used online, after the installation of CDF player. The program code may be downloaded at the same address [10]. The results obtained in Example 11.2 may be tested online at the same address [10].

11.7 Concluding Remarks

By applying the generalization of the well known notions and by applying the combination of the synthesis function method and the method of intersection of the best response graphs, the conditions for the Pareto–Nash solutions existence are highlighted with the aim to apply it in the case of the dyadic two-criterion mixed-strategy games.

The method for determining Pareto–Nash Equilibrium Set in dyadic two-criterion mixed-strategy games is exposed. The Wolfram Language Program is published on the Wolfram Demonstration Project [10].

The illustrative examples are presented.

Since, the investigated problems have an exponential complexity, a further development of the method for games with large dimensions is welcomed.

References

1. Sagaidac, M., and V. Ungureanu. 2004. *Operational Research*. Chişinău: CEP USM, 296 pp. (in Romanian).
2. Ungureanu, V. 2006. Nash equilibrium set computing in finite extended games. *Computer Science Journal of Moldova* 14 (3(42)): 345–365.
3. Ungureanu, V. 2008. Solution principles for simultaneous and sequential games mixture. *ROMAI Journal* 4 (1): 225–242.
4. Selten, R. 1975. Reexamination of the perfectness concept for equilibrium points in extensive games. *International Journal of Game Theory* 4 (1): 22–55.
5. Borm, P., F. Van Megen, and S. Tijs. 1999. A perfectness concept for multicriteria games. *Mathematical Methods of Operations Research* 49 (3): 401–412.
6. Yu, H. 2003. Weak Pareto equilibria for multiobjective constrained games. *Applied Mathematics Letters* 16: 773–776.
7. Podinovskii, V.V., and V.D. Nogin. 1982. *Pareto-optimal solutions of the multi-criteria problems*. Moscow: Nauka, 255 pp. (in Russian).
8. Jahn, J. 2004. *Vector Optimization: Theory, Applications and Extensions*, Series Operations Research. Berlin: Springer, XV+481 pp.
9. Lozan, V., and V. Ungureanu. 2009. Principles of Pareto–Nash equilibrium. *Studia Universitatis* 7: 52–56.
10. Lozan, V., and V. Ungureanu. 2011. Pareto–Nash equilibria in bicriterial dyadic games with mixed strategies, Wolfram Demonstrations Project. http://demonstrations.wolfram.com/ParetoNashEquilibriaInBicriterialDyadicGamesWithMixedStrateg/. Accessed 13 Oct 2011.

Chapter 12
Taxonomy of Strategic Games with Information Leaks and Corruption of Simultaneity

Abstract In this chapter, pseudo-simultaneous normal form games are considered, i.e. strategic games with rules violated by information leaks and the corruption of simultaneity by Ungureanu (Comput Sci J Moldova 24, 1: 83–105, 2016) [1]. A classification of such games is provided. The taxonomy is constructed based on applicable fundamental solution principles. Existence conditions are highlighted, formulated and analysed.

12.1 Introduction

Strategic or normal form game constitutes an abstract mathematical model of decision processes with two or more decision makers (players) [2, 3]. An important supposition of the game is that all the players choose their strategies simultaneously and confidentially, and that everyone determines his gain on the resulting profile. Reality is somewhat diverse. The rules of the games may be broken. Some players may cheat and know the choices of the other players. So, the rule of confidentiality and simultaneity is not respected. Is the essence of initial normal form game change in such games? Is still the Nash equilibrium principle applicable? Do we need other solution principles and other interpretations? How many types of games appear and may they be classified? Can we construct a taxonomy (classification) of these games? Answers to these and other related questions are the objective of this chapter.

Usually, the traditional research approach to games of such types relies on consideration of all possible players' best response mappings and analysis of all possible profiles [4] or to use an approach of the maximum guaranteed result as in hierarchical games with uncertain factors [5–9]. There is a stable opinion about a high complexity of their analysis and solving [4].

We initiate an approach which sets rules of all possible games with information leaks and highlights their specific characteristics [1]. The approach relies on knowledge vectors of the players and game knowledge net. A taxonomy (classification) of all possible games is done on the bases of the applicable solution principles. The name of every taxon (class) reflects the principle used for including respective games in the same taxon.

V. Ungureanu, *Pareto-Nash-Stackelberg Game and Control Theory*,
Smart Innovation, Systems and Technologies 89,
https://doi.org/10.1007/978-3-319-75151-1_12

As a result of the taxonomy construction and establishing strict characteristics and rules for any taxon, we reveals a simplicity of analysis and solving the games. It is an unexpectedly and impressive result.

For the beginning, let us recall that a Nash equilibrium sample and the entire Nash equilibrium set may be determined via intersection of the graphs of best response mappings—a method considered earlier in works [10–15]. The approach proves to be expedient as well for strategic games with information leaks and broken simultaneity [1]. Initially, we expose the results for bimatrix games with different levels of knowledge. Then, we expose results for the general polymatrix games. It is a useful approach both for simplicity of exposition, and for understanding the ideas and results.

12.1.1 Normal Form Game and Axioms

Consider finite strategic (normal form) game

$$\Gamma = \left\langle N, \{S_p\}_{p \in N}, \{a_{\mathbf{s}}^p = a_{s_1 s_2 \ldots s_n}^p\}_{p \in N} \right\rangle,$$

where

$N = \{1, 2, ..., n\}$ is a set of players,
$S_p = \{1, 2, \ldots, m_p\}$ is a set of strategies of the player $p \in N$,
$s_p \in S_p$ is a strategy of $p \in N$ player,
$\# S_p = m_p < +\infty$, $p \in N$, is a finiteness constraint,
$a_{\mathbf{s}}^p = a_{s_1 s_2 \ldots s_n}^p$ is a player's $p \in N$ pay-off function defined on the Cartesian product $S = \underset{p \in N}{\times} S_p$, i.e. for every player $p \in N$ an n dimensional pay-off matrix $A^p[m_1 \times m_2 \times \cdots \times m_n]$ is defined.

For the normal form game a system of axioms is stated.

Axiom 12.1.1 Rationality. The players behave rationally. The rationality means that every rational player optimizes the value of his pay-off function.

Axiom 12.1.2 Knowledge. The players know the set of players, the strategy sets and the payoff functions.

Axiom 12.1.3 Simultaneity. The players choose their strategies simultaneously and confidentially in a single-act (single stage) without knowing the chosen strategies of the others players.

Axiom 12.1.4 Payoff. After all strategy selection, the players compute their payoffs as the values of their payoff functions on the resulting profile.

It is clear, (simultaneous) Nash games [2, 3] are based on these four axioms.

In Stackelberg game, the axiom of simultaneity is replaced by the axiom of hierarchy.

Axiom 12.1.5 Hierarchy. The players chose their strategies in a known order, e.g. the first player (leader) chooses his strategy and communicates it to the second player. The second player (follower) knows the strategy of the leader, chooses his strategy and communicates it to the third player and so on. The last player knows all the strategies of the precedent players and chooses his strategy at the last.

Both the Nash game, and the Stackelberg game, are based commonly on the axioms of rationality, knowledge and payoff. Additionally and distinctive, the Nash game [3] is based on the axiom of simultaneity, while the Stackelberg game [16, 17] is based on the axiom of hierarchy.

Finally, we can deduce that in the Nash game all the players choose their strategies simultaneously and every player determines his payoff as the value of his payoff function on the resulting profile. But distinctively, in the Stackelberg game the players choose their strategies sequentially, in a known order and knowing the strategies chosen by the precedent players, and every player determines his gain as the value of his payoff function on the resulting profile.

12.1.2 Axiom of Simultaneity and Its Corruption

Both for Nash games, and for Stackelberg games, their own solution principles exist. If the axioms of the games are respected, these solution principles may be applied. Actually, the names of the games are chosen to reflect the solution concept which is applicable.

But, what does it happen when the axioms are violated by the corruption of some of their elements, e.g. some players may know chosen strategies of the other players in Nash games? May such games be examined by applying the same general solution principles (Nash and Stackeberg equilibria) or must new solution concepts be defined and applied?

To respond to these questions, we need to avoid the ambiguity. So, let us examine more exactly the process of decision making in conditions of information leaks, by establishing the axioms of the corrupt games. It is very convenient to consider in such case an abstract manager of the games—a person (or persons) which organizes and manages the decision process in the games. Thereby, we can describe exactly the process of decision making in corrupt games, knowing the source of corruption.[1]

At the **first pseudo-stage** of the decision process, the manager declares the players must choose their strategies. After players choose their strategies (intentions), the manager dishonestly from the viewpoint of the rules of the strategy game may submit to some players information about chosen strategies (corruption and information leaks). With additional information, some of the players may intend to change their strategies and they can do this at the second pseudo-stage.

[1]Corruption: "the abuse of entrusted power to private gain" (Transparency International); "dishonest or fraudulent conduct by those in power, typically involving bribery" (Oxford Dictionary, 2014).

At the **second pseudo-stage**, the manager declares that the players must submit immediately their choices. At this moment, the players may change their initial decisions. After possible changes in their intentions and choices, the players submit definitely their chosen strategies.

For an honest player, the decision process looks like an one stage process. Only for the dishonest manager and for the players which obtain additional information, the decision process looks like a two stage process.

As an axiom of the such game, the axiom of information leak may be stated as

Axiom 12.1.6 Information Leak. The decision process has two pseudo-stages.

At the **first pseudo-stage**, the information leak about player chosen strategies may occur.

At the **second pseudo-stage**, the players chose their strategies, some of them knowing eventually the strategies chosen by the other players.

Definition 12.1 A game with information leak or a corrupt game is a game for which four axioms are fulfil: rationality, knowledge, payoff and information leak.

Remark 12.1 Let us observe that three axioms of rationality, knowledge and payoff, are common for the Nash game, Stackelberg game and corrupt game (game with information leak).

Remark 12.2 Generally, the game with information leak is only a particular case of the corrupt game. We will use interchangeably this name unless we will define a more general context of a corrupt game.

Remark 12.3 The game with information leak, as it is defined above, actually includes different types of games.

To respond to the variate questions which appear in the context of corrupt games, we consider further the taxonomy (classification) of all possible types of games, principles of solutions, solution existence conditions and algorithms for solutions determining. The exposition will start with bimatrix games, but first of all, we must highlight shortly the essence of so called theory of moves [18] in this context.

12.1.3 Theory of Moves

We must observe that the games we consider in this work may be related to theory of moves [18]. Nevertheless, it is an important difference—we consider only the two pseudo-stages of the decision making process. While, theory of moves does not limit the number of moves to one fixed number. More the more, theory of moves has initial axioms which defined in a strict manner as the process of decision making, as the end condition. Additionally, those axioms differ from that accepted for games with information leaks.

Theory of moves is based, on the concepts of thinking ahead, stable outcomes, outcomes induced when one player has "*moving power*", incomplete information, non-myopic concept of equilibrium, etc. The non-myopic equilibrium depends on some parameters, such as, e.g., initial state from which the process of moving starts and who moves the first. It is essential that all the games have at least one non-myopic equilibrium. In the games we consider, there are different solution concepts and it is not guaranteed that the solutions exist. These reasons are exposed in continuation.

12.2 Taxonomy of Bimatrix Games with Information Leaks

For bimatrix strategic games, we suppose the process of decision making occurs in two pseudo-stages, on an account of possible information leaks. On the first pseudo-stage, the players choose their strategies and by corruption, it is possible either for one of them, or for both players, to know the chosen (intention) strategy of the opponent. On the second pseudo-stage, the players use obtained information, choose their strategies and no more corruption is possible.

First, we can distinguish **simultaneous** and **sequential** bimatrix games for such processes of decision making.

Second, simultaneous bimatrix games may obtain some features of sequential games taking into consideration obtained information/knowledge ($\gamma\nu\omega\sigma\eta$) possessed by each player in the process of realizing the game.

Remark 12.4 We suppose initially, when the players begin the strategy selections, they start playing a Nash game, but in the process of strategy selections information leak may occur, the Nash game may degenerate and may change its essence.

Remark 12.5 In order to distinguish players without their numbers, we will refer them as the player and his opponent. So, if the first player is referred simply as the player, then the second player is referred as the opponent, and vice versa.

12.2.1 Knowledge and Types of Games

The knowledge of the players is associated with their knowledge vectors γ^A and γ^B.

12.2.1.1 Knowledge Vectors

Essentially, the knowledge vectors have an infinite number of components $\gamma^A = (\gamma_0^A, \gamma_1^A, \dots)$ and $\gamma^B = (\gamma_0^B, \gamma_1^B, \dots)$, with components defined and interpreted as follows.

- **Player's knowledge of the components of the normal form.** γ_0^A and γ_0^B are reserved to knowledge about the normal form of the game. The values $\gamma_0^A = 1$ and $\gamma_0^B = 1$ mean that the players have full information about the strategy sets and payoff functions. It is the case we consider in this work, i.e. mutually $\gamma_0^A = 1$ and $\gamma_0^B = 1$, and these components of the knowledge vectors are simply omitted.
- **Player's knowledge of the opponent's chosen strategy.** $\gamma_1^A = 0$ and $\gamma_1^B = 0$ mean for each player, correspondingly, that he doesn't know the opponent's strategy. $\gamma_1^A = 1$ and $\gamma_1^B = 1$ mean for each player, correspondingly, that he knows the opponent's strategy. The combined cases are possible too.
- **Player's knowledge of the opponent's knowledge of the player's chosen strategy.** $\gamma_2^A = 0$ and $\gamma_2^B = 0$ mean for each player, correspondingly, that he knows that the opponent doesn't know player's strategy. $\gamma_2^A = 1$ and $\gamma_2^B = 1$ mean for each player, correspondingly, that he knows that the opponent knows player's strategy. Evidently, the combined cases are possible too. Remark that these components may be thought rather as the players beliefs, because such type of knowledge may be as true, as false. In this context it must be observed, that the values of γ_2^A and γ_2^B represent the knowledge/belief about the values of γ_1^B and γ_1^A, correspondingly.
- **The next components** $\gamma_3^A, \gamma_4^A, \ldots$ and $\gamma_3^B, \gamma_4^B, \ldots$ **of the knowledge vectors are omitted, initially.** Nevertheless, it must be remarked that the values of γ_i^A and γ_i^B represent the knowledge/belief about the values of γ_{i-1}^B and γ_{i-1}^A, correspondingly.

We distinguish the games with l levels of knowledge, for which all components of the knowledge vectors with indices greater than l are equal to 0.

Remark 12.6 Remark, once again, that there are two pseudo-level of decision making process. Information leaks may occur only on the first pseudo-level. The knowledge vectors may have any number $l \geq 1$ of components (levels of knowledge).

12.2.1.2 Types of Games

Depending on the values of the knowledge vectors, different types of games may be considered.

Proposition 12.1 *There are 4^l possible types of games $\Gamma_{\gamma^A\gamma^B}$ with l levels of knowledge.*

Proof It is enough to emphasize the components of the knowledge vectors $\gamma^A = (\gamma_1^A, \ldots, \gamma_l^A)$ and $\gamma^B = (\gamma_1^B, \ldots, \gamma_l^B)$ and their possible values as 0 or 1. Accordingly, there are 4^l possible pairs of such vectors, i.e. 4^l possible games. □

12.2.1.3 Knowledge Net

Knowledge net is defined as:

$$G = (V, E),$$

where $V = I \cup J \cup \gamma_A \cup \gamma_B$, $E \subseteq V \times V$.

For the moment, we will limit ourselves to knowledge vectors.

12.2.2 *Taxonomy Elements*

If the information leaks only on the first 2 levels, there are $4^2 = 16$ possible kinds of games with information leaks with $\gamma^A = \left(\gamma_1^A, \gamma_2^A\right)$ and $\gamma^B = \left(\gamma_1^B, \gamma_2^B\right)$, according to the above. From the solution principle perspective, some of them are similar and they may be included in common taxa (classes, families, sets).

Let us highlight the possible kinds of such games by the values of low index, where the first two digits are the values for knowledge vector components of the first player, and the following two digits are the values for knowledge vector components of the second player. We obtain the following taxonomy for two matrix games with information leaks and two levels of knowledge:

1. Nash taxon: $\mathrm{NT} = \{\Gamma_{00\,00}, \Gamma_{11\,11}, \Gamma_{00\,11}, \Gamma_{11\,00}\}$,
2. Stackelberg taxon: $\mathrm{ST} = \{\Gamma_{01\,10}, \Gamma_{01\,11}, \Gamma_{10\,01}, \Gamma_{11\,01}\}$,
3. Maximin taxon: $\mathrm{MT} = \{\Gamma_{01\,01}\}$,
4. Maximin-Nash taxon: $\mathrm{MNT} = \{\Gamma_{00\,01}, \Gamma_{01\,00}\}$,
5. Optimum taxon: $\mathrm{OT} = \{\Gamma_{10\,10}\}$,
6. Optimum-Nash taxon: $\mathrm{ONT} = \{\Gamma_{00\,10}, \Gamma_{10\,00}\}$,
7. Optimum-Stackelberg taxon: $\mathrm{OST} = \{\Gamma_{10\,11}, \Gamma_{11\,10}\}$.

The generic name of each taxon is selected on the base of the correspondent solution principles applied by the players: Nash equilibrium, Stackelberg equilibrium, Maximin principle, Optimum principle or two of them, together. Even though the taxon may include some games, the name reflects solution principles or applicable principles in all the games of the taxon. If the taxon is formed only by one element, its name is the same as for the game.

Remark 12.7 We choose the term taxon (plural—taxa) to name the set of related games in order to highlight additionally their acquired pseudo-dynamics [16, 17, 19] and to avoid confusion with mathematically overcharged or too used terms of class, cluster, family, group or set.

Let us investigate the solution principles for all these taxa.

12.3 Solution Principles for Bimatrix Games with Information Leak on Two Levels of Knowledge

Consider a bimatrix $m \times n$ game Γ with matrices

$$A = (a_{ij}),\ B = (b_{ij}),\ i \in I,\ j \in J,$$

where $I = \{1, 2, \dots, m\}$ is the set of strategies of the first player, and $J = \{1, 2, \dots, n\}$ is the set of strategies of the second player.

We consider the games based on the four axioms of rationality, knowledge, payoff and information leak (Axioms 12.1.1, 12.1.2, 12.1.4, 12.1.6). The players choose simultaneously their strategies and the information leak may occur before submitting the results off their selections. One or both of them may know the intention of the opponent. Let us suppose in such case, they may change only once their strategy accordingly to leaked information. So, the strategic games may be transformed by acquiring additional information into two stage games. At first stage, they choose strategies but do not submit them because of additional information acquiring. At the second stage, they may change initial strategies and submit definitely new strategies adjusted to obtained information accordingly to leaked information. After a such submission, the games end and both the players determine the values of their payoff functions.

Evidently, other types of games with information leaks may be considered. First, we will limit ourselves only to such two-pseudo-stage games with information leaks on two levels of knowledge.

For any taxon, we will firstly define it and after that we will argue its consistency.

12.3.1 Nash Taxon

Let us argue that all elements of NT $= \{\Gamma_{00\,00}, \Gamma_{11\,11}, \Gamma_{00\,11}, \Gamma_{11\,00}\}$ are Nash games, i.e. Axioms 12.1.1–12.1.4 are typical for these games and for them the Nash equilibrium principle may be applied as a common solution principle.

First, let us recall that the process of decision making in the Nash game, denoted by NΓ, is described in the following way. Simultaneously and confidentially, the first player selects the lines i^* of the matrices A and B, and the second player selects columns j^* of the same matrices. The first player gains $a_{i^*j^*}$, and the second player gains $b_{i^*j^*}$.

Evidently, $\Gamma_{00\,00}$ is a pure Nash games, i.e. $\Gamma_{00\,00} =$ N. But, it is not difficult to understand that $\Gamma_{11\,11}$, $\Gamma_{00\,11}$, $\Gamma_{11\,00}$, are Nash games too. So, the taxon (group) is formed by the four Nash games, differing only by the knowledge/belief of the players.

If we call the player which apply a Nash equilibrium strategy as an atom (a Nash atom, Nash atomic player) and denote him as N, then the two-player Nash game may be denoted as N_2 (Nash game, Nash molecular game).

Remark 12.8 We will call and denote in the same manner other types of players and games.

12.3.1.1 Nash Equilibrium

The pair of strategies (i^*, j^*) forms a Nash equilibrium if

$$a_{i^*j^*} \geq a_{ij^*}, \forall i \in I,$$
$$b_{i^*j^*} \geq b_{i^*j}, \forall j \in J.$$

12.3.1.2 Set of Nash Equilibria

An equivalent Nash equilibrium definition may be formulated in terms of graphs of best response (optimal reaction) applications (mappings).

Let

$$\mathrm{Gr_A} = \left\{(i, j) : j \in J,\ i \in \operatorname*{Argmax}_{k \in I} a_{kj}\right\},$$

be the graph of best response application of the first player, and

$$\mathrm{Gr_B} = \left\{(i, j) : i \in I,\ j \in \operatorname*{Argmax}_{k \in J} b_{ik}\right\}.$$

be the graph of best response application of the second player.

$$\mathrm{NE} = \mathrm{Gr_A} \cap \mathrm{Gr_B}$$

forms the set of Nash equilibria.

12.3.1.3 Nash Equilibrium Existence

Proposition 12.2 *There are Nash games which do not have a Nash equilibrium.*

Proof Examples of games which do not have a Nash equilibrium are commonly known. □

Remark, the games we consider are pure strategy games. It is a largely known result that every poly-matrix strategic game has Nash equilibria in mixed strategies. In this work we consider only pure strategy games.

12.3.2 *Stackelberg Taxon*

The Stackelberg Taxon is defined as

$$\mathrm{ST} = \{\Gamma_{01\,10}, \Gamma_{01\,11}, \Gamma_{10\,01}, \Gamma_{11\,01}\}.$$

To argue the inclusion of each element in the Stackelberg Taxon, let us recall the essence of the decision making process in the Stackelberg game.

A Stackelberg two-player game has two stages, from the beginning, and for the Stackeberg game the Axioms 12.1.1, 12.1.2, 12.1.4 and 12.1.5 are typical. At the first stage, the first player (leader) selects the lines i^* of the matrices A and B, and communicates his choice two the second player (follower). At the second stage, the second player (follower) knows the choice of the first player (leader) and selects the columns j^* of the matrices A and B. The first player gains $a_{i^*j^*}$, and the second player gains $b_{i^*j^*}$. If the players change their roles as the leader and the follower, an another Stackelberg game is defined.

The Stackelberg game is denoted by SG_{12} if the first player is the leader and by SG_{21} if the second player is the leader.

$\Gamma_{01\,10}$ is a pure Stackelberg game S_{12}, i.e. $\Gamma_{01\,10} = \mathrm{S}_{12}$, and Γ_{1001} is a pure Stackelberg game $\mathrm{S}\Gamma_{21}$, i.e. $\Gamma_{1001} = \mathrm{S}\Gamma_{21}$. It is clear that $\Gamma_{01\,11} = \mathrm{S}\Gamma_{12}$ and $\Gamma_{11\,01} = \mathrm{S}\Gamma_{21}$.

12.3.2.1 Stackelberg Equilibrium

The pair of strategies $(i^*, j^*) \in \mathrm{Gr_B}$ forms a Stackelberg equilibrium if

$$a_{i^*j^*} \geq a_{ij}, \forall (i, j) \in \mathrm{Gr_B}.$$

If the players change their roles and the second player is the leader, then the pair of strategies $(i^*, j^*) \in \mathrm{Gr_A}$ forms a Stackelberg equilibrium if

$$b_{i^*j^*} \geq b_{ij}, \forall (i, j) \in \mathrm{Gr_A}.$$

12.3.2.2 Set of Stackelberg Equilibria

The sets of Stackelberg equilibria are generally different for Stackelberg games $\mathrm{S}\Gamma_{12}$ and $\mathrm{S}\Gamma_{21}$.

$$\mathrm{SE}_{12} = \underset{(i,j)\in \mathrm{Gr_B}}{\mathrm{Argmax}}\ a_{ij}$$

forms the set of Stackelberg equilibria in a Stackelberg game $\mathrm{S}\Gamma_{12} = \mathrm{S}_{12}$.

$$\mathrm{SE}_{21} = \operatorname*{Argmax}_{(i,j)\in \mathrm{Gr_A}} b_{ij}$$

forms the set of Stackelberg equilibria in a Stackelberg game $S\Gamma_{21} = \mathrm{S}_{21}$.

It is evident that the notions of Nash and Stackeberg equilibria are not identical. The respective sets of equilibria may have common elements, but the sets generally differ.

12.3.2.3 Stackelberg Equilibrium Existence

Proposition 12.3 *Every finite Stackelberg game has a Stackelberg equilibrium.*

Proof The proof follows from the Stackelberg equilibrium definition and the finiteness of the player strategy sets. □

12.3.3 Maximin Taxon

Maximin Taxon contains only one element $\mathrm{MT} = \{\Gamma_{01\,01}\}$.

The decision making process in the Maximin game $\mathrm{M} = \mathrm{M}_2$ respects the Axioms 12.1.1–12.1.4 as for the Nash Game. Simultaneously and secretly, as in the Nash game, the first player selects the lines i^* of the matrices A and B, and the second player selects the columns j^* of the same matrices. Unlike the Nash Game, every player suspects that the opponent may know his choice, i.e. distinction of the Maximin Game consists in the player attitudes.

12.3.3.1 Maximin Solution Principle

Players compute the set of their pessimistic strategies. The set

$$\mathrm{MS_A} = \mathrm{Arg}\max_{i\in I}\min_{j\in J} a_{ij}$$

forms the set of pessimistic strategies of the first player. The set

$$\mathrm{MS_B} = \mathrm{Arg}\max_{j\in J}\min_{i\in I} b_{ij}$$

forms the set of pessimistic strategies of the second player.

Every element of the Cartesian product $\mathrm{MS} = \mathrm{MS_A} \times \mathrm{MS_B}$ forms a maximin solution of Maximin Game $\mathrm{M}\Gamma = \mathrm{M}_2$.

12.3.3.2 Set of Maximin Solutions

The set of Maximin Solutions of the Maximin Game is

$$\mathrm{MS} = \mathrm{M_A} \times \mathrm{M_B}.$$

Proposition 12.4 *The sets* NE, SE_{12}, SE_{21} *and* MS *are generally not identical for the matrices A and B.*

Proof It is enough to mention that every Stackelberg game has Stackelberg equilibria and every Maximin game has the maximin solution, but the Nash game with the same matrices may do not have Nash equilibria. Even though the Nash game has equilibria, simple examples may be constructed which illustrate that Nash equilibrium is not identical with Stackelberg equilibrium and Maximin solution. □

12.3.3.3 Maximin Solution Existence

Proposition 12.5 *Every finite Maximin Game has maximin solutions.*

Proof The proof follows from the finiteness of the strategy sets. □

12.3.4 Maximin-Nash Taxon

Maximin-Nash Taxon contains two elements $\mathrm{MNT} = \{\Gamma_{00\,01}, \Gamma_{01\,00}\}$.

Let us suppose, without loss of generality, the players choose their strategies without knowing the opponent choice. However, one of them (and only one) has the belief that there is information leak about the chosen strategy. Let us denote such a game by $\mathrm{MN}\Gamma = \mathrm{MN}$ or $\mathrm{NM}\Gamma = \mathrm{NM}$.

12.3.4.1 Maximin-Nash Solution Principle

For defining the solution concept of such games, we can observe that they may be seen as a constrained Nash Game $\Gamma_{00\,00}$ in which the Maximin principle must be applied additionally for the pessimistic player which suspects the corruption. So, for $\Gamma_{00\,01}$ we can define as the solution any element from:

$$\mathrm{NMS} = \mathrm{NE} \cap (\mathrm{I} \times \mathrm{MS_B}),$$

For the game $\Gamma_{01\,00}$, the solution is any element from the set

$$\mathrm{MNS} = \mathrm{NE} \cap (\mathrm{MS_A} \times \mathrm{J}).$$

From the above definitions, it follows that a Maximin-Nash Solution is a Nash Equilibrium for which one of its components (corresponding to the player which suspect corruption) is a Maximin strategy too.

12.3.4.2 Set of Maximin-Nash Solutions

NMS is the set of solutions in game NM $= \Gamma_{00\,01}$, and MNS is the set of solutions in game MN $= \Gamma_{01\,00}$.

12.3.4.3 Maximin-Nash Solution Existence

Proposition 12.6 *If the Maximin-Nash Game* MN *has a solution, then the Nash Game has a Nash equilibrium.*

Proof The proof follows from the definition of the Maximin vs Nash solution. □

Generally, the reciprocal proposition is not true.

12.3.5 Optimum Taxon

Optimum Taxon is formed only by one element OT $= \{\Gamma_{10\,10}\}$.

The player strategies are selected as follows. Let us suppose the both players declare that they play the Nash game, but everyone cheats and (by corruption and information leak) knows the choice of the opponent. Such a game is denoted by $\mathrm{O} = \mathrm{O}_2$.

To formalize this game, we must highlight two pseudo-stages of the game. At the first pseudo-stage, the players choose initially their strategies (i_0, j_0). At the second pseudo-stage, the players choose their final strategies (i_1, j_1) knowing (i_0, j_0).

12.3.5.1 Optimum Profile

As everyone does not suspect opponent of cheating, but both of them cheat, they play as followers, i.e. in the game O_2 the players act as followers.

The resulting profile is (i_1, j_1), where

$$i_1 \in \underset{i \in I}{\mathrm{Argmax}}\ a_{i j_0}$$

and

$$j_1 \in \underset{j \in J}{\mathrm{Argmax}}\ b_{i_0 j}.$$

As both i_1 and j_1 correspond to j_0 and i_0, respectively, the pair (i_1, j_1) is not a solution concept. It is a simple profile—an Optimum Profile.

12.3.5.2 Set of Optimum Profiles

For this game we can define only the set of Optimum Profiles:

$$\mathrm{O_2P}(i_0, j_0) = \left(\operatorname*{Argmax}_{i \in I} a_{i j_0}, \operatorname*{Argmax}_{j \in J} b_{i_0 j} \right).$$

12.3.5.3 Optimum Profiles Existence

We mentioned above that the OT taxon is based on the Optimum Profile, which generally is not a solution concept. Nevertheless, we may conclude that the Optimum Profile exists for any finite game because of strategy set finiteness.

12.3.6 Optimum-Nash Taxon

This Taxon has two symmetric elements $\mathrm{ONT} = \{\Gamma_{00\,10}, \Gamma_{10\,00}\}$.

Suppose the players declare that they play the Nash game, but one of them cheats and (by corruption and information leak) knows the choice of the opponent. We denote this game by ON = ON or NO = NO.

To formalize this game, we highlight two stages of the game as in the precedent case. At the first stage, the players choose initially their strategies (i_0, j_0). At the second stage, the cheater changes his strategy as an optimal response to the opponent strategy. So, at the second stage, or (i_0, j_1), or (i_1, j_0) is realised.

12.3.6.1 Optimum-Nash Profile Principle

As in the case of the Maximin vs Nash Game, for defining the solution concept we can observe first if they play Nash Game $\Gamma_{00\,00}$, i.e. they choose to play a Nash Equilibrium, the cheating is not convenient. For such games, Nash Equilibrium is the solution principle to apply. If the honest player does not play Nash Equilibrium Strategy, he may lose out comparably with the Nash Equilibrium. So, he plays Nash Equilibrium. In such case, for the cheater is convenient to play a Nash Equilibrium strategy.

As a conclusion, this type of game may be thought as a Nash Game if the game has Nash equilibrium. If the game doesn't have Nash Equilibrium or it have many Nash Equilibria, the principle of the Optimum-Nash profile is applied. One of them chooses his strategy as in Nash game (leader). He can apply the maximin or the Stackelberg strategy of the leader. The opponent chooses his strategy as the last player in Stackelberg game (follower).

12.3.6.2 Set of Optimum-Nash Profiles

Evidently, if the honest player chooses the Nash Equilibrium Strategy, the set of solutions is identical to NES.

If the honest player chooses maximin strategy, e.g. the first player chooses one of the elements of $\text{MS}_\text{A} = \text{Arg} \max\limits_{i \in I} \min\limits_{j \in J} a_{ij}$, the opponent chooses every element from $J^* = \text{Arg} \max\limits_{j \in J} b_{ij}$.

If the honest player chooses Stackelberg leader strategy, the opponent chooses the follower strategy. In such case, the ON Profile is a Stackelberg equilibrium.

12.3.6.3 Optimum-Nash Profile Existence

Based on the above, ON Profile exists for any ON game. It may be NE, SE, or a simple Maximin-Optimum Profile.

12.3.7 Optimum-Stackelberg Taxon

Optimum-Stackelberg Taxon contains two symmetric elements

$$\text{OST} = \{\Gamma_{10\,11},\ \Gamma_{11\,10}\}.$$

Let us suppose that each player knows the opponent's chosen strategy and only one of them knows additionally that the opponent knows his chosen strategy. So, the one which does not know that the opponent knows his chosen strategy will simply select his strategy as an optimal response to the opponent's strategy (he will play as an unconscious leader in a Stackelberg game), but the other (which knows additionally that the opponent knows his chosen strategy; player with the value of knowledge vector equal to '11') will know the opponent's reaction and will play as a follower in a Stackelberg game.

Remark 12.9 If every player knows a priory what information leak he will use (he knows the values of his respective knowledge vector), then the player with the value of knowledge vector equal to '11' will play as a leader, and his opponent will play as a follower.

It is not the case we consider.

12.3.7.1 Optimum-Stackelberg Solution Principle

If the first player does not suspect of information leak to the second player ($\Gamma_{10\,11}$), but he knows the strategy j selected by the second player, then he chooses his strategy as an optimal response to j, i.e.

$$i^* \in I^* = \underset{i \in I}{\text{Argmax}}\, a_{ij}.$$

Let us suppose $\#I^* = 1$. The second player knows that for his selected strategy j the first player will select i^*. He must select his strategy as an optimal response to the i^*, i.e.

$$j^* \in J^* = \underset{j \in J}{\text{Argmax}}\, b_{i^* j}.$$

So, the solution of the game $\Gamma_{10\,11}$ is (i^*, j^*).

Analogically, we can define the solution concept for $\Gamma_{11\,10}$. If the second player does not suspect of information leak to the first player, but he knows the strategy i selected by the first player, then he chooses his strategy as an optimal response to i, i.e. $j^* \in J^* = \underset{j \in J}{\text{Argmax}}\, b_{ij}$. Let us suppose $\#J^* = 1$. The first player knows that for his selected strategy i the second player will select j^*. He must select his strategy as an optimal response to the j^*, i.e. $i^* \in I^* = \underset{i \in I}{\text{Argmax}}\, a_{i j^*}$. So, the solution of $\Gamma_{11\,10}$ is (i^*, j^*).

Let us denote such a game by OS = OS. The symmetric game is denoted by SO = SO.

12.3.7.2 Set of Optimum-Stackelberg Solutions

Let us recall that to define solution concept we impose the cardinality of sets I^* and J^* to be 1. To define the set of solutions we must exclude this supposition. So, for $\Gamma_{10\,11}$ the set $I^* = \underset{i \in I}{\text{Argmax}}\, a_{ij}$ represents all optimal responses to the strategy j of the second player. The second player knows/calculates this optimal response set. On its base, he defines his set of Maximin Response $J^* = \underset{j \in J}{\text{Argmax}}\underset{i \in I^*}{\min}\, b_{ij}$ by applying the Maximin Principle. So, the set of solutions of $\Gamma_{10\,11}$ is $I^* \times J^*$.

Analogically, for $\Gamma_{11\,10}$ the set $J^* = \underset{j \in J}{\text{Argmax}}\, a_{ij}$ represents all optimal responses to the strategy i of the first player. The first player knows/calculates this optimal response set. On its base, he defines his set of Maximin Response $I^* = \underset{i \in I}{\text{Argmax}}\underset{j \in J^*}{\min}\, a_{ij}$ by applying Maximin Principle. So, the set of solutions of $\Gamma_{11\,10}$ is $I^* \times J^*$.

12.3.7.3 Optimum-Stackelberg Solution Existence

Proposition 12.7 *Every finite Optimum-Stackelberg Game* OS *has an Optimum-Stackeberg solution.*

Proof The proof follows from the definition of the Optimum-Stackeberg Solution and the finiteness of the strategy sets. □

12.4 Taxonomy of Bimatrix Games with Information Leak and More Than Two Levels of Knowledge

According to the above results, there are $4^3 = 64$ possible kinds of games with information leak $\Gamma_{\gamma^A\gamma^B}$ in the case when the vectors of knowledge have three components $\gamma^A = (\gamma_1^A, \gamma_2^A, \gamma_3^A)$ and $\gamma^B = (\gamma_1^B, \gamma_2^B, \gamma_3^B)$ (information leak may occur on 3 levels).

In this case and in the general case, is it enough to examine only seven taxa of games as for games with two level of knowledge or the number of taxa increases?

Theorem 12.1 *The number of taxa for bimatrix games with information leaks with a number of knowledge levels $l \geq 2$ does not depend on the number of knowledge levels l.*

Proof First, let us observe that the abstract maximal number of possible taxa depends on the number of solution principle applied by two players. In our case, we apply only four solution principles: Nash equilibrium, Stackelberg equilibrium, Maximin principle, and Optimum principle. So, the maximal number of taxa may be equal to 16. But, the rules of the games and knowledge possessed by players in the case of two levels of knowledge make up possible only seven taxa.

By induction, it is provable that these number of taxa remains unchanged for $l \geq 3$. □

12.5 Repeated Bimatrix Games with Information Leaks

If the games described above are considered as molecular games, then we can examine a series of molecular games on every stage of which a molecular game is played. Evidently, such games are a simple consequence of the games of the seven types, corresponding to seven taxa highlighted above.

12.6 Taxonomy of Polymatrix Games with Information Leaks and More Than Two Levels of Knowledge

In the case of three and more players, we can adjust the molecular approach and we can denote the games by their atoms (atom players). Evidently, the number of taxa for such games may increase. Can we present a taxonomy of such games? Can we present a scheme or a table of elementary or molecular games? These questions have to be answered by the future work.

12.7 Conclusions

Normal form games pretend to be a mathematical model of situations often met in reality. Actually, they formalize an essential part of real decision making situations and processes, but not ultimate. Real decision making situations are influenced by different factors, which may change the essence of the games and the solution principles applicable for their solving. It follows that the initial mathematical models must be modified, at least.

This chapter may be seen as a **work in progress**. It presents a taxonomy of normal form games with information leaks. Every taxon contains the games solvable on the base of the same solution principles, highlighted in the name.

The games with arbitrary pseudo-levels and levels of knowledge, and the games with bribe are the subject of following works. So, we can admit that the results of this chapter establish a foundation for a new domain of research in game theory.

References

1. Ungureanu V. 2016. Taxonomy of Strategic Games with Information Leaks and Corruption of Simultaneity. *Computer Science Journal of Moldova* 24, 1(70): 83–105.
2. Von Neumann, J., and O. Morgenstern. 1944. *Theory of Games and Economic Behavior*, 2nd ed., 674 pp. Princeton, New Jersey: Annals Princeton University Press.
3. Nash, J. (1951). Noncooperative games. *Annals of Mathematics*, *54*(2), 280–295.
4. Kukushkin, N.S., and V.V. Morozov. 1984. *Theory of Non-Antagonistic Games*, 2nd ed., 105 pp. Moscow State University. (in Russian).
5. Gorelov, M.A. 2016. Hierarchical games with uncertain factors. *Upr. Bol?sh. Sist.* 59: 6–22. (in Russian).
6. Vatel', I.A., and N.S. Kuku u skin. 1973. Optimal conduct of a player having the right to the first move when his knowledge of partner's interests is inaccurate. *u Z. Vy u cislsl. Mat. i Mat. Fiz.* 13: 303–310. (in Russian).
7. Kuku u skin, N.S. 1973. A certain game with incomplete information. *u Z. Vy u cislsl. Mat. i Mat. Fiz.* 13: 210–216. (in Russian).
8. Kononenko, A.F. 1973. The rôle of information on the opponent's objective function in two-person games with a fixed sequence of moves. *u Z. Vy u cislsl. Mat. i Mat. Fiz.* 13: 311–317. (in Russian).

9. Ere u sko, F.I., and A.F. Kononenko. 1973. Solution of the game with the right to precedence in case of incomplete information on the partner's intention. *u Z. Vy u cisl. Mat. i Mat. Fiz.* 13: 217–221. (in Russian).
10. Ungureanu, V. 2006. Nash equilibrium set computing in finite extended games. *Computer Science Journal of Moldova* 14, 3 (42): 345–365.
11. Ungureanu, V. (2008). Solution principles for simultaneous and sequential games mixture. *ROMAI Journal*, *4*(1), 225–242.
12. Ungureanu V. 2013. Linear discrete-time Pareto-Nash-Stackelberg control problem and principles for its solving. *Computer Science Journal of Moldova* 21, 1(61): 65–85.
13. Ungureanu, V., & Lozan, V. (2013). Linear discrete-time set-valued Pareto-Nash-Stackelberg control processes and their principles. *ROMAI Journal*, *9*(1), 185–198.
14. Ungureanu, V. 2015. Mathematical Theory of Pareto-Nash-Stackelberg Game-Control Models. *Workshop on Applied Optimization Models and Computation, Indian Statistical Institute*, 55–66, Delhi Centre, Jan 28–30, Program and Abstracts.
15. Ungureanu, V. 2015. Linear Discrete Optimal and Stackelberg Control Processes with Echoes and Retroactive Future. In *CSCS20-2015, 2015 20th International Conference on Control Systems and Computer Science*, 5–9, May 27–29, 2015, Bucharest, Romania, Editors: Ioan Dumitrache, Adina Magda Florea, Florin Pop, Alexandru Dumitra c scu, Proceedings, vol. 1, CSCS20 Main Track.
16. Von Stackelberg, H. 1934. Marktform und Gleichgewicht (Market Structure and Equilibrium), XIV+134 pp. Vienna: Springer. (in German).
17. Leitmann, G. (1978). On generalized Stackelberg strategies. *Journal of Optimization Theory and Applications*, *26*, 637–648.
18. Brams, S.J. 1994. *Theory of Moves*, XII+248 pp. Cambridge: Cambridge University Press.
19. Bellman, R. (1957). *Dynamic Programming*. New Jersey: Princeton University Press. 365 pp.

Part III
Pareto-Nash-Stackelberg Game and Control Processes

The third part of the monograph is dedicated to the Pareto-Nash-Stackelberg control of the systems governed by linear discrete-time laws.

Chapter 13
Linear Discrete-Time Pareto-Nash-Stackelberg Control and Its Principles

Abstract A direct-straightforward method for solving linear discrete-time optimal control problem is applied to solve the control problem of a linear discrete-time system as a mixture of multi-criteria Stackelberg and Nash games. For simplicity, the exposition starts with the simplest case of linear discrete-time optimal control problem and, by sequential considering of more general cases, investigation finalizes with the highlighted Pareto-Nash-Stackelberg and set valued control problems. Different solution principles are compared and their equivalence is proved. We need to remark that there are other possible title variants of the considered models like, e.g., a multi-agent control problem of the Pareto-Nash-Stackelberg type. There is an appropriate approach in (Leitmann, Pickl and Wang Dynamic Games in Economics, Springer, Berlin, 205–217, 2014), [1]. A more simple and largely used title for such games is dynamic games, see e.g. (Başar, Olsder, Society for Industrial and Applied Mathematics, Philadelphia, 536, 1999), [2], (Long, A Survey of Dynamic Games in Economics, World Scientific, New Jersey, XIV–275, 2010), [3]. We insist on the above title in order to highlight both the game and control natures of the modelled real situations and processes. More the more we can refer in this context a Zaslavski's recent monograph (Zaslavski, Discrete-Time Optimal Control and Games on Large Intervals, Springer, Switzerland, X+398, 2017, [4]) that uses an appropriate approach to the names of considered mathematical models.

13.1 Introduction

Optimal control theory, which appeared in a great measure due to Lev Pontryagin [5] and Richard Bellman [6] as a natural extension of the calculus of variations, often does not satisfy all requirements and needs for modelling and solving problems of real dynamic systems and processes. A situation of this type occurs for the problem of a linear discrete-time system control [7] by a decision process that evolves as a Pareto-Nash-Stackelberg game with constraints — a mixture of hierarchical and simultaneous games [8–12]. For such systems, the optimal control concept evolves

V. Ungureanu, *Pareto-Nash-Stackelberg Game and Control Theory*,
Smart Innovation, Systems and Technologies 89,
https://doi.org/10.1007/978-3-319-75151-1_13

naturally to a concept of Pareto-Nash-Stackelberg control and to a natural principle for solving the highlighted problem by applying the concept of Pareto-Nash-Stackelberg equilibrium [12] with a direct-straightforward principle for solving. It is important to mention that the investigation of the problems that mix game and control processes appeared almost simultaneously with game theory and optimal control theory due to Rufus Isaacs [13]. It is still a modern, very dynamic and expanding domain of research, see e.g. [2, 14, 15].

A direct method and principle for solving linear discrete-time optimal control problem is extended to a control problem of a linear system in discrete time as a mixture of multi-criteria Stackelberg and Nash games [12].

The exposition starts with the investigation of the simplest case of linear discrete-time optimal control problems [16] and, by sequential considering of more general cases, finalizes with the Pareto-Nash-Stackelberg and set valued control problems.

The maximum principle of Pontryagin [5] is formulated and proved for all the considered problems. Its equivalence with the direct-straightforward solution principle is established.

13.2 Linear Discrete-Time Optimal Control Problem

Consider the following problem [16][1]:

$$\begin{aligned} f(x,u) = \sum_{t=1}^{T}(c^t x^t + b^t u^t) &\to \max, \\ x^t = A^{t-1}x^{t-1} + B^t u^t&,\ t = 1,\dots,T, \\ D^t u^t \le d^t&,\ t = 1,\dots,T, \end{aligned} \tag{13.1}$$

where $x^0,\ x^t,\ c^t \in \mathbb{R}^n,\ u^t,\ b^t \in \mathbb{R}^m,\ A^{t-1} \in \mathbb{R}^{n\times n},\ B^t \in \mathbb{R}^{n\times m},\ d^t \in \mathbb{R}^k,\ D^t \in \mathbb{R}^{k\times n}, c^t x^t = \langle c^t, x^t\rangle, b^t u^t = \langle b^t, u^t\rangle, t = 1,\dots,T, u = (u^1,\dots,u^T), n, m, k, T \in \mathbb{N}$.

Theorem 13.1 *Let problem (13.1) be solvable. The sequence $\bar{u}^1, \bar{u}^2, \dots, \bar{u}^T$ forms an optimal control if and only if $\bar{u}^t$ is the solution of the linear programming problem*

$$\begin{aligned} &\left(c^t B^t + c^{t+1}A^t B^t + \dots + c^T A^{T-1}A^{T-2}\dots A^t B^t + b^t\right) u^t \to \max, \\ &D^t u^t \le d^t, \end{aligned}$$

for $t = 1,\dots,T$.

[1] Symbol T means discrete time horizon in this chapter. The symbol of matrix translation is omitted. Left and right matrix multiplications are largely used. The reader is asked to understand by himself when column or row vector are used.

Proof Problem (13.1) may be represented in the form:

$$\begin{array}{llllllll}
Ex^1 & & & -B^1u^1 & & & & = A^0x^0, \\
-A^1x^1 + Ex^2 & & & & -B^2u^2 & & & = 0, \\
\dots & & & & \dots & & & \dots \\
& -A^{T-1}x^{T-1} + Ex^T & & & & & -B^Tu^T & = 0, \\
& & & D^1u^1 & & & & \le d^1, \\
& & & & D^2u^2 & & & \le d^2, \\
& & & & \dots & & & \dots \\
& & & & & & D^Tu^T & \le d^T, \\
c^1x^1 + c^2x^2 & + \dots + & & b^1u^1 & + b^2u^2 + & \dots + & b^Tu^t & \to \max .
\end{array}$$

Its dual problem is

$$\begin{array}{lllllll}
p^1 & -p^2A^1 & & & & & = c^1, \\
& p^2 - p^3A^2 & & & & & = c^2, \\
& & \dots & & & & \dots \\
& & p^T & & & & = c^T, \\
-p^1B^1 & & & +q^1D^1 & & & = b^1, \\
& -p^2B^2 & & & +q^2D^2 & & = b^2, \\
& & & & \dots & & \dots \\
& & -p^TB^T & & & +q^TD^T & = b^T, \\
& & & q^1 \ge 0, q^2 \ge 0, & & q^T \ge 0, & \\
p^1A^0x^0 & & & +q^1d^1 & +q^2d^2 + & \dots + q^Td^T & \to \min .
\end{array}$$

From the constraints of dual problem, it follows that the values of dual variables $p^1, p^2, \dots, p^T$ are calculated on the bases of a recurrence relation:

$$\begin{aligned}
p^T &= c^T, \\
p^t &= p^{t+1}A^t + c^t,\ t = T-1, \dots, 1.
\end{aligned} \tag{13.2}$$

So, the dual problem is equivalent to:

$$\begin{aligned}
q^1D^1 &= p^1B^1 + b^1, \\
q^2D^2 &= p^2B^2 + b^2, \\
&\dots \\
q^TD^T &= p^TB^T + b^T, \\
q^t &\ge 0, t = 1, \dots, T, \\
p^1A^0x^0 + q^1d^1 + q^2d^2 + \dots + q^Td^T &\to \min .
\end{aligned}$$

The dual of the last problem is:

$$
\begin{aligned}
D^1u^1 &\le d^1,\\
D^2u^2 &\le d^2,\\
&\dots\\
D^Tu^T &\le d^T,
\end{aligned}
\tag{13.3}
$$
$$
\sum_{t=1}^{T}\langle u^t, p^t B^T + b^t\rangle \to \max .
$$

The solution of problem (13.3) may be found by solving T linear programming problems

$$
\begin{aligned}
D^tu^t &\le d^t,\\
\langle u^t, p^t B^T + b^t\rangle &\to \max,
\end{aligned}
$$

for $t = 1, \dots, T$. So, the solution of initial control problem (13.1) is identical with a sequence of the solutions of T linear programming problems.

Similar results may be obtained by performing direct transformations of problem (13.1):

$$
\begin{aligned}
x^1 &= A^0x^0 + B^1u^1,\\
x^2 &= A^1x^1 + B^2u^2 = A^1(A^0x^0 + B^1u^1) + B^2u^2 =\\
&= A^1A^0x^0 + A^1B^1u^1 + B^2u^2,\\
x^3 &= A^2x^2 + B^3u^3 = A^2(A^1A^0x^0 + A^1B^1u^1 + B^2u^2) + B^3u^3 =\\
&= A^2A^1A^0x^0 + A^2A^1B^1u^1 + A^2B^2u^2 + B^3u^3,\\
&\dots\\
x^T &= A^{T-1}x^{T-1} + B^Tu^T =\\
&= \prod_{t=0}^{T-1} A^tx^0 + \prod_{t=1}^{T-1} A^tB^1u^1 + \prod_{t=2}^{T-1} A^tB^2u^2 +\\
&\quad + \dots + \prod_{t=T-1}^{T-1} A^tB^{T-1}u^{T-1} + B^Tu^T,
\end{aligned}
$$

and by subsequent substitution of these expressions in the objective function:

$$
\begin{aligned}
f(x,u) = {} & c^1\left(A^0x^0 + B^1u^1\right) + c^2\left(A^1A^0x^0 + A^1B^1u^1 + B^2u^2\right) +\\
& + c^3\left(A^2A^1A^0x^0 + A^2A^1B^1u^1 + A^2B^2u^2 + B^3u^3\right) + \dots +\\
& + c^T\Bigg(\prod_{t=0}^{T-1} A^tx^0 + \prod_{t=1}^{T-1} A^tB^1u^1 + \prod_{t=2}^{T-1} A^tB^2u^2 +\\
& + \dots + \prod_{t=T-1}^{T-1} A^tB^{T-1}u^{T-1} + B^Tu^T\Bigg) +\\
& + b^1u^1 + b^2u^2 + \dots + b^Tu^T =
\end{aligned}
$$

$$
\begin{aligned}
&= \left(c^1 + c^2 A^1 + c^3 A^2 A^1 + \cdots + c^T A^{T-1} A^{T-2} \dots A^1\right) A^0 x^0 + \\
&+ \left(c^1 B^1 + c^2 A^1 B^1 + c^3 A^2 A^1 B^1 + \cdots + \right.\\
&\left. + c^T A^{T-1} A^{T-2} \dots A^1 B^1 + b^1\right) u^1 + \\
&+ \left(c^2 B^2 + c^3 A^2 B^2 + c^4 A^3 A^2 B^2 + \cdots + \right.\\
&\left. + c^T A^{T-1} A^{T-2} \dots A^2 B^2 + b^2\right) u^2 + \\
&+ \cdots + \left(c^T B^T + b^T\right) u^T.
\end{aligned}
$$

Finally, the problem obtains the form

$$
\begin{aligned}
&f(u) = \\
&= \left(c^1 + c^2 A^1 + c^3 A^2 A^1 + \cdots + c^T A^{T-1} A^{T-2} \dots A^1\right) A^0 x^0 + \\
&+ \left(c^1 B^1 + c^2 A^1 B^1 + c^3 A^2 A^1 B^1 + \cdots + \right.\\
&\left. + c^T A^{T-1} A^{T-2} \dots A^1 B^1 + b^1\right) u^1 + \\
&+ \left(c^2 B^2 + c^3 A^2 B^2 + c^4 A^3 A^2 B^2 + \cdots + \right.\\
&\left. + c^T A^{T-1} A^{T-2} \dots A^2 B^2 + b^2\right) u^2 + \\
&+ \cdots + \left(c^T B^T + b^T\right) u^T \rightarrow \max, \\
&D^t u^t \leq d^t, t = 1, \dots, T.
\end{aligned}
\tag{13.4}
$$

Obviously, (13.3) and (13.4) are identical. So, the solution of problem (13.4) is obtained as a sequence of solutions of T linear programming problems. Apparently, the method complexity is polynomial, but really it has pseudo-polynomial complexity because of the possible exponential value of T on n. □

Various particular cases may be considered for problem (13.1) and the optimality conditions follow as a consequence of Theorem 13.1.

Theorem 13.2 *If*

$$
\begin{aligned}
A^0 &= A^1 = \cdots = A^{T-1} = A, \\
B^1 &= B^2 = \cdots = B^T = B,
\end{aligned}
$$

and problem (13.1) is solvable, then the sequence $\bar{u}^1, \bar{u}^2, \dots, \bar{u}^T$ *forms an optimal control if and only if* $\bar{u}^t$ *is the solution of the linear programming problem*

$$
\begin{aligned}
&\left(c^t B + c^{t+1} A B + c^{t+2} (A)^2 B + \cdots + c^T (A)^{T-t} B + b^t\right) u^t \rightarrow \max, \\
&D^t u^t \leq d^t,
\end{aligned}
$$

for $t = 1, \dots, T$.

Theorem 13.1 establishes a principle for solving (13.1).

By considering the Hamiltonian functions

$$
H_t(u^t) = \left\langle p^t B^t + b^t, u^t \right\rangle, t = T, \dots, 1,
$$

where p^t, $t = T, \dots, 1$ are defined by (13.2), as it is conjectured in [16] and proved above by two ways, the maximum principle of Pontryagin [5] holds.

Theorem 13.3 *Let (13.1) be solvable. The sequence* $\bar{u}^1, \bar{u}^2, \ldots, \bar{u}^T$ *forms an optimal control if and only if*

$$H_t(\bar{u}^t) = \max_{u^t: D^t u^t \le d^t} H_t(u^t), t = T, \ldots, 1.$$

Evidently, Theorems 13.1 and 13.3 are equivalent.

13.3 Linear Discrete-Time Stackelberg Control Problem

Let us modify Problem (13.1) by considering the control of Stackelberg type [10], that is a Stackelberg game with T players [8–11]. In such game, at each stage t ($t = 1, \ldots, T$) the player t selects his strategy and communicates his and all precedent selected strategies to the following $t+1$ player. After all stage strategy selections, all the players compute their gains on the resulting profile.

Let us call such type of system control Stackelberg control, and the corresponding problem — linear discrete-time Stackelberg control problem.

The decision process described above may be formalized as follows:

$$\begin{aligned}
&f_1(x, u) = \sum_{t=1}^{T} \left(c^{1t}x^t + b^{1t}u^t\right) \underset{u^1}{\longrightarrow} \max,\\
&f_2(x, u) = \sum_{t=1}^{T} \left(c^{2t}x^t + b^{2t}u^t\right) \underset{u^2}{\longrightarrow} \max,\\
&\ldots\\
&f_T(x, u) = \sum_{t=1}^{T} \left(c^{Tt}x^t + b^{Tt}u^t\right) \underset{u^T}{\longrightarrow} \max,\\
&x^t = A^{t-1}x^{t-1} + B^t u^t, t = 1, \ldots, T,\\
&D^t u^t \le d^t, t = 1, \ldots, T,
\end{aligned} \tag{13.5}$$

where $x^0, x^t, c^{\pi t} \in \mathbb{R}^n$, $u^t, b^{\pi t} \in \mathbb{R}^m$, $A^{t-1} \in \mathbb{R}^{n\times n}$, $B^t \in \mathbb{R}^{n\times m}$, $d^t \in \mathbb{R}^k$, $D^t \in \mathbb{R}^{k\times n}$, $c^{\pi t}x^t = \langle c^{\pi t}, x^t\rangle$, $b^{\pi t}u^t = \langle b^{\pi t}, u^t\rangle$, $t, \pi = 1, \ldots, T$.

Formally, the set of strategies of the player π ($\pi = 1, 2, \ldots, T$) is determined only by admissible solutions of the next problem:

$$\begin{aligned}
&f_\pi\left(x, u^\pi || u^{-\pi}\right) = \sum_{t=1}^{T} \left(c^{\pi t}x^t + b^{\pi t}u^t\right) \underset{u^\pi}{\longrightarrow} \max,\\
&x^\pi = A^{\pi-1}x^{\pi-1} + B^\pi u^\pi,\\
&D^\pi u^\pi \le d^\pi.
\end{aligned}$$

Actually, as we can find out, the strategy sets of the players are interconnected and the game is not a simple normal form game. A situation similar with that in optimization theory may be established – there are problems without constraints and with constraints. So, the strategy (normal form) game may be called strategy game without constraints. Game which contains common constraints on strategies may be called strategy game with constraints.

Theorem 13.4 *Let (13.5) be solvable. The sequence* $\bar{u}^1, \bar{u}^2, \ldots, \bar{u}^T$ *forms a Stackelberg equilibrium control in (13.5) if and only if* $\bar{u}^\pi$ *is optimal solution of linear programming problem*

$$\begin{array}{l} f_\pi(u^\pi) = \left(c^{\pi\pi} B^\pi + c^{\pi\pi+1} A^\pi B^\pi + c^{\pi\pi+2} A^{\pi+1} A^\pi B^\pi + \cdots + \right. \\ \left. + c^{\pi T} A^{T-1} A^{T-2} \ldots A^\pi B^\pi + b^{\pi\pi}\right) u^\pi \underset{u^\pi}{\rightarrow} \max, \\ D^\pi u^\pi \leq d^\pi, \end{array}$$

for $\pi = 1, \ldots, T$.

Proof Player π ($\pi = 1, 2, \ldots, T$) decision problem is defined by linear programming problem (13.5). Since the controlled system is one for all players, by performing the direct transformations as above, problem (13.5) is transformed into

$$\begin{array}{l} f_\pi\left(u^\pi \| u^{-\pi}\right) = \\ = \left(c^{\pi 1} + c^{\pi 2} A^1 + c^{\pi 3} A^2 A^1 + \cdots + \right. \\ \left. + c^{\pi T} A^{T-1} A^{T-2} \ldots A^1\right) A^0 x^0 + \\ + \left(c^{\pi 1} B^1 + c^{\pi 2} A^1 B^1 + c^{\pi 3} A^2 A^1 B^1 + \cdots + \right. \\ \left. + c^{\pi T} A^{T-1} A^{T-2} \ldots A^1 B^1 + b^{\pi 1}\right) u^1 + \\ + \left(c^{\pi 2} B^2 + c^{\pi 3} A^2 B^2 + c^{\pi 4} A^3 A^2 B^2 + \cdots + \right. \\ \left. + c^{\pi T} A^{T-1} A^{T-2} \ldots A^2 B^2 + b^{\pi 2}\right) u^2 + \\ + \cdots + \left(c^{\pi T} B^T + b^{\pi T}\right) u^T \underset{u^\pi}{\rightarrow} \max, \pi = 1, \ldots, T, \\ D^t u^t \leq d^t, t = 1, \ldots, T. \end{array} \tag{13.6}$$

From the equivalence of (13.5) and (13.6), the proof of Theorem 13.4 follows. □

There are various particular cases of (13.5) and Theorem 13.4. Theorem 13.5 presents one of such cases.

Theorem 13.5 *If*

$$\begin{array}{l} A^0 = A^1 = \cdots = A^{T-1} = A, \\ B^1 = B^2 = \cdots = B^T = B, \end{array}$$

and problem (13.5) is solvable, then the sequence $\bar{u}^1, \bar{u}^2, \ldots, \bar{u}^T$ *forms a Stackelberg equilibrium control if and only if* $\bar{u}^\pi$ *is the solution of linear programming problem*

$$\begin{array}{l} \left(c^{\pi\pi} B + c^{\pi\pi+1} AB + \cdots + c^{\pi T} (A)^{T-\pi} B + b^{\pi\pi}\right) u^\pi \underset{u^\pi}{\rightarrow} \max, \\ D^\pi u^\pi \leq d^\pi, \end{array}$$

for $\pi = 1, \ldots, T$.

Theorem 13.4 establishes a principle for solving (13.5).

The maximum principle of Pontryagin may be applied for solving problem (13.5) too.

Let us consider the following recurrence relations

$$\begin{aligned} p^{\pi T} &= c^{\pi T}, \\ p^{\pi t} &= p^{\pi t+1} A^t + c^{\pi t}, \; t = T-1, \ldots, 1, \end{aligned} \tag{13.7}$$

where $\pi = 1, \ldots, T$. Hamiltonian functions are defined as

$$H_{\pi t}\left(u^t\right) = \left\langle p^{\pi t} B^t + b^{\pi t}, u^t \right\rangle, t = T, \ldots, 1, \pi = 1, \ldots, T,$$

where $p^{\pi t}$, $t = T, \ldots, 1, \pi = 1, \ldots, T$, are defined by (13.7).

Theorem 13.6 *Let (13.5) be solvable. The sequence of controls $\bar{u}^1, \ldots, \bar{u}^T$ forms a Stackelberg equilibrium control if and only if*

$$H_{\pi\pi}\left(\bar{u}^{\pi}\right) = \max_{u^{\pi}: D^{\pi} u^{\pi} \leq d^{\pi}} H_{\pi\pi}\left(u^{\pi}\right),$$

for $\pi = 1, \ldots, T$.

Proof The proof of Theorem 13.6 may be provided by direct substitution of relations (13.7) in the Hamiltonian functions defined above and by comparing the final results with linear programming problems from Theorem 13.4. □

Obviously, Theorems 13.4 and 13.6 are equivalent.

From computational point of view, method for solving problem (13.5) established by Theorem 13.4 looks more preferable than the method established by Theorem 13.6.

13.4 Linear Discrete-Time Pareto-Stackelberg Control Problem

Let us modify problem (13.5) by considering a control of Pareto-Stackelberg type. At each stage, a single player makes decision. Every player selects on his stage his strategy according to his criteria and communicates his choice and the precedent player choices to the following player. At the end stage, after all stage strategy selections, the players compute their gains. Such type of control is called Pareto-Stackelberg control, and the corresponding problem is called a linear discrete-time Pareto-Stackelberg control problem.

The described decision process may be formalized in a following manner:

$$\begin{aligned}
&f_1(x,u)=\sum_{t=1}^{T}\left(c^{1t}x^t+b^{1t}u^t\right)\underset{u^1}{\longrightarrow}\max,\\
&f_2(x,u)=\sum_{t=1}^{T}\left(c^{2t}x^t+b^{2t}u^t\right)\underset{u^2}{\longrightarrow}\max,\\
&\dots\\
&f_T(x,u)=\sum_{t=1}^{T}\left(c^{Tt}x^t+b^{Tt}u^t\right)\underset{u^T}{\longrightarrow}\max,\\
&x^t=A^{t-1}x^{t-1}+B^tu^t,\ t=1,\dots,T,\\
&D^tu^t\le d^t,\ t=1,\dots,T,
\end{aligned}\tag{13.8}$$

where $x^0, x^t\in\mathbb{R}^n$, $c^{\pi t}\in\mathbb{R}^{k_\pi\times n}$, $u^t\in\mathbb{R}^m$, $b^{\pi t}\in\mathbb{R}^{k_\pi\times m}$, $A^{t-1}\in\mathbb{R}^{n\times n}$, $B^t\in\mathbb{R}^{n\times m}$, $d^t\in\mathbb{R}^k$, $D^t\in\mathbb{R}^{k\times n}$, $t,\pi=1,\dots,T$.

Theorem 13.7 *Let (13.8) be solvable. The sequence $\bar u^1,\bar u^2,\dots,\bar u^T$ forms a Pareto-Stackelberg equilibrium control in (13.8) if and only if $\bar u^\pi$ is an efficient solution of the multi-criteria linear programming problem*

$$\begin{aligned}
&f_\pi(u^\pi)=\left(c^{\pi\pi}B^\pi+c^{\pi\pi+1}A^\pi B^\pi+c^{\pi\pi+2}A^{\pi+1}A^\pi B^\pi+\dots+\right.\\
&\left.+c^{\pi T}A^{T-1}A^{T-2}\dots A^\pi B^\pi+b^{\pi\pi}\right)u^\pi\underset{u^\pi}{\to}\max,\\
&D^\pi u^\pi\le d^\pi,
\end{aligned}$$

for $\pi=1,\dots,T$.

Proof The set of strategies of the player π ($\pi=1,\dots,T$) is defined formally by the problem

$$\begin{aligned}
&f_\pi\left(x,u^\pi\|u^{-\pi}\right)=\sum_{t=1}^{T}\left(c^{\pi t}x^t+b^{\pi t}u^t\right)\underset{u^\pi}{\longrightarrow}\max,\\
&x^\pi=A^{\pi-1}x^{\pi-1}+B^\pi u^\pi,\\
&D^\pi u^\pi\le d^\pi.
\end{aligned}$$

By performing direct transformations as above, problem (13.8) is transformed into

$$\begin{aligned}
&f_\pi\left(u^\pi\|u^{-\pi}\right)=\\
&=\left(c^{\pi 1}+c^{\pi 2}A^1+c^{\pi 3}A^2A^1+\dots+\right.\\
&\left.+c^{\pi T}A^{T-1}A^{T-2}\dots A^1\right)A^0x^0+\\
&+\left(c^{\pi 1}B^1+c^{\pi 2}A^1B^1+c^{\pi 3}A^2A^1B^1+\dots+\right.\\
&\left.+c^{\pi T}A^{T-1}A^{T-2}\dots A^1B^1+b^{\pi 1}\right)u^1+\\
&+\left(c^{\pi 2}B^2+c^{\pi 3}A^2B^2+c^{\pi 4}A^3A^2B^2+\dots+\right.\\
&\left.+c^{\pi T}A^{T-1}A^{T-2}\dots A^2B^2+b^{\pi 2}\right)u^2+\\
&+\dots+\left(c^{\pi T}B^T+b^{\pi T}\right)u^{\pi T}\underset{u^\pi}{\to}\max,\ \pi=1,\dots,T,\\
&D^tu^t\le d^t,\ t=1,\dots,T.
\end{aligned}\tag{13.9}$$

Equivalence of (13.8) and (13.9) proves Theorem 13.7. □

As above, a particular case of (13.8) is considered by Theorem 13.8.

Theorem 13.8 *If*

$$A^0 = A^1 = \cdots = A^{T-1} = A,$$
$$B^1 = B^2 = \cdots = B^T = B,$$

and problem (13.8) is solvable, then the sequence $\bar{u}^1, \bar{u}^2, \ldots, \bar{u}^T$ forms a Pareto-Stackelberg equilibrium control if and only if $\bar{u}^\pi$ is an efficient solution of the multi-criteria linear programming problem

$$\left(c^{\pi\pi} B + c^{\pi\pi+1} AB + \cdots + c^{\pi T} (A)^{T-\pi} B + b^{\pi\pi}\right) u^\pi \underset{u^\pi}{\rightarrow} \max,$$
$$D^\pi u^\pi \leq d^\pi,$$

for $\pi = 1, \ldots, T$.

The Pontryagin maximum principle may be extended on problem (13.8).
Let us consider the following recurrence relations

$$\begin{aligned} p^{\pi T} &= c^{\pi T}, \\ p^{\pi t} &= p^{\pi t+1} A^t + c^{\pi t},\ t = T-1, \ldots, 1, \end{aligned} \tag{13.10}$$

where $\pi = 1, \ldots, T$. The Hamiltonian vector-functions are

$$H_{\pi t}\left(u^t\right) = \left\langle p^{\pi t} B^t + b^{\pi t}, u^t\right\rangle, t = T, \ldots, 1, \pi = 1, \ldots, T,$$

where $p^{\pi t}$, $t = T, \ldots, 1, \pi = 1, \ldots, T$ are defined by (13.10).

Theorem 13.9 *Let problem (13.8) be solvable. The sequence of controls $\bar{u}^1, \ldots, \bar{u}^T$ forms a Pareto-Stackelberg equilibrium control if and only if*

$$\bar{u}^\pi \in \underset{u^\pi : D^\pi u^\pi \leq d^\pi}{Arg\ max} \quad H_{\pi\pi}\left(u^\pi\right),$$

for $\pi = 1, \ldots, T$.

Proof By direct substitution of relations (13.10) in the Hamiltonian functions and by comparing the final results with the multi-criteria linear programming problems from Theorem 13.7, the truth of Theorem 13.9 arises. □

Theorems 13.7 and 13.9 are equivalent.

Remark 13.1 The method of Pareto-Stackelberg equilibrium control computing, established by Theorems 13.7–13.9, needs solutions of the multi-criteria linear programming problems.

13.5 Linear Discrete-Time Nash-Stackelberg Control Problem

Let us modify problem (13.5) by considering the control of Nash-Stackelberg type with T stages and $\nu_1 + \nu_2 + \cdots + \nu_T$ players, where $\nu_1, \nu_2, \ldots, \nu_T$ are the numbers of players on the stages $1, 2, \ldots, T$. Every player is identified by two numbers/indices (τ, π), where τ is the number of the stage on which the player selects his strategy and $\pi \in \{1, 2, \ldots, \nu_\tau\}$ is his number at the stage τ.

In this game, at each stage τ, the players $1, 2, \ldots, \nu_\tau$ play a Nash game by selecting simultaneously their strategies and by communicating their and all precedent selected strategies to the following $\tau + 1$ stage players. After all stage strategy selections, on the resulting profile, all the players compute their gains. Such type of control is called Nash-Stackelberg control, and the corresponding problem — a linear discrete-time Nash-Stackelberg control problem.

The described decision process may be modelled as follows:

$$
\begin{aligned}
& f_{\tau\pi}(x, u^{\tau\pi} || u^{-\tau\pi}) = \sum_{t=1}^{T} \left(c^{\tau\pi t} x^t + \sum_{\mu=1}^{\nu_t} b^{\tau\pi t\mu} u^{t\mu} \right) \underset{u^{\tau\pi}}{\longrightarrow} \max, \\
& \tau = 1, \ldots, T, \ \pi = 1, \ldots, \nu_\tau, \\
& x^t = A^{t-1} x^{t-1} + \sum_{\pi=1}^{\nu_t} B^{t\pi} u^{t\pi}, \ t = 1, \ldots, T, \\
& D^{t\pi} u^{t\pi} \le d^{t\pi}, \ t = 1, \ldots, T, \ \pi = 1, \ldots, \nu_t,
\end{aligned}
\tag{13.11}
$$

where $x^0, x^t, c^{\tau\pi t} \in \mathbb{R}^n, u^t, b^{\tau\pi t\mu} \in \mathbb{R}^m, A^{t-1} \in \mathbb{R}^{n\times n}, \ B^{\tau\pi} \in \mathbb{R}^{n\times m}, \ d^{\tau\pi} \in \mathbb{R}^k$, $D^{\tau\pi} \in \mathbb{R}^{k\times n}, t, \tau = 1, \ldots, T, \pi = 1, \ldots, \nu_\tau, \mu = 1, \ldots, \nu_t$.

Theorem 13.10 *Let problem (13.11) be solvable. The sequence $\bar{u}^{11}, \bar{u}^{12}, \ldots, \bar{u}^{T\nu_T}$ forms a Nash-Stackelberg equilibrium control in problem (13.11) if and only if $\bar{u}^{\tau\pi}$ is optimal in linear programming problem*

$$
\begin{aligned}
& f(u^{\tau\pi} || u^{-\tau\pi}) = \\
& = \left(c^{\tau\pi\tau} B^{\tau\pi} + c^{\tau\pi\tau+1} A^{\tau} B^{\tau\pi} + c^{\tau\pi\tau+2} A^{\tau+1} A^{\tau} B^{\tau\pi} + \cdots + \right. \\
& \left. + c^{\tau\pi T} A^{T-1} A^{T-2} \ldots A^{\tau} B^{\tau\pi} + b^{\tau\pi\tau\pi} \right) u^{\tau\pi} \underset{u^{\tau\pi}}{\rightarrow} \max, \\
& D^{\tau\pi} u^{\tau\pi} \le d^{\tau\pi},
\end{aligned}
\tag{13.12}
$$

for $\tau = 1, \ldots, T, \pi = 1, \ldots, \nu_\tau$.

Proof By performing direct transformations

$$
\begin{aligned}
x^1 &= A^0x^0 + \sum_{\pi=1}^{\nu_1} B^{1\pi}u^{1\pi},\\
x^2 &= A^1x^1 + \sum_{\pi=1}^{\nu_2} B^{2\pi}u^{2\pi} =\\
&= A^1\left(A^0x^0 + \sum_{\pi=1}^{\nu_1} B^{1\pi}u^{1\pi}\right) + \sum_{\pi=1}^{\nu_2} B^{2\pi}u^{2\pi} =\\
&= A^1A^0x^0 + A^1\sum_{\pi=1}^{\nu_1} B^{1\pi}u^{1\pi} + \sum_{\pi=1}^{\nu_2} B^{2\pi}u^{2\pi},\\
x^3 &= A^2x^2 + \sum_{\pi=1}^{\nu_3} B^{3\pi}u^{3\pi} =
\end{aligned}
$$

$$
\begin{aligned}
&= A^2\left(A^1A^0x^0 + A^1\sum_{\pi=1}^{\nu_1} B^{1\pi}u^{1\pi} + \sum_{\pi=1}^{\nu_2} B^{2\pi}u^{2\pi}\right) +\\
&\quad + \sum_{\pi=1}^{\nu_3} B^{3\pi}u^{3\pi} =\\
&= A^2A^1A^0x^0 + A^2A^1\sum_{\pi=1}^{\nu_1} B^{1\pi}u^{1\pi} + A^2\sum_{\pi=1}^{\nu_2} B^{2\pi}u^{2\pi} +\\
&\quad + \sum_{\pi=1}^{\nu_3} B^{3\pi}u^{3\pi},\\
&\dots\\
x^T &= A^{T-1}x^{T-1} + \sum_{\pi=1}^{\nu_T} B^{T\pi}u^{T\pi} =\\
&= \prod_{t=0}^{T-1} A^t x^0 + \prod_{t=1}^{T-1} A^t \sum_{\pi=1}^{\nu_1} B^{1\pi}u^{1\pi} + \prod_{t=2}^{T-1} A^t \sum_{\pi=1}^{\nu_2} B^{2\pi}u^{2\pi} +\\
&\quad + \dots + \prod_{t=T-1}^{T-1} A^t \sum_{\pi=1}^{\nu_{T-1}} B^{T-1\pi}u^{T-1\pi} + \sum_{\pi=1}^{\nu_T} B^{T\pi}u^{T\pi},
\end{aligned}
$$

and, by subsequent substitution in the objective/cost functions of (13.11), problem (13.11) is reduced to

$$\begin{aligned}
&f(u^{\tau\pi}||u^{-\tau\pi}) = \\
&= \left(c^{\tau\pi 1} + c^{\tau\pi 2}A^1 + c^{\tau\pi 3}A^2A^1 + \cdots + \right.\\
&\left.+ c^{\tau\pi T}A^{T-1}A^{T-2}\dots A^1\right)A^0x^0 + \\
&+ \left(c^{\tau\pi 1}B^{11} + c^{\tau\pi 2}A^1B^{11} + c^{\tau\pi 3}A^2A^1B^{11} + \cdots + \right.\\
&\left.+ c^{\tau\pi T}A^{T-1}A^{T-2}\dots A^1B^{11} + b^{\tau\pi 11}\right)u^{11} + \\
&+ \left(c^{\tau\pi 1}B^{12} + c^{\tau\pi 2}A^1B^{12} + c^{\tau\pi 3}A^2A^1B^{12} + \cdots + \right.\\
&\left.+ c^{\tau\pi T}A^{T-1}A^{T-2}\dots A^1B^{12} + b^{\tau\pi 12}\right)u^{12} + \cdots + \\
&+ \left(c^{\tau\pi 1}B^{1\nu_1} + c^{\tau\pi 2}A^1B^{1\nu_1} + c^{\tau\pi 3}A^2A^1B^{1\nu_1} + \cdots + \right.\\
&\left.+ c^{\tau\pi T}A^{T-1}A^{T-2}\dots A^1B^{1\nu_1} + b^{\tau\pi 1\nu_1}\right)u^{1\nu_1} + \\
&+ \left(c^{\tau\pi 2}B^{21} + c^{\tau\pi 3}A^2B^{21} + c^{\tau\pi 4}A^3A^2B^{21} + \cdots + \right.\\
&\left.+ c^{\tau\pi T}A^{T-1}A^{T-2}\dots A^2B^{21} + b^{\tau\pi 21}\right)u^{21} + \\
&+ \left(c^{\tau\pi 2}B^{22} + c^{\tau\pi 3}A^2B^{22} + c^{\tau\pi 4}A^3A^2B^{22} + \cdots + \right.\\
&\left.+ c^{\tau\pi T}A^{T-1}A^{T-2}\dots A^2B^{22} + b^{\tau\pi 22}\right)u^{22} + \cdots + \\
&+ \left(c^{\tau\pi 2}B^{2\nu_2} + c^{\tau\pi 3}A^2B^{2\nu_2} + c^{\tau\pi 4}A^3A^2B^{2\nu_2} + \cdots + \right.\\
&\left.+ c^{\tau\pi T}A^{T-1}A^{T-2}\dots A^2B^{2\nu_2} + b^{\tau\pi 2\nu_2}\right)u^{2\nu_2} + \\
&+ \cdots + \left(c^{\tau\pi T}B^{T\nu_T} + b^{\tau\pi T\nu_T}\right)u^{T\nu_T} \underset{u^{\tau\pi}}{\rightarrow} \max, \\
&\tau = 1, \dots, T, \pi = 1, \dots, \nu_\tau, \\
&D^{\tau\pi}u^{\tau\pi} \le d^{\tau\pi}, \tau = 1, \dots, T, \pi = 1, \dots, \nu_\tau.
\end{aligned} \tag{13.13}$$

Evidently, problem (13.13) defines a strategic game for which the Nash-Stackelberg equilibrium is also a Nash equilibrium and it is simply computed as a sequence of solutions of $\nu_1 + \cdots + \nu_T$ linear programming problems

$$\begin{aligned}
&f(u^{\tau\pi}||u^{-\tau\pi}) = \\
&= \left(c^{\tau\pi\tau}B^{\tau\pi} + c^{\tau\pi\tau+1}A^\tau B^{\tau\pi} + c^{\tau\pi\tau+2}A^{\tau+1}A^\tau B^{\tau\pi} + \cdots + \right.\\
&\left.+ c^{\tau\pi T}A^{T-1}A^{T-2}\dots A^\tau B^{\tau\pi} + b^{\tau\pi\tau\pi}\right)u^{\tau\pi} \underset{u^{\tau\pi}}{\rightarrow} \max, \\
&D^{\tau\pi}u^{\tau\pi} \le d^{\tau\pi},
\end{aligned} \tag{13.14}$$

$\tau = 1, \dots, T, \pi = 1, \dots, \nu_\tau.$

The truth of Theorem 13.10 follows from the equivalence of problems (13.14) with (13.11) and (13.12). □

Various particular cases of problem (13.11) and Theorem 13.10 may be examined, e.g. Theorem 13.11 considers one of the such cases.

Theorem 13.11 *If*

$$\begin{aligned}
A^0 &= A^1 = \cdots = A^{T-1} = A, \\
B^{11} &= B^{12} = \cdots = B^{T\nu_T} = B,
\end{aligned}$$

and problem (13.11) is solvable, then the sequence $\bar{u}^{11}, \bar{u}^{12}, \dots, \bar{u}^{T\nu_T}$ forms a Nash-Stackelberg equilibrium control if and only if $\bar{u}^{\tau\pi}$ is optimal in the linear programming problem

$$
\begin{aligned}
& f(u^{\tau\pi} || u^{-\tau\pi}) = \\
& = \left(c^{\tau\pi\tau} B + c^{\tau\pi\tau+1} AB + c^{\tau\pi\tau+2} (A)^2 B + \cdots + \right. \\
& \left. + c^{\tau\pi T} (A)^{T-\tau} B + b^{\tau\pi\tau\pi}\right) u^{\tau\pi} \underset{u^{\tau\pi}}{\rightarrow} \max, \\
& D^{\tau\pi} u^{\tau\pi} \leq d^{\tau\pi},
\end{aligned}
$$

for $\tau = 1, \ldots, T,\ \pi = 1, \ldots, \nu_\tau$.

Let us extend Pontryagin maximum principle on problem (13.11).
Consider the following recurrence relations

$$
\begin{aligned}
p^{\tau\pi T} &= c^{\tau\pi T}, \\
p^{\tau\pi t} &= p^{\tau\pi t+1} A^t + c^{\tau\pi t},\ t = T-1, \ldots, 1,
\end{aligned}
\tag{13.15}
$$

where $\tau = 1, \ldots, T,\ \pi = 1, \ldots, \nu_\tau$. The Hamiltonian functions are

$$
H_{\tau\pi t}(u^{\tau\pi}) = \left\langle p^{\tau\pi t} B^{\tau\pi} + b^{\tau\pi\tau\pi}, u^{\tau\pi} \right\rangle, t = T, \ldots, 1,
$$

where $\tau = 1, \ldots, T,\ \pi = 1, \ldots, \nu_\tau$ and $p^{\tau\pi t},\ t = T, \ldots, 1, \tau = 1, \ldots, T,\ \pi = 1, \ldots, \nu_\tau$ are defined by (13.15).

Theorem 13.12 *Let problem (13.11) be solvable. The sequence* $\bar{u}^{11}, \bar{u}^{12}, \ldots, \bar{u}^{T\nu_T}$ *forms a Nash-Stackelberg equilibrium control if and only if*

$$
H_{\tau\pi t}(\bar{u}^{\tau\pi}) = \max_{u^{\tau\pi}: D^{\tau\pi} u^{\tau\pi} \leq d^{\tau\pi}} H_{\tau\pi t}(u^{\tau\pi}),
$$

for $t = T, \ldots, 1,\ \tau = 1, \ldots, T,\ \pi = 1, \ldots, \nu_\tau$.

Proof It is sufficient to make direct substitutions above and to compare obtained results with the expressions in Theorem 13.10. It is evident that Theorems 13.10 and 13.12 are equivalent. □

13.6 Linear Discrete-Time Pareto-Nash-Stackelberg Control Problem

Let us unite problems (13.8) and (13.11) by considering the control of Pareto-Nash-Stackelberg type with T stages and $\nu_1 + \cdots + \nu_T$ players, where $\nu_1, \ldots, \nu_T$ are the correspondent numbers of players on the stages $1, \ldots, T$.

Every player is identified by two numbers as above in Nash-Stackelberg control: τ is the stage on which the player selects his strategy and π is player's number at the stage τ. In such game, at each stage τ, the players $1, 2, \ldots, \nu_\tau$ play a Pareto-Nash game by selecting simultaneously their strategies according to their criteria ($k_{\tau 1}, k_{\tau 2}, \ldots, k_{\tau\nu_\tau}$ are the numbers of criteria of respective players) and by communicating their and all precedent selected strategies to the following $\tau + 1$ stage players. After the all stage strategy selections, all the players compute their gains on the resulting profile. Such type of control is called Pareto-Nash-Stackelberg control, and the corresponding problem — a linear discrete-time Pareto-Nash-Stackelberg control problem.

The decision control process may be modelled as:

$$\begin{aligned}
& f_{\tau\pi}(x, u^{\tau\pi} \| u^{-\tau\pi}) = \sum_{t=1}^{T} \left(c^{\tau\pi t} x^t + \sum_{\mu=1}^{\nu_t} b^{\tau\pi t\mu} u^{t\mu} \right) \underset{u^{\tau\pi}}{\longrightarrow} \max, \\
& \tau = 1, \ldots, T, \ \pi = 1, \ldots, \nu_\tau, \\
& x^t = A^{t-1} x^{t-1} + \sum_{\pi=1}^{\nu_t} B^{t\pi} u^{t\pi}, \ t = 1, \ldots, T, \\
& D^{t\pi} u^{t\pi} \leq d^{t\pi}, \ t = 1, \ldots, T, \ \pi = 1, \ldots, \nu_t,
\end{aligned} \tag{13.16}$$

where $x^0,\ x^t \in \mathbb{R}^n,\ c^{\tau\pi t\mu} \in \mathbb{R}^{k_{tp} \times n},\ u^{\tau\pi} \in \mathbb{R}^m,\ b^{\tau\pi t\mu} \in \mathbb{R}^{k_{tp} \times n},$ $A^{t-1} \in \mathbb{R}^{n \times n},\ B^{\tau\pi} \in \mathbb{R}^{n \times m},\ d^{\tau\pi} \in \mathbb{R}^k,\ D^{\tau\pi} \in \mathbb{R}^{k \times n},\ t,\ \tau = 1, \ldots, T,\ \pi = 1, \ldots, \nu_\tau,\ \mu = 1, \ldots, \nu_t.$

Theorem 13.13 *Let (13.16) be solvable. The control sequence* $\bar{u}^{11}, \bar{u}^{12}, \ldots, \bar{u}^{T\nu_T}$ *forms a Pareto-Nash-Stackelberg equilibrium control in (13.16) if and only if* $\bar{u}^{\tau\pi}$ *is an efficient solution of the multi-criteria linear programming problem*

$$\begin{aligned}
& f(u^{\tau\pi} \| u^{-\tau\pi}) = \\
& = \left(c^{\tau\pi\tau} B^{\tau\pi} + c^{\tau\pi\tau+1} A^\tau B^{\tau\pi} + c^{\tau\pi\tau+2} A^{\tau+1} A^\tau B^{\tau\pi} + \cdots + \right. \\
& \left. + c^{\tau\pi T} A^{T-1} A^{T-2} \ldots A^\tau B^{\tau\pi} + b^{\tau\pi\tau\pi} \right) u^{\tau\pi} \underset{u^{\tau\pi}}{\rightarrow} \max, \\
& D^{\tau\pi} u^{\tau\pi} \leq d^{\tau\pi},
\end{aligned} \tag{13.17}$$

for $\tau = 1, \ldots, T, \pi = 1, \ldots, \nu_\tau$.

Proof By performing similar direct transformation as above, Problems (13.16) are reduced to a sequence of multi-criteria linear programming problems (13.17).

Equivalence of (13.16) and (13.17) proves Theorem 13.13. □

As a corollary, Theorem 13.14 follows.

Theorem 13.14 *If*

$$\begin{aligned}
A^0 = A^1 = \cdots = A^{T-1} = A, \\
B^{11} = B^{12} = \cdots = B^{T\nu_T} = B,
\end{aligned}$$

and Problem (13.16) is solvable, then the sequence $\bar{u}^{11}, \bar{u}^{12}, \ldots, \bar{u}^{T\nu_T}$ *forms a Pareto-Nash-Stackelberg equilibrium control if and only if* $\bar{u}^{\tau\pi}$ *is an efficient solution of the multi-criteria linear programming problem*

$$
\begin{aligned}
& f(u^{\tau\pi} || u^{-\tau\pi}) = \\
& = \left(c^{\tau\pi\tau} B + c^{\tau\pi\tau+1} AB + c^{\tau\pi\tau+2} (A)^2 B + \cdots + \right. \\
& \left. + c^{\tau\pi T} (A)^{T-\tau} B + b^{\tau\pi\tau\pi}\right) u^{\tau\pi} \underset{u^{\tau\pi}}{\rightarrow} \max, \\
& D^{\tau\pi} u^{\tau\pi} \leq d^{\tau\pi},
\end{aligned}
$$

for $\tau = 1, \ldots, T,\ \pi = 1, \ldots, \nu_\tau$.

Pontryagin maximum principle may be generalized for (13.16).
By considering recurrence relations

$$
\begin{aligned}
p^{\tau\pi T} &= c^{\tau\pi T}, \\
p^{\tau\pi t} &= p^{\tau\pi t+1} A^t + c^{\tau\pi t},\ t = T - 1, \ldots, 1,
\end{aligned} \tag{13.18}
$$

where $\tau = 1, \ldots, T, \pi = 1, \ldots, \nu_\tau$, Hamiltonian vector-functions are defined as

$$
H_{\tau\pi t} (u^{\tau\pi}) = \left\langle p^{\tau\pi t} B^{\tau\pi} + b^{\tau\pi\tau\pi}, u^{\tau\pi} \right\rangle, t = T, \ldots, 1,
$$

where $\tau = 1, \ldots, T, \pi = 1, \ldots, \nu_\tau$ and $p^{\tau\pi t}, t = T, \ldots, 1, \tau = 1, \ldots, T,\ \pi = 1, \ldots, \nu_\tau$. Remark the vector nature of (13.18) via (13.15).

Theorem 13.15 *Let (13.16) be solvable. The control sequence* $\bar{u}^{11}, \bar{u}^{12}, \ldots, \bar{u}^{T\nu_T}$ *forms a Pareto-Nash-Stackelberg equilibrium control if and only if*

$$
\bar{u}^{\tau\pi} \in \underset{u^{\tau\pi}: D^{\tau\pi} u^{\tau\pi} \leq d^{\tau\pi}}{Arg\ max} H_{\tau\pi t} (u^{\tau\pi}),
$$

for $t = T, \ldots, 1, \tau = 1, \ldots, T, \pi = 1, \ldots, \nu_\tau$.

Proof The proof is providing by direct substitutions and comparing the final expressions with them in Theorem 13.13. Theorems 13.13 and 13.15 are equivalent. □

13.7 Linear Discrete-Time Set-Valued Optimal Control Problem

The state of controlled system is described above by a point of a vector space. Indeed, real systems may be treated as an n-dimension body, the state of which is described by a set of points in every time moment. Evidently, the initial state of the system is described by the initial set $X^0 \subset \mathbb{R}^n$. Naturally, the following problem arises

$$
\begin{aligned}
F(X, U) = \sum_{t=1}^{T} (c^t X^t + b^t U^t) &\rightarrow \max, \\
X^t = A^{t-1} X^{t-1} + B^t U^t,\ t &= 1, \ldots, T, \\
D^t U^t \leq d^t,\ t &= 1, \ldots, T,
\end{aligned} \tag{13.19}
$$

where $X^0, X^t \subset \mathbb{R}^n$, $c^t \in \mathbb{R}^n$, $U^t \subset \mathbb{R}^m$, $b^t \in \mathbb{R}^m$, $A^{t-1} \in \mathbb{R}^{n\times n}$, $B^t \in \mathbb{R}^{n\times m}$, $d^t \in \mathbb{R}^k$, $D^t \in \mathbb{R}^{k\times n}$, $c^t X^t = \langle c^t, X^t \rangle$, $b^t U^t = \langle b^t, U^t \rangle$, $t = 1, \dots, T$. Linear set operations in (13.19) are defined obviously (see, e.g., [17]): $AX = \{Ax : x \in X\}$, for all $X \subset \mathbb{R}^n$, for all $A \in \mathbb{R}^{n\times n}$.

The objective set-valued mapping $F : X \times Y \multimap R$, $F(X, Y) \subset R$, represents a summation of intervals. That is, the optimization of the objective mapping in problem (13.19) needs an interval arithmetic treatment.

Theorem 13.16 *Let problem (13.19) be solvable. The control sequence*

$$\bar{u}^1, \bar{u}^2, \dots, \bar{u}^T$$

forms an optimal control if and only if $\bar{u}^t$ *is the solution of linear programming problem*

$$\begin{aligned}
&\left(c^t B^t + c^{t+1} A^t B^t + \dots + c^T A^{T-1} A^{T-2} \dots A^t B^t + b^t\right) u^t \to \max,\\
&D^t u^t \le d^t,
\end{aligned}$$

for $t = 1, \dots, T$.

Proof By direct transformations, (13.19) is transformed into

$$\begin{aligned}
&F(U) =\\
&= \left(c^1 + c^2 A^1 + c^3 A^2 A^1 + \dots +\right.\\
&\left.+ c^T A^{T-1} A^{T-2} \dots A^1\right) A^0 X^0 +\\
&+ \left(c^1 B^1 + c^2 A^1 B^1 + c^3 A^2 A^1 B^1 + \dots +\right.\\
&\left.+\, c^T A^{T-1} A^{T-2} \dots A^1 B^1 + b^1\right) U^1 +\\
&+ \left(c^2 B^2 + c^3 A^2 B^2 + c^4 A^3 A^2 B^2 + \dots +\right.\\
&\left.+\, c^T A^{T-1} A^{T-2} \dots A^2 B^2 + b^2\right) U^2 +\\
&+ \dots + \left(c^T B^T + b^T\right) U^T \to \max,\\
&D^t U^t \le d^t, t = 1, \dots, T.
\end{aligned} \tag{13.20}$$

The equivalence of problems (13.19) and (13.20) as well as the form of the objective mapping proves that the optimal solution does not depend of initial point. The cardinality of every control set $U^1, \dots, U^T$ is equal to one. Thus, the truth of Theorem 13.16 for problem (13.19) is established. □

Analogue conclusions are true for all the problems and theorems considered above. These facts are the subject of the following chapters.

13.8 Concluding Remarks

There are different types of control: optimal control, Stackelberg control, Pareto-Stackelberg control, Nash-Stackelberg control, Pareto-Nash-Stackelberg control, etc.

The direct-straightforward, dual and classical principles (Pontryagin and Bellman) may be applied for determining the desired control of dynamic processes. These principles are the bases for pseudo-polynomial methods, which are exposed as a consequence of theorems for the linear discrete-time Pareto-Nash-Stackelberg control problems.

The direct-straightforward principle is applicable for solving the problem of determining the optimal control of the set-valued linear discrete-time processes. Pseudo-polynomial solution method is constructed.

References

1. Leitmann, G., S. Pickl, and Z. Wang. 2014. Multi-agent Optimal Control Problems and Variational Inequality Based Reformulations. In *Dynamic Games in Economics*, ed. J. Haunschmied, V.M. Veliov, and S. Wrzaczek, 205–217. Berlin: Springer.
2. Başar, T., and G.J. Olsder. 1999. *Dynamic Noncooperative Game Theory, 536*. Philadelphia: Society for Industrial and Applied Mathematics.
3. Long, N.V. 2010. *A Survey of Dynamic Games in Economics*. New Jersey: World Scientific, XIV–275 pp.
4. Zaslavski, A.J. 2017. *Discrete-Time Optimal Control and Games on Large Intervals*. Springer, Switzerland, X+398 pp.
5. Pontryagin, L.S., V.G. Boltyanskii, R.V. Gamkrelidze and E.F. Mishchenko. 1961. *Mathematical theory of optimal processes*. Moscow: Nauka, 393 pp. (in Russian).
6. Bellman, R. 1957. *Dynamic Programming*, 365. New Jersey: Princeton University Press.
7. Vignaly, R., and M. Prandini. 1996. Minimum resource commitment for reachability specifications in a discrete time linear setting. *IEEE Transactions on Automatic Control* 62 (6): 3021–3028.
8. Von Neumann, J., and O. Morgenstern. 1944. *Theory of Games and Economic Behavior*. Princeton, New Jersey: Annals Princeton University Press, (2nd edn., 1947), 674 pp.
9. Nash, J. 1951. Noncooperative games. *Annals of Mathematics* 54 (2): 280–295.
10. Von Stackelberg, H. *Marktform und Gleichgewicht (Market Structure and Equilibrium)*, Vienna: Springer Verlag, 1934, XIV+134 pp. (in German).
11. Leitmann, G. 1978. On Generalized Stackelberg Strategies. *Journal of Optimization Theory and Applications* 26: 637–648.
12. Ungureanu, V. 2008. Solution principles for simultaneous and sequential games mixture. *ROMAI Journal* 4 (1): 225–242.
13. Isaacs, R. 1965. *Differential games: a mathematical theory with applications to warfare and pursuit, control and optimization*. Berlin: Wiley, XXIII+385 pp.
14. Chikriy, A.A. 1992. *Conflict-controlled processes*. Kiev: Naukova Dumka, 384 pp. (in Russian).
15. Lin, Ch.-S., Ch.-T. Chen, F.-Sh. Chen and W.-Zh. Hung. 2014. A Novel Multiperson Game Approach for Linguistic Multicriteria Decision Making Problems. *Mathematical Problems in Engineering*, Article ID 592326, 20 pp.
16. Ashmanov, S.A., and A.V. Timohov. 1991. *The Optimization Theory in Problems and Exercises, 142–143*. Moscow: Nauka.
17. Rockafellar, T. 1970. *Convex Analysis*, 468. Princeton: Princeton University Press.

Chapter 14
Linear Discrete-Time Set-Valued Pareto-Nash-Stackelberg Control and Its Principles

Abstract In this chapter, game and control concepts that are examined in the precedent chapters are unified in unique mathematical models based on a foundation of set-valued mappings. New solutions concepts are inherent and they are introduced. After concept definitions and problem formulations, methods for their computing are examined.

14.1 Introduction

Pareto-Nash-Stackelberg control processes, examined in [1] as extension and integration of optimal multi-objective control processes [2, 3] with simultaneous and sequential games [4–8], are generalized by considering the set-valued multi-criteria control processes of a system with discrete-time dynamics described by a system of set-valued linear equations. The Pareto-Nash-Stackelberg set-valued control problems of linear discrete-time system are solved by applying a straightforward principle [1]. The characteristics and properties of Set-Valued Algebra [9] together with Interval Analysis [10] serve as foundation for obtained results.

The exposition starts with the simplest case of linear discrete-time set-valued optimal control problem and, by sequential considering of more general cases, finalizes with the Pareto-Nash-Stackelberg set-valued control problem. The maximum principle of Pontryagin [1, 2] is extended, formulated and proved for all the considered problems too. Its equivalence with the straightforward direct principle is established.

14.2 Linear Discrete-Time Set-Valued Optimal Control Problem

The system may be imagined as an n-dimension dynamic body the state of which is described by the set of points in every time moment. So, the initial state is described by the initial set $X^0 \subset \mathbb{R}^n$. An optimal control problem naturally arises:

V. Ungureanu, *Pareto-Nash-Stackelberg Game and Control Theory*,
Smart Innovation, Systems and Technologies 89,
https://doi.org/10.1007/978-3-319-75151-1_14

$$
\begin{aligned}
F(X, U) = \sum_{t=1}^{T} (c^t X^t + b^t U^t) &\to \max, \\
X^t &= A^{t-1} X^{t-1} + B^t X^t,\ t = 1, \ldots, T, \\
D^t U^t &\le d^t, t = 1, \ldots, T,
\end{aligned}
\tag{14.1}
$$

where

$$
\begin{aligned}
&X^0,\ X^t \subset \mathbb{R}^n, c^t \in \mathbb{R}^n, U^t \subset \mathbb{R}^m, b^t \in \mathbb{R}^m, \\
&A^{t-1} \in \mathbb{R}^{n \times n}, B^t \in \mathbb{R}^{n \times m}, d^t \in \mathbb{R}^k, D^t \in \mathbb{R}^{k \times n}, \\
&c^t X^t = \langle c^t, X^t \rangle, b^t U^t = \langle b^t, U^t \rangle, t = 1, \ldots, T, \\
&U = (U^1, U^2, \ldots, U^T), X = (X^0, X^1, \ldots, X^T).
\end{aligned}
$$

Set operations in (14.1) are defined obviously [9]:

$$
AX = \{Ax : x \in X\}, \text{for all } X \subset \mathbb{R}^n,\ \ A \subset \mathbb{R}^{n \times n}.
$$

Remark 14.1 The objective set-valued mapping

$$
F : X \times U \multimap \mathbb{R},\ F(X, U) \subset \mathbb{R},
$$

represents a summation of intervals. So, the further application of the interval arithmetic [10] is intrinsic.

Theorem 14.1 *Let problem (14.1) be solvable. The control sequence $\bar{u}^1, \bar{u}^2, \ldots, \bar{u}^T$, is optimal if and only if $\bar{u}^t$ is a solution of the linear programming problem*

$$
\begin{aligned}
(c^t B^t + c^{t+1} A^t B^t + \cdots + c^T A^{T-1} A^{T-2} \ldots A^t B^t + b^t) u^t &\to \max, \\
D^t u^t &\le d^t,
\end{aligned}
$$

for $t = 1, \ldots, T$.

Proof By performing direct substitutions in (14.1):

$$
\begin{aligned}
X^1 &= A^0 X^0 + B^1 U^1, \\
X^2 &= A^1 X^1 + B^2 U^2 = A^1 (A^0 X^0 + B^1 U^1) + B^2 U^2 = \\
&= A^1 A^0 X^0 + A^1 B^1 U^1 + B^2 U^2, \\
X^3 &= A^2 X^2 + B^3 U^3 = A^2 (A^1 A^0 X^0 + A^1 B^1 U^1 + B^2 U^2) + B^3 U^3 = \\
&= A^2 A^1 A^0 X^0 + A^2 A^1 B^1 U^1 + A^2 B^2 U^2 + B^3 U^3, \\
&\ldots \\
X^T &= A^{T-1} X^{T-1} + B^T U^T = \\
&= \prod_{t=0}^{T-1} A^t X^t + \prod_{t=1}^{T-1} A^t B^1 U^1 + \prod_{t=2}^{T-1} A^t B^2 U^2 + \cdots + \\
&\quad + A^{T-1} B^{T-1} U^{T-1} + B^T U^T,
\end{aligned}
$$

and by subsequent substitution of the resulting relations in the objective mapping:

$$
\begin{aligned}
&F(X,U)=\\
&=c^1(A^0X^0+B^1U^1)+c^2(A^1A^0X^0+A^1B^1U^1+B^2U^2)+\\
&\quad+c^3(A^2A^1A^0X^0+A^2A^1B^1U^1+A^2B^2U^2+B^3U^3)+\\
&\quad+\dots+c^T(\prod_{t=0}^{T-1}A^tX^t+\prod_{t=1}^{T-1}A^tB^1U^1+\prod_{t=2}^{T-1}A^tB^2U^2+\\
&\quad+\dots+A^{T-1}B^{T-1}U^{T-1}+B^TU^T)+\\
&\quad+b^1U^1+b^2U^2+\dots+b^TU^T=\\
&=(c^1+c^2A^1+c^3A^2A^1+\dots+c^TA^{T-1}A^{T-2}\dots A^1)A^0X^0+\\
&\quad+(c^1B^1+c^2A^1B^1+c^3A^2A^1B^1+\dots+\\
&\quad+c^TA^{T-1}A^{T-2}\dots A^1B^1+b^1)U^1+\\
&\quad+(c^2B^2+c^3A^2B^2+c^4A^3A^2B^2+\dots+\\
&\quad+c^TA^{T-1}A^{T-2}\dots A^2B^2+b^2)U^2+\dots+\\
&\quad+(c^TB^T+b^T)U^T,
\end{aligned}
$$

problem (14.1) is transformed into:

$$
\begin{aligned}
F(U)=\;&(c^1+c^2A^1+c^3A^2A^1+\dots+\\
&+c^TA^{T-1}A^{T-2}\dots A^1)A^0X^0+\\
&+(c^1B^1+c^2A^1B^1+c^3A^2A^1B^1+\dots+\\
&+c^TA^{T-1}A^{T-2}\dots A^1B^1+b^1)U^1+\\
&+(c^2B^2+c^3A^2B^2+c^4A^3A^2B^2+\dots+\\
&+c^TA^{T-1}A^{T-2}\dots A^2B^2+b^2)U^2+\dots+\\
&+(c^TB^T+b^T)U^T\to\max,\\
D^tU^t\le d^t,&\ t=1,\dots,T.
\end{aligned}
\tag{14.2}
$$

Obviously, (14.1) and (14.2) are equivalent.

The form of the objective mapping $F(U)$ in (14.2) establishes that the optimal control does not depend of initial state X^0.

By applying the specific interval arithmetic properties of the linear set-valued programming problems, we can conclude that the solution set of problem (14.2) is equivalent with the solution set of the traditional point-valued linear programming problem, that is we can consider that, in general, the cardinality of every control set $U^1, U^2, \dots, U^T$ is equal to 1. So, the solution of problem (14.2) may be obtained as a sequence of solutions of T linear programming problems. □

Apparently, we constructed above a polynomial method of solving (14.1). Actually, the method has a pseudo-polynomial complexity because of possible exponential value of T on n.

Following Theorem 14.2 is an important particular corollary of the precedent theorem, but it has its own independent importance.

Theorem 14.2 *If*

$$A^0 = A^1 = \cdots = A^{T-1} = A,$$
$$B^1 = B^2 = \cdots = B^T = B,$$

and problem (14.1) is solvable, then the control sequence $\bar{u}^1, \bar{u}^2, \ldots, \bar{u}^T$, forms an optimal control if and only if $\bar{u}^t$ is a solution of the linear programming problem

$$(c^t B + c^{t+1} AB + c^{t+2}(A)^2 B + \cdots + c^T (A)^{T-t} B + b^t)u^t \rightarrow \max,$$
$$D^t u^t \leq d^t,$$

for $t = 1, \ldots, T$.

Theorem 14.1 establishes a principle for solving problem (14.1). The maximum principle of Pontryagin may be applied for solving (14.1) too. As the cardinality of every control set $U^1, U^2, \ldots, U^T$ is equal to 1, let us consider the following recurrent relations:

$$\begin{aligned} p^T &= c^T \\ p^t &= p^{t+1} A^t + c^t,\ t = T-1, \ldots, 1. \end{aligned} \tag{14.3}$$

Hamiltonian functions are defined on the basis of (14.3) as

$$H_t(u^t) = \langle p^t B^t + b^t, u^t \rangle,\ t = T, \ldots, 1.$$

Theorem 14.3 *Let problem (14.1) be solvable. The control sequence $\bar{u}^1, \bar{u}^2, \ldots, \bar{u}^T$, is optimal if and only if*

$$H_t(\bar{u}^t) = \max_{u^t : D^t u^t \leq d^t} H_t(u^t),\ t = T, \ldots, 1.$$

It's obvious that Theorems 14.1 and 14.3 are equivalent. Only the means that we used to formulate and prove them differ. For more details on the theoreme proving, the proofs of the precedent chapter's theorems may be consulted.

14.3 Linear Discrete-Time Set-Valued Stackelberg Control Problem

Let us modify problem (14.1) by considering the control of Stackelberg type, that is the Stackelberg game with T players [1, 4, 5, 8]. In such game, at each stage t ($t = 1, \ldots, T$) the player t selects his strategy and communicates his and all precedent selected strategies to the following $t+1$ player. After all stage strategy selections, all the players compute their gains on the resulting profile.

Let us call a Stackelberg control this type of system control and the corresponding problem — a linear discrete-time set-valued Stackelberg control problem. The decision process described above may be formalized as

$$\begin{aligned} F_1(X,U) &= \sum_{t=1}^{T}(c^{1t}X^t + b^{1t}U^t) \underset{U^1}{\longrightarrow} \max, \\ F_2(X,U) &= \sum_{t=1}^{T}(c^{2t}X^t + b^{2t}U^t) \underset{U^2}{\longrightarrow} \max, \\ &\dots \\ F_T(X,U) &= \sum_{t=1}^{T}(c^{Tt}X^t + b^{Tt}U^t) \underset{U^T}{\longrightarrow} \max, \\ X^t &= X^{t-1}A^{t-1} + B^tX^t, t = 1,\dots,T, \\ D^tU^t &\le d^t, t = 1,\dots,T, \end{aligned} \tag{14.4}$$

where

$$\begin{aligned} &X^0, X^t \subset \mathbb{R}^n, c^{\pi t} \in \mathbb{R}^n, U^t \subset \mathbb{R}^m, \\ &b^{\pi t} \in \mathbb{R}^m, A^{t-1} \in \mathbb{R}^{n\times n}, B^t \in \mathbb{R}^{n\times m}, \\ &d^t \in \mathbb{R}^k, D^t \in \mathbb{R}^{k\times n}, \\ &c^tX^t = \langle c^t, X^t\rangle, \\ &b^tU^t = \langle b^t, U^t\rangle, \\ &t, \pi = 1,\dots,T. \end{aligned}$$

Theorem 14.4 *Let problem (14.4) be solvable. The control sequence $\bar{u}^1, \bar{u}^2, \dots, \bar{u}^T$, forms a Stackelberg equilibrium control if and only if $\bar{u}^\pi$ is an optimal solution of the problem*

$$\begin{aligned} &(c^{\pi\pi}B^\pi + c^{\pi\pi+1}A^\pi B^\pi + \dots + c^{\pi T}A^{T-1}A^{T-2}\dots A^\pi B^\pi + b^{\pi\pi})u^\pi \underset{u^\pi}{\longrightarrow} \max, \\ &D^\pi u^\pi \le d^\pi, \end{aligned}$$

for every $\pi = 1,\dots,T$.

Proof The strategy set of the player π ($\pi = 1, 2, \dots, T$) is equivalent to the admissible set of the problem:

$$\begin{aligned} F_\pi(X, U^\pi||U^{-\pi}) &= \sum_{t=1}^{T}(c^{\pi t}X^t + b^{\pi t}U^t) \underset{U^\pi}{\longrightarrow} \max, \\ X^\pi &= X^{\pi-1}A^{\pi-1} + B^\pi X^\pi, \\ D^\pi U^\pi &\le d^\pi. \end{aligned}$$

The decision problem of the player π ($\pi = 1, 2, \dots, T$) is defined by the precedent linear set-valued programming problem. Since the controlled system is one for all the players, by performing the direct substitutions as above, problem (14.4) is transformed into the problem

$$\begin{aligned} F_\pi(U^\pi || U^{-\pi}) = &(c^{\pi 1} + c^{\pi 2}A^1 + c^{\pi 3}A^2A^1 + \dots + \\ &+ c^{\pi T}A^{T-1}A^{T-2}\dots A^1)A^0X^0 + \\ &+ (c^{\pi 1}B^1 + c^{\pi 2}A^1B^1 + c^{\pi 3}A^2A^1B^1 + \dots + \\ &+ c^{\pi T}A^{T-1}A^{T-2}\dots A^1B^1 + b^{\pi 1})U^1 + \\ &+ (c^{\pi 2}B^2 + c^{\pi 3}A^2B^2 + c^{\pi 4}A^3A^2B^2 + \dots + \\ &+ c^{\pi T}A^{T-1}A^{T-2}\dots A^2B^2 + b^{\pi 2})U^2 + \dots + \\ &+ (c^{\pi T}B^T + b^{\pi T})U^T \underset{U^\pi}{\longrightarrow} \max, \pi = 1, \cdots, T, \\ D^tU^t \leq d^t,\ &t = 1, \dots, T. \end{aligned} \tag{14.5}$$

As in the precedent optimal control case, the cardinality of each Stackelberg control set $U^1, U^2, \dots, U^T$ may be reduced to "1", i.e. the solution set of the traditional linear programming problem defines the control. From the equivalence of problems (14.4) and (14.5), the proof of the theorem follows. □

Following Theorem 14.5 is an important particular case of Theorem 14.4.

Theorem 14.5 *If $A^0 = A^1 = \dots = A^{T-1} = A$, $B^1 = B^2 = \dots = B^T = B$ and (14.4) is solvable, then the sequence $\bar{u}^1, \bar{u}^2, \dots, \bar{u}^T$, forms a Stackelberg equilibrium control if and only if $\bar{u}^\pi$ is the solution of linear programming problem*

$$(c^{\pi\pi}B + c^{\pi\pi+1}AB + c^{\pi\pi+2}(A)^2B + \dots + c^{\pi T}(A)^{T-\pi}B + b^{\pi\pi})u^\pi \underset{u^\pi}{\longrightarrow} \max,$$
$$D^\pi u^\pi \leq d^\pi,$$

for $\pi = 1, \dots, T$.

Theorem 14.4 establishes a principle for solving problem (14.4).

The maximum principle of Pontryagin may be applied for solving (14.4) too. Let us consider the following recurrence relations

$$\begin{aligned} p^{\pi T} &= c^{\pi T}, \\ p^{\pi t} &= p^{\pi t+1}A^t + c^{\pi t},\ t = T-1, \dots, 1, \end{aligned} \tag{14.6}$$

where $\pi = 1, \dots, T$. Hamiltonian functions are defined on the bases of relations (14.6) as

$$H_{\pi t}(u^t) = \langle p^{\pi t}B^t + b^{\pi t}, u^t \rangle,\ t = T, \dots, 1,\ \pi = 1, \dots, T.$$

Theorem 14.6 *Let problem (14.4) be solvable. The sequence* $\bar{u}^1, \bar{u}^2, \ldots, \bar{u}^T$, *forms a Stackelberg equilibrium control if and only if*

$$H_{\pi\pi}(\bar{u}^\pi) = \max_{u^\pi : D^\pi u^\pi \le d^\pi} H_{\pi\pi}(u^\pi),$$

for $\pi = 1, \ldots, T$.

Proof The proof is provided by direct substitution of relations (14.6) in the Hamiltonian functions and by comparing the final results with the linear programming problems from Theorem 14.4. Obviously, Theorems 14.4 and 14.6 are equivalent. □

From computational point of view, the method for solving problem (14.4) established by Theorem 14.4 seems to be more preferable than the method established by Theorem 14.4.

14.4 Linear Discrete-Time Set-Valued Pareto-Stackelberg Control Problem

Let us modify problem (14.4) by considering a control of Pareto-Stackelberg type. At each stage a single player makes decision. Every player selects his strategy (control) on his stage by considering his criteria and communicates his choice and precedent players choices to the following player. At last stage, after all stage strategy selections, the players compute their gains. Such type of control is called Pareto-Stackelberg control, and the corresponding problem is called a linear discrete-time set-valued Pareto-Stackelberg control problem.

The decision process is formalized as follows:

$$\begin{aligned}
F_1(X, U) &= \sum_{t=1}^{T} (c^{1t} X^t + b^{1t} U^t) \underset{U^1}{\longrightarrow} \max, \\
F_2(X, U) &= \sum_{t=1}^{T} (c^{2t} X^t + b^{2t} U^t) \underset{U^2}{\longrightarrow} \max, \\
&\ldots \\
F_T(X, U) &= \sum_{t=1}^{T} (c^{Tt} X^t + b^{Tt} U^t) \underset{U^T}{\longrightarrow} \max, \\
X^t &= A^{t-1} X^{t-1} + B^t X^t,\ t = 1, \ldots, T, \\
D^t U^t &\le d^t,\ t = 1, \ldots, T,
\end{aligned} \tag{14.7}$$

where $X^0, X^t \subset \mathbb{R}^n, c^{\pi t} \in \mathbb{R}^{k_\pi \times n}, U^t \subset \mathbb{R}^m, b^{\pi t} \in \mathbb{R}^{k_\pi \times m}, A^{t-1} \in \mathbb{R}^{n \times n}, B^t \in \mathbb{R}^{n \times m}$, $d^t \in \mathbb{R}^k, D^t \in \mathbb{R}^{k \times n},\ t, \pi = 1, \ldots, T$.

Theorem 14.7 *Let problem (14.7) be solvable. The control sequence $\bar{u}^1, \bar{u}^2, \ldots, \bar{u}^T$, forms a Pareto-Stackelberg equilibrium control if and only if $\bar{u}^\pi$ is an efficient solution of the multi-criteria linear programming problem*

$$\begin{aligned}
&(c^{\pi\pi} B^\pi + c^{\pi\pi+1} A^\pi B^\pi + \cdots + \\
&+ c^{\pi T} A^{T-1} A^{T-2} \ldots A^\pi B^\pi + b^{\pi\pi}) u^\pi \underset{u^\pi}{\longrightarrow} \max, \\
&D^\pi u^\pi \leq d^\pi,
\end{aligned}$$

for $\pi = 1, \ldots, T$.

Proof The strategy set of the player π ($\pi = 1, 2, \ldots, T$) is defined formally by the problem:

$$\begin{aligned}
F_\pi(X, U^\pi || U^{-\pi}) = \sum_{t=1}^{T} (c^{\pi t} X^t + b^{\pi t} U^t) \underset{U^\pi}{\longrightarrow} \max, \\
X^\pi = X^{\pi-1} A^{\pi-1} + B^\pi X^\pi, \\
D^\pi U^\pi \leq d^\pi.
\end{aligned}$$

By performing the direct transformations as above, problem (14.7) is transformed into

$$\begin{aligned}
F_\pi(U^\pi || U^{-\pi}) = &(c^{\pi 1} + c^{\pi 2} A^1 + c^{\pi 3} A^2 A^1 + \cdots + \\
&+ c^{\pi T} A^{T-1} A^{T-2} \ldots A^1) A^0 X^0 + \\
&+ (c^{\pi 1} B^1 + c^{\pi 2} A^1 B^1 + c^{\pi 3} A^2 A^1 B^1 + \cdots + \\
&+ c^{\pi T} A^{T-1} A^{T-2} \ldots A^1 B^1 + b^{\pi 1}) U^1 + \\
&+ (c^{\pi 2} B^2 + c^{\pi 3} A^2 B^2 + c^{\pi 4} A^3 A^2 B^2 + \cdots + \\
&+ c^{\pi T} A^{T-1} A^{T-2} \ldots A^2 B^2 + b^{\pi 2}) U^2 + \cdots + \\
&+ (c^{\pi T} B^T + b^{\pi T}) U^T \underset{U^\pi}{\longrightarrow} \max, \\
&\pi = 1, \ldots, T, \\
D^t U^t \leq d^t, &\ t = 1, \ldots, T.
\end{aligned} \tag{14.8}$$

On the basis of interval arithmetic properties, we can conclude that (14.8) is equivalent with a simple multi-criteria linear programming problem. Additionally, from the equivalence of (14.7) and (14.8), the truth of the theorem follows. □

As above, Theorem 14.8 is dedicated to a particular case of problem (14.7).

Theorem 14.8 *If*

$$\begin{aligned}
A^0 = A^1 = \cdots = A^{T-1} = A, \\
B^1 = B^2 = \cdots = B^T = B,
\end{aligned}$$

and problem (14.7) is solvable, then the sequence $\bar{u}^1, \bar{u}^2, \ldots, \bar{u}^T$, forms a Pareto-Stackelberg equilibrium control if and only if $\bar{u}^\pi$ is an efficient solution of the multi-

criteria linear programming problem

$$
\begin{aligned}
&(c^{\pi\pi} B + c^{\pi\pi+1} AB + c^{\pi\pi+2} (A)^2 B + \cdots + \\
&+ c^{\pi T} (A)^{T-\pi} B + b^{\pi\pi}) u^{\pi} \underset{u^{\pi}}{\longrightarrow} \max, \\
&D^{\pi} u^{\pi} \leq d^{\pi},
\end{aligned}
$$

for $\pi = 1, \ldots, T$.

Let us extend the Pontryagin maximum principle on problem (14.7). By considering the recurrence relations

$$
\begin{aligned}
p^{\pi T} &= c^{\pi T}, \\
p^{\pi t} &= p^{\pi t+1} A^t + c^{\pi t}, \ t = T-1, \ldots, 1,
\end{aligned}
\tag{14.9}
$$

where $\pi = 1, \ldots, T$. The Hamiltonian vector-functions may be defined on the basis of (14.7) and (14.9) as

$$
H_{\pi t}(u^t) = \langle p^{\pi t} B^t + b^{\pi t}, u^t \rangle, t = T, \ldots, 1, \pi = 1, \ldots, T.
$$

Theorem 14.9 *Let problem (14.7) be solvable. The sequence* $\bar{u}^1, \bar{u}^2, \ldots, \bar{u}^T$, *forms a Pareto-Stackelberg equilibrium control if and only if*

$$
\bar{u}^{\pi} \in \underset{u^{\pi}: D^{\pi} u^{\pi} \leq d^{\pi}}{\text{Arg max}} \ H_{\pi\pi}(u^{\pi}),
$$

for $\pi = 1, \ldots, T$.

Proof By direct substitution of (14.9) in Hamiltonian functions and by comparing the final results with multi-criteria linear programming problems from Theorem 14.7 the truth of the theorem arises. □

Theorems 14.7 and 14.9 are equivalent.

It can be remarked especially that the method of Pareto-Stackelberg control determining, established by Theorems 14.7 and 14.9, needs the solutions of multi-criteria linear programming problems.

14.5 Linear Discrete-Time Set-Valued Nash-Stackelberg Control Problem

Let us modify problem (14.4) by considering the control of Nash-Stackelberg type with T stages and $\nu_1 + \nu_2 + \cdots + \nu_T$ players, where $\nu_1, \nu_2, \ldots, \nu_T$ are the numbers of players at the stages $1, 2, \ldots, T$. Each player is identified by the two numbers/indices (τ, π), where τ is the stage number on which the player selects his strategy and $\pi \in \{1, 2, \dots, \nu_\tau\}$ is his number at the stage τ. In such game, at each stage τ the players $1, 2, \dots, \nu_\tau$ play a Nash game by selecting simultaneously their strategies and by communicating their and all strategies selected by precedent players to the following $\tau + 1$ stage players. After the all stage strategy selections, all the players compute their gains on the resulting profile. Such type of control is called Nash-Stackelberg control and the corresponding problem — a linear discrete-time set-valued Nash-Stackelberg control problem.

The decision process may be modelled as

$$
\begin{aligned}
&F_{\tau\pi}(X, U^{\tau\pi}||U^{-\tau\pi}) = \\
&= \sum_{t=1}^{T}(c^{\tau\pi t}X^t + \sum_{\mu=1}^{\nu_t} b^{\tau\pi t\mu}U^{t\mu}) \underset{U^{\tau\pi}}{\longrightarrow} \max, \\
&\tau = 1, \dots, T,\ \pi = 1, \dots, \nu_\tau, \\
X^t &= X^{t-1}A^{t-1} + \sum_{\pi=1}^{\nu_t} B^{t\pi}U^{t\pi},\ t = 1, \dots, T, \\
D^{t\pi}U^{t\pi} &\le d^{t\pi},\ t = 1, \dots, T,\ \pi = 1, \dots, \nu_t,
\end{aligned}
\tag{14.10}
$$

where $X^0, X^t \subset \mathbb{R}^n, c^{\tau\pi t} \in \mathbb{R}^n, U^{\tau\pi} \subset \mathbb{R}^m, b^{\tau\pi t\mu} \in \mathbb{R}^m, A^{t-1} \in \mathbb{R}^{n\times n}, B^{\tau\pi} \in \mathbb{R}^{n\times m}$, $d^{\tau\pi} \in \mathbb{R}^k$, $D^{\tau\pi} \in \mathbb{R}^{k\times n}, t, \tau = 1, \dots, T, \pi = 1, \dots, \nu_\tau, \mu = 1, \dots, \nu_t$.

Theorem 14.10 *Let problem (14.10) be solvable. The sequence $\bar{u}^{11}, \bar{u}^{12}, \dots, \bar{u}^{T\nu_T}$, forms a Nash-Stackelberg equilibrium control if and only if the control $\bar{u}^{\tau\pi}$ is an optimal solution of the linear programming problem*

$$
\begin{aligned}
f_{\tau\pi}(u^{\tau\pi}||u^{-\tau\pi}) = &(c^{\tau\pi\tau}B^{\tau\pi} + c^{\tau\pi\tau+1}A^{\tau}B^{\tau\pi} + \\
&+c^{\tau\pi\tau+2}A^{\tau+1}A^{\tau}B^{\tau\pi} + \dots + \\
&+c^{\tau\pi T}A^{T-1}A^{T-2}\dots A^{\tau}B^{\tau\pi} + \\
&+b^{\tau\pi\tau\pi})u^{\tau\pi} \underset{u^{\tau\pi}}{\longrightarrow} \max, \\
D^{\tau\pi}u^{\tau\pi} \le d^{\tau\pi}&, \tau = 1, \dots, T, \pi = 1, \dots, \nu_\tau.
\end{aligned}
\tag{14.11}
$$

for $\tau = 1, \dots, T, \pi = 1, \dots, \nu_\tau$.

Proof By performing direct transformations

$$
\begin{aligned}
X^1 &= A^0X^0 + \sum_{\pi=1}^{\nu_1} B^{1\pi}U^{1\pi}, \\
X^2 &= A^1X^1 + \sum_{\pi=1}^{\nu_2} B^{2\pi}U^{2\pi} = \\
&= A^1\left(A^0X^0 + \sum_{\pi=1}^{\nu_1} B^{1\pi}U^{1\pi}\right) + \sum_{\pi=1}^{\nu_2} B^{2\pi}U^{2\pi} = \\
&= A^1A^0X^0 + A^1\sum_{\pi=1}^{\nu_1} B^{1\pi}U^{1\pi} + \sum_{\pi=1}^{\nu_2} B^{2\pi}U^{2\pi},
\end{aligned}
$$

$$
\begin{aligned}
X^3 &= A^2X^2 + \sum_{\pi=1}^{\nu_3} B^{3\pi}U^{3\pi} = \\
&= A^2\left(A^1A^0X^0 + A^1\sum_{\pi=1}^{\nu_1} B^{1\pi}U^{1\pi} + \sum_{\pi=1}^{\nu_2} B^{2\pi}U^{2\pi}\right) + \\
&\quad + \sum_{\pi=1}^{\nu_3} B^{3\pi}U^{3\pi} = \\
&= A^2A^1A^0X^0 + A^2A^1\sum_{\pi=1}^{\nu_1} B^{1\pi}U^{1\pi} + A^2\sum_{\pi=1}^{\nu_2} B^{2\pi}U^{2\pi} + \\
&\quad + \sum_{\pi=1}^{\nu_3} B^{3\pi}U^{3\pi},
\end{aligned}
$$

$$\ldots$$

$$
\begin{aligned}
X^T &= A^{T-1}X^{T-1} + \sum_{\pi=1}^{\nu_T} B^{T\pi}U^{T\pi} = \\
&= \prod_{t=0}^{T-1} A^tX^t + \prod_{t=1}^{T-1} A^t\sum_{\pi=1}^{\nu_1} B^{1\pi}U^{1\pi} + \prod_{t=2}^{T-1} A^t\sum_{\pi=1}^{\nu_2} B^{2\pi}U^{2\pi} + \cdots + \\
&\quad + A^{T-1}\sum_{\pi=1}^{\nu_{T-1}} B^{T-1\pi}U^{T-1\pi} + \sum_{\pi=1}^{\nu_T} B^{T\pi}U^{T\pi},
\end{aligned}
$$

and by subsequent substitution in the objective/payoff functions, problem (14.10) is reduced to

$$
\begin{aligned}
&F_{\tau\pi}(U^{\tau\pi}||U^{-\tau\pi}) = \\
= &(c^{\tau\pi 1} + c^{\tau\pi 2}A^1 + c^{\tau\pi 3}A^2A^1 + \cdots + \\
&+c^{\tau\pi T}A^{T-1}A^{T-2}\dots A^1)A^0X^0 + \\
&+(c^{\tau\pi 1}B^{11} + c^{\tau\pi 2}A^1B^{11} + c^{\tau\pi 3}A^2A^1B^{11} + \cdots + \\
&+c^{\tau\pi T}A^{T-1}A^{T-2}\dots A^1B^{11} + b^{\tau\pi 11})U^{11} + \\
&+(c^{\tau\pi 1}B^{12} + c^{\tau\pi 2}A^1B^{12} + c^{\tau\pi 3}A^2A^1B^{12} + \cdots + \\
&+c^{\tau\pi T}A^{T-1}A^{T-2}\dots A^1B^{12} + b^{\tau\pi 12})U^{12} + \\
&+\cdots+ \\
&+(c^{\tau\pi 1}B^{1\nu_1} + c^{\tau\pi 2}A^1B^{1\nu_1} + c^{\tau\pi 3}A^2A^1B^{1\nu_1} + \cdots + \\
&+c^{\tau\pi T}A^{T-1}A^{T-2}\dots A^1B^{1\nu_1} + b^{\tau\pi 1\nu_1})U^{1\nu_1} + \\
&+(c^{\tau\pi 2}B^{21} + c^{\tau\pi 3}A^2B^{21} + c^{\tau\pi 4}A^3A^2B^{21} + \cdots + \\
&+c^{\tau\pi T}A^{T-1}A^{T-2}\dots A^2B^{21} + b^{\tau\pi 21})U^{21} + \cdots + \\
&+(c^{\tau\pi 2}B^{22} + c^{\tau\pi 3}A^2B^{22} + c^{\tau\pi 4}A^3A^2B^{22} + \cdots + \\
&+c^{\tau\pi T}A^{T-1}A^{T-2}\dots A^2B^{22} + b^{\tau\pi 22})U^{22} + \cdots + \\
&+\cdots+ \\
&+(c^{\tau\pi 2}B^{2\nu_2} + c^{\tau\pi 3}A^2B^{2\nu_2} + c^{\tau\pi 4}A^3A^2B^{2\nu_2} + \cdots + \\
&+c^{\tau\pi T}A^{T-1}A^{T-2}\dots A^2B^{2\nu_2} + b^{\tau\pi 2\nu_2})U^{2\nu_2} + \cdots + \\
&+\cdots+ \\
&+(c^{\tau\pi T}B^{T\nu_T} + b^{\tau\pi T\nu_T})U^{T\nu_T} \underset{U^{\tau\pi}}{\longrightarrow} \max, \\
&\tau = 1, \dots, T, \pi = 1, \dots, \nu_\tau, \\
&D^{\tau\pi}U^{\tau\pi} \leq d^{\tau\pi}, \tau = 1, \dots, T, \pi = 1, \dots, \nu_\tau.
\end{aligned}
\tag{14.12}
$$

Problem (14.12) is equivalent to a point-valued problem. So, the control sets $U^1, U^2, \dots, U^T$ may be identified with the sets of cardinality 1. Evidently, model (14.11) defines a strategic game for which any Nash-Stackelberg equilibrium is also a Nash equilibrium and it is simply computed as a sequence of solutions of the problem

$$
\begin{aligned}
f_{\tau\pi}(u^{\tau\pi}||u^{-\tau\pi}) = &(c^{\tau\pi\tau}B^{\tau\pi} + c^{\tau\pi\tau+1}A^\tau B^{\tau\pi} + \\
&+c^{\tau\pi\tau+2}A^{\tau+1}A^\tau B^{\tau\pi} + \cdots + \\
&+c^{\tau\pi T}A^{T-1}A^{T-2}\dots A^\tau B^{\tau\pi} + \\
&+b^{\tau\pi\tau\pi})u^{\tau\pi} \underset{u^{\tau\pi}}{\longrightarrow} \max, \\
D^{\tau\pi}u^{\tau\pi} \leq d^{\tau\pi}, &\tau = 1, \dots, T, \pi = 1, \dots, \nu_\tau.
\end{aligned}
$$

Equivalence of (14.10) with the last problem proves Theorem 14.10. □

An important particular case of (14.10) and Theorem 14.10 is considered by following Theorem 14.11 which is an analogous of that ones from the precedent sections.

Theorem 14.11 *If*

$$A^0 = A^1 = \cdots = A^{T-1} = A,$$
$$B^{11} = B^{12} = \cdots = B^{T\nu_T} = B,$$

and problem (14.10) is solvable, then the sequence $\bar{u}^{11}, \bar{u}^{12}, \ldots, \bar{u}^{T\nu_T}$, *forms a Nash-Stackelberg equilibrium control if and only if the control* $\bar{u}^{\tau\pi}$ *is an optimal solution of the linear programming problem*

$$\begin{aligned} f_{\tau\pi}(u^{\tau\pi}||u^{-\tau\pi}) &= (c^{\tau\pi\tau}B + c^{\tau\pi\tau+1}AB + c^{\tau\pi\tau+2}(A)^2B + \cdots + \\ &\quad + c^{\tau\pi T}(A)^{T-\tau}B + b^{\tau\pi\tau\pi})u^{\tau\pi} \underset{u^{\tau\pi}}{\longrightarrow} \max, \\ D^{\tau\pi}u^{\tau\pi} &\le d^{\tau\pi}, \end{aligned}$$

for $\tau = 1, \ldots, T, \pi = 1, \ldots, \nu_\tau$.

Let us provide an extension of the Pontryagin maximum principle on problem (14.10).

Consider the following recurrence relations

$$\begin{aligned} p^{\tau\pi T} &= c^{\tau\pi T}, \\ p^{\tau\pi t} &= p^{\tau\pi t+1}A^t + c^{\tau\pi t}, \; t = T-1, \ldots, 1, \end{aligned} \tag{14.13}$$

where $\tau = 1, \ldots, T, \pi = 1, \ldots, \nu_\tau$. Hamiltonian functions are defined as

$$H_{\tau\pi t}(u^{\tau\pi}) = \langle p^{\tau\pi t}B^{\tau\pi} + b^{\tau\pi\tau\pi}, u^{\tau\pi}\rangle, \; t = T, \ldots, 1,$$

where $\tau = 1, \ldots, T, \pi = 1, \ldots, \nu_\tau$, and $p^{\tau\pi t}$, $t = T, \ldots, 1$, $\tau = 1, \ldots, T$, $\pi = 1, \ldots, \nu_\tau$, are defined by (14.13).

Theorem 14.12 *Let problem (14.10) be solvable. The sequence* $\bar{u}^{11}, \bar{u}^{12}, \ldots, \bar{u}^{T\nu_T}$, *forms a Nash-Stackelberg equilibrium control if and only if*

$$H_{\tau\pi t}(\bar{u}^{\tau\pi}) = \max_{u^{\tau\pi}: D^{\tau\pi}u^{\tau\pi} \le d^{\tau\pi}} H_{\tau\pi t}(u^{\tau\pi}),$$

for $t = T, \ldots, 1, \tau = 1, \ldots, T, \pi = 1, \ldots, \nu_\tau$.

Proof The proof is provided by direct transformation and comparing the obtained results with that from Theorem 14.10. Their equivalence proves the theorem. □

Theorems 14.10 and 14.12 are equivalent.

14.6 Linear Discrete-Time Set-Valued Pareto-Nash-Stackelberg Control Problem

Let us unify problems (14.7) and (14.10) by considering the control of Pareto-Nash-Stackelberg type with T stages and $\nu_1 + \nu_2 + \dots + \nu_T$ players, where the values $\nu_1, \nu_2, \dots, \nu_T$, are the correspondent numbers of players on the stages $1, 2, \dots, T$. Every player is identified by two numbers as above in the Nash-Stackelberg control: τ is the number of stage on which player selects his strategy and π is the player number at the stage τ. In such game, at each stage τ the players $1, 2, \dots, \nu_\tau$ play a Pareto-Nash game by selecting simultaneously their strategies accordingly to their criteria ($k_{\tau 1}, k_{\tau 2}, \dots, k_{\tau \nu_\tau}$ are the numbers of criteria of respective players) and by communicating their and all the precedent selected strategies to the following $\tau + 1$ stage players. After the all stage strategy selections, all the players compute their payoffs on the resulting profile. Such type of control is called a Pareto-Nash-Stackelberg control and the corresponding problem is called a linear discrete-time set-valued Pareto-Nash-Stackelberg control problem.

The mathematical model of the decision control process may be formalised as the problem

$$
\begin{gathered}
F_{\tau\pi}(X, U^{\tau\pi} || U^{-\tau\pi}) = \\
= \sum_{t=1}^{T} (c^{\tau\pi t} X^t + \sum_{\mu=1}^{\nu_t} b^{\tau\pi t\mu} U^{t\mu}) \underset{U^{\tau\pi}}{\longrightarrow} \max, \\
\tau = 1, \dots, T, \ \pi = 1, \dots, \nu_\tau, \\
X^t = A^{t-1} X^{t-1} + \sum_{\pi=1}^{\nu_t} B^{t\pi} U^{t\pi}, \ t = 1, \dots, T, \\
D^{t\pi} U^{t\pi} \leq d^{t\pi}, \ t = 1, \dots, T, \ \pi = 1, \dots, \nu_t,
\end{gathered} \tag{14.14}
$$

where $X^0, \ X^t \subset \mathbb{R}^n, c^{\tau\pi t} \in \mathbb{R}^{k_{tp} \times n}, U^{\tau\pi} \subset \mathbb{R}^m, b^{\tau\pi t\mu} \in \mathbb{R}^{k_{tp} \times m}, A^{t-1} \in \mathbb{R}^{n \times n}, B^{\tau\pi} \in \mathbb{R}^{n \times m}, d^{\tau\pi} \in \mathbb{R}^k, D^{\tau\pi} \in \mathbb{R}^{k \times n}, t, \tau = 1, \dots, T, \pi = 1, \dots, \nu_\tau, \mu = 1, \dots, \nu_t$.

Theorem 14.13 *Let problem (14.14) be solvable. The sequence $\bar{u}^{11}, \bar{u}^{12}, \dots, \bar{u}^{T\nu_T}$, forms a Pareto-Nash-Stackelberg equilibrium control in problem (14.14) if and only if the control $\bar{u}^{\tau\pi}$ is an efficient solution of the multi-criteria linear programming problem*

$$
\begin{aligned}
f_{\tau\pi}(u^{\tau\pi} || u^{-\tau\pi}) = {} & (c^{\tau\pi\tau} B^{\tau\pi} + c^{\tau\pi\tau+1} A^\tau B^{\tau\pi} + \\
& + c^{\tau\pi\tau+2} A^{\tau+1} A^\tau B^{\tau\pi} + \dots + \\
& + c^{\tau\pi T} A^{T-1} A^{T-2} \dots A^\tau B^{\tau\pi} + \\
& + b^{\tau\pi\tau\pi}) u^{\tau\pi} \underset{u^{\tau\pi}}{\longrightarrow} \max, \\
D^{\tau\pi} u^{\tau\pi} \leq {} & d^{\tau\pi}, \tau = 1, \dots, T, \pi = 1, \dots, \nu_\tau.
\end{aligned} \tag{14.15}
$$

for $\tau = 1, \dots, T, \pi = 1, \dots, \nu_\tau$.

Proof By performing similar transformation as above, problem (14.14) is reduced to a sequence of multi-criteria linear programming problems (14.15).

Equivalence of (14.14) and (14.15) proves Theorem 14.13. □

As a corollary follows Theorem 14.14.

Theorem 14.14 *If*

$$A^0 = A^1 = \cdots = A^{T-1} = A,$$
$$B^{11} = B^{12} = \cdots = B^{T\nu_T} = B,$$

and problem (14.14) is solvable, then the sequence $\bar{u}^{11}, \bar{u}^{12}, \ldots, \bar{u}^{T\nu_T}$, *forms a Pareto-Nash-Stackelberg equilibrium control if and only if* $\bar{u}^{\tau\pi}$ *is an efficient solution of the multi-criteria linear programming problem*

$$\begin{aligned} f_{\tau\pi}(u^{\tau\pi} \| u^{-\tau\pi}) = (c^{\tau\pi\tau} B + c^{\tau\pi\tau+1} AB + c^{\tau\pi\tau+2}(A)^2 B + \cdots + \\ + c^{\tau\pi T}(A)^{T-\tau} B + b^{\tau\pi\tau\pi}) u^{\tau\pi} \underset{u^{\tau\pi}}{\longrightarrow} \max, \\ D^{\tau\pi} u^{\tau\pi} \leq d^{\tau\pi}, \end{aligned}$$

for $\tau = 1, \ldots, T$, $\pi = 1, \ldots, \nu_\tau$.

The maximum principle of Pontryagin may be also generalized for problem (14.14). By considering the recurrence relations

$$\begin{aligned} p^{\tau\pi T} &= c^{\tau\pi T}, \\ p^{\tau\pi t} &= p^{\tau\pi t+1} A^t + c^{\tau\pi t}, \ t = T-1, \ldots, 1, \end{aligned} \tag{14.16}$$

where $\tau = 1, \ldots, T$, $\pi = 1, \ldots, \nu_\tau$. Hamiltonian vector-functions are defined on (14.16) as

$$H_{\tau\pi t}(u^{\tau\pi}) = \langle p^{\tau\pi t} B^{\tau\pi} + b^{\tau\pi\tau\pi}, u^{\tau\pi} \rangle, \ t = T, \ldots, 1.$$

Remark, the vector nature of (14.16).

Theorem 14.15 *Let problem (14.14) be solvable. The sequence* $\bar{u}^{11}, \bar{u}^{12}, \ldots, \bar{u}^{T\nu_T}$, *forms a Pareto-Nash-Stackelberg equilibrium control if and only if*

$$\bar{u}^{\tau\pi} \in \underset{u^{\tau\pi}:\, D^{\tau\pi} u^{\tau\pi} \leq d^{\tau\pi}}{\operatorname{Arg\,max}} H_{\tau\pi t}(u^{\tau\pi}),$$

for $t = T, \ldots, 1$, $\tau = 1, \ldots, T$, $\pi = 1, \ldots, \nu_\tau$.

Theorems 14.13 and 14.15 are equivalent.

14.7 Concluding Remarks

As it has been mentioned above, various types of control processes may appear in real life: optimal control, Stackelberg control, Pareto-Stackelberg control, Nash-Stackelberg control, Pareto-Nash-Stackelberg control, etc. Traditionally, a single valued control is studied. Nevertheless, the control may have a set valued nature too. For such type of control processes the mathematical models and solving principles were exposed.

The direct-straightforward and the classical Pontryagin principles are applied for determining the desired control of set-valued dynamic processes. These principles are the bases for pseudo-polynomial methods, which are exposed as consequences of theorems for the set-valued linear discrete-time Pareto-Nash-Stackelberg control problems.

The task to obtain results for various types of set-valued non-linear control processes with discrete and continuous time is a subject for a follow-up research.

References

1. Ungureanu, V. 2013. Linear discrete-time Pareto-Nash-Stackelberg control problem and principles for its solving. *Computer Science Journal of Moldova* 21 (1 (61)): 65–85.
2. Pontryagin, L.S., V.G. Boltyanskii, R.V. Gamkrelidze, and E.F. Mishchenko. 1961. *Mathematical Theory of Optimal Processes*. Moscow: Nauka, 393 pp. (in Russian).
3. Blackwell, D. 1956. An analog of the minimax theorem for vector payoffs. *Pacific Journal of Mathematics* 6: 1–8.
4. Ungureanu, V. 2008. Solution principles for simultaneous and sequential games mixture. *ROMAI Journal* 4 (1): 225–242.
5. Von Stackelberg, H. Marktform und Gleichgewicht (Market Structure and Equilibrium), Vienna: Springer Verlag, 1934, XIV+134 pp. (in German).
6. Nash, J. 1951. Noncooperative games. *Annals of Mathematics* 54 (2): 280–295.
7. Von Neumann, J., and O. Morgenstern. 1944. *Theory of Games and Economic Behavior*. Princeton: Annals Princeton University Press; 2nd ed., 1947, 674 pp.
8. Leitmann, G. 1978. On generalized stackelberg strategies. *Journal of Optimization Theory and Applications* 26: 637–648.
9. Rockafellar, T. 1970. *Convex Analysis*. Princeton: Princeton University Press, 468 pp.
10. Moore, R.E., R.B. Kearfott, and M.J. Cloud. 2009. *Introduction to Interval Analysis*. Philadelphia: SIAM, 234 pp.

Chapter 15
Linear Discrete Pareto-Nash-Stackelberg Control Processes with Echoes and Retroactive Future

Abstract Mathematical models of linear discrete Pareto-Nash-Stackelberg processes with echoes and retroactive future develop the mathematical models of Pareto-Nash-Stackelberg control exposed in the precedent chapters and introduced initially in (Ungureanu, Computer Science Journal of Moldova, 21, No. 1:(61), 65–85,2013) [1]. Applications of a straightforward method along with Pontryagin's principle produce important theoretical and practical results for the investigated models. Software benchmarking confirms and illustrates the value of the obtained results.

15.1 Introduction

Mathematical models of linear discrete-time Pareto-Nash-Stackelberg processes with echoes and retroactive future develop general models of Pareto-Nash-Stackelberg control processes [1], based on a mixture of controls and games of simultaneous and sequential types [1–9]. Such models reflect real economic and technical control processes that are governed by linear systems and criteria, see e.g. [10–13].

If echoes of precedent stages phenomena influence system state and control, new mathematical models appear. Demographic phenomena are real examples of this type — social phenomena from the past (e.g., wars) influence the present.

In considered models, the future influences on the present admit an interpretation from the perspective of a future final state and control as an objective, plan or program. Evidently, such a point of view generates new kinds of models too.

In order to solve considered models we need to define solution concepts and to apply a straightforward principle or the maximum principle of Pontryagin [1, 4].

The exposition starts with particular models of optimal control and develops them consecutively, from particular to general models, throw Stackelberg, Pareto-Stackelberg, Nash-Stackelberg, to the final Pareto-Nash-Stackelberg models. For simplicity's sake, we adopt a unique approach for the proofs.

Remark 15.1 Considered problems could and would have bang-bang solutions because at every moment the control consists of a mixture of some player strategies.

V. Ungureanu, *Pareto-Nash-Stackelberg Game and Control Theory*,
Smart Innovation, Systems and Technologies 89,
https://doi.org/10.1007/978-3-319-75151-1_15

Remark 15.2 Obtained results may be seen finally as a special type of decomposition of the initial control problems to a series of linear programming problems.

15.2 Optimal Control of Linear Discrete-Time Processes with Periodic Echoes

Let us consider a dynamic system with the state evolution and control described and governed by the following mathematical model:

$$
\begin{aligned}
f(X,U) &= \sum_{t=1}^{T}(c^t x^t + d^t u^t) \to \max, \\
x^t &= \begin{cases} A^{t-1}x^{t-1} + B^t u^t, & t \notin [k\tau, k\tau+\omega), \\ A^{t-1}x^{t-1} + B^t u^t + E^t u^{t-\gamma}, & t \in [k\tau, k\tau+\omega), \end{cases} \\
G^t u^t &\le g^t, \\
& \quad t = 1, \ldots, T, \\
& \quad k = 1, 2, \ldots, \left\lfloor \tfrac{T}{\tau} \right\rfloor,
\end{aligned}
\tag{15.1}
$$

where $x^0,\ x^t \in \mathbb{R}^n,\ u^t \in \mathbb{R}^m,\ c^t \in \mathbb{R}^n,\ d^t \in \mathbb{R}^m,\ A^{t-1} \in \mathbb{R}^{n\times n},\ B^t,\ E^t \in \mathbb{R}^{n\times m},\ g^t \in \mathbb{R}^r,\ G^t \in \mathbb{R}^{r\times m},\ t = 1, \ldots, T,\ r, n, m \in \mathbb{N}^*$. The evolution parameters τ, γ and ω are fixed and have an obvious interpretation:

- $\tau \in \mathbb{N}$ is a time period on which echoes start to be active and to influence on the system,
- $\omega \in \mathbb{N}$ is a length of the time interval for which echoes are active,
- $\gamma \in \mathbb{N}$ is the time span between the control applying and its echo producing.

Some controls may not have echoes.

Without loss of generality, we can assume that $\gamma,\ \omega \le \tau$. Evidently, x^0 is an initial state of the system, the tuple $X = (x^0, \ldots, x^T)$ represents the system's trajectory, the tuple $U = (u^1, \ldots, u^T)$ forms the system's trajectory control.

Notation 15.2.1 By defining $E^t = 0 \in \mathbb{R}^{n\times m}, t \notin [k\tau, k\tau+\omega), k = 1, 2, \ldots, \left\lfloor \frac{T}{\tau} \right\rfloor$, the system of equations in (15.1) may be written as

$$x^t = A^{t-1}x^{t-1} + B^t u^t + E^t u^{t-\gamma}, \quad t = 1, 2, \ldots, T. \tag{15.2}$$

The echo matrix E^t is a zero/null matrix if there are no echoes at the moment t. Let us consider $A^t = I$ for $t \ge T$, i.e. A^t is an identity matrix.

Definition 15.1 Any solution (X, U) of problem (15.1) forms an optimal trajectory X and control U.

Theorem 15.1 *Let problem (15.1) be solvable. The control*

$$(\bar{u}^1, \bar{u}^2, \ldots, \bar{u}^T)$$

is optimal if and only if the component $\bar{u}^t$ is an optimal solution of the linear programming problem

$$\left(\sum_{l=t}^{T} c^l \left(B^t_{l-1} + E^t_{l-1}\right) + d^t\right) u^t \to \max, \qquad (15.3)$$
$$G^t u^t \leq g^t,$$

for $t = T, T-1, ..., 1$, where B^t_{l-1}, $E^t_{l-1} \in \mathbb{R}^{n\times m}$, and

$$\begin{aligned} B^t_l &= \begin{cases} A^l A^{l-1} \dots A^t B^t, \ l = T-1, T-2, \dots, t, \\ B^t, \ l = t-1, \end{cases} \\ E^t_l &= \begin{cases} A^{l+\gamma} A^{l-1+\gamma} \dots A^{t+\gamma} E^{t+\gamma}, \ l = T-1, T-2, \dots, t, \\ E^{t+\gamma}, \ l = t-1. \end{cases} \end{aligned} \qquad (15.4)$$

Proof Let us perform direct substitutions in (15.1).

For $k = 0,\ t < \tau$, we have

$$\begin{aligned} x^1 &= A^0 x^0 + B^1 u^1, \\ x^2 &= A^1 x^1 + B^2 u^2 = A^1(A^0 x^0 + B^1 u^1) + B^2 u^2 \\ &= A^1 A^0 x^0 + A^1 B^1 u^1 + B^2 u^2, \\ x^3 &= A^2 x^2 + B^3 u^3 = A^2(A^1 A^0 x^0 + A^1 B^1 u^1 + B^2 u^2) + B^3 u^3 \\ &= A^2 A^1 A^0 x^0 + A^2 A^1 B^1 u^1 + A^2 B^2 u^2 + B^3 u^3, \\ &\dots \\ x^t &= A^{t-1} x^{t-1} + B^t u^t \\ &= x^0_{t-1} + \textstyle\sum_{l=1}^{t} B^l_{t-1} u^l, \end{aligned}$$

where $B^l_{t-1} = \begin{cases} A^{t-1} A^{t-2} \dots A^l B^l, \ l = t-1, t-2, \dots, 1, \\ B^t, \ l = t, \end{cases}$

$x^0_{t-1} = A^{t-1} A^{t-2} \dots A^0 x^0.$

For $k = 1,\ \tau \leq t < \tau + \omega$, we obtain

$$\begin{aligned} x^\tau &= A^{\tau-1} x^{\tau-1} + B^\tau u^\tau + E^\tau u^{\tau-\gamma} \\ &= x^0_{\tau-1} + \textstyle\sum_{l=1}^{\tau} B^l_{\tau-1} u^l + E^\tau u^{\tau-\gamma}, \end{aligned}$$

$$\begin{aligned} x^{\tau+1} &= A^\tau x^\tau + B^{\tau+1} u^{\tau+1} + E^{\tau+1} u^{\tau+1-\gamma} \\ &= x^0_\tau + \textstyle\sum_{l=1}^{\tau+1} B^l_\tau u^l + A^\tau E^\tau u^{\tau-\gamma} + E^{\tau+1} u^{\tau+1-\gamma}, \end{aligned}$$

$$\begin{aligned} x^{\tau+2} &= A^{\tau+1} x^{\tau+1} + B^{\tau+2} u^{\tau+2} + E^{\tau+2} u^{\tau+2-\gamma} \\ &= x^0_{\tau+1} + \textstyle\sum_{l=1}^{\tau+1} B^l_\tau u^l \\ &\quad + A^{\tau+1} A^\tau E^\tau u^{\tau-\gamma} + A^{\tau+1} E^{\tau+1} u^{\tau+1-\gamma} + E^{\tau+2} u^{\tau+2-\gamma}, \\ &\dots \end{aligned}$$

$$\begin{aligned} x^t &= A^{t-1} x^{t-1} + B^t u^t + E^t u^{t-\gamma} \\ &= x^0_{t-1} + \sum_{l=1}^{t} B^l_{t-1} u^l + \sum_{l=\tau}^{t-1} A^{t-1} A^{t-2} \dots A^l E^l u^{t-\gamma} + E^t u^{t-\gamma} \\ &= x^0_{t-1} + \sum_{l=1}^{t} B^l_{t-1} u^l + \sum_{l=\tau-\gamma}^{t-1-\gamma} A^{t-1} A^{t-2} \dots A^{l+\gamma} E^{l+\gamma} u^l + E^t u^{t-\gamma}. \end{aligned}$$

Considering (15.2) and the values of E^t, $t = 1, \dots, T + \gamma$, we can deduce the general formula

$$x^t = x^0_{t-1} + \sum_{l=1}^{t} \left(B^l_{t-1} + E^l_{t-1}\right) u^l, \tag{15.5}$$

for $t = 1, 2, \dots, T$.

By subsequent substitutions of these relations in the objective function we obtain

$$\begin{aligned}
f(X,U) &= \sum_{t=1}^{T} (c^t x^t + d^t u^t) \\
&= \sum_{t=1}^{T} \left(c^t \left(x^0_{t-1} + \sum_{l=1}^{t} \left(B^l_{t-1} + E^l_{t-1}\right) u^l \right) + d^t u^t \right) \\
&= \sum_{t=1}^{T} c^t x^0_{t-1} + \sum_{t=1}^{T} c^t \sum_{l=1}^{t} \left(B^l_{t-1} + E^l_{t-1}\right) u^l + \sum_{t=1}^{T} d^t u^t \\
= \sum_{t=1}^{T} c^t x^0_{t-1} & \\
&+ (c^T B^1_{T-1} + c^{T-1} B^1_{T-2} + \cdots + c^2 B^1_1 + c^1 B^1_0 \\
&+ c^T E^1_{T-1} + c^{T-1} E^1_{T-2} + \cdots + c^2 E^1_1 + c^1 E^1_0 + d^1) u^1 \\
&+ (c^T B^2_{T-1} + c^{T-1} B^2_{T-2} + \cdots + c^3 B^2_2 + c^2 B^2_1 \\
&+ c^T E^2_{T-1} + c^{T-1} E^2_{T-2} + \cdots + c^3 E^2_2 + c^2 E^2_1 + d^2) u^2 \\
&+ \cdots + \\
&+ (c^T B^T_{T-1} + c^T E^T_{T-1} + d^T) u^T \\
= \sum_{t=1}^{T} c^t x^0_{t-1} &+ \sum_{t=1}^{T} \left(\sum_{l=t}^{T} c^l \left(B^t_{l-1} + E^t_{l-1}\right) + d^t \right) u^t .
\end{aligned}$$

So, problem (15.1) is transformed into

$$\begin{gathered}
\sum_{t=1}^{T} c^t x^0_{t-1} + \sum_{t=1}^{T} \left(\sum_{l=t}^{T} c^l \left(B^t_{l-1} + E^t_{l-1}\right) + d^t \right) u^t \to \max, \\
G^t u^t \le g^t, \ t = 1, \dots, T.
\end{gathered} \tag{15.6}$$

Obviously, problems (15.1) and (15.6) are equivalent.

As the objective function in (15.6) is separable, problem (15.1) is transformed into T linear programming problems (15.3). □

Corollary 15.1 *Optimal control in problem (15.1) does not depend on the initial state* x^0.

Proof It is sufficient to observe the form of the objective function and constraints in (15.6). □

Notation 15.2.2 Notation x^0_{t-1}, B^t_{l-1}, E^t_{l-1} are for images of x^0, B^t, $E^{t+\gamma}$, obtained by applying mappings

- $A^{t-1}A^{t-2}\ldots A^0$ to x^0,
- $A^{l-1}A^{l-2}\ldots A^t$ to B^t
- and
- $A^{l-1+\gamma}A^{l-2+\gamma}\ldots A^{t+\gamma}$ to $E^{t+\gamma}$.

Remark 15.3 The solution of problem (15.1) is composed of a solution sequence of T linear programming problems (15.3). Apparently, this result establishes a polynomial time method for solving (15.1) as the coefficients of objective function (15.3) may be computed in about $O(Tn^3)$ time and there are polynomial time algorithms for solving linear programming problems. Actually, the method has a pseudo-polynomial complexity because T may be expressed exponentially on n.

Remark 15.4 Problems (15.3) may be solved independently, but it is better to solve them retrospectively from T to 1 because the coefficients of the objective function may be represented in the form of two triangles for any value of t

$$\begin{aligned}
B^t_{t-1} &= B^t,\\
B^t_t &= A^t B^t,\\
B^t_{t+1} &= A^{t+1}A^t B^t,\\
B^t_{t+2} &= A^{t+2}A^{t+1}A^t B^t,\\
&\ldots\\
B^t_{T-2} &= A^{T-2}\ldots A^{t+2}A^{t+1}A^t B^t,\\
B^t_{T-1} &= A^{T-1}A^{T-2}\ldots A^{t+2}A^{t+1}A^t B^t,
\end{aligned}$$

and

$$\begin{aligned}
E^t_{t-1} &= E^{t+\gamma},\\
E^t_t &= A^{t+\gamma}E^{t+\gamma},\\
E^t_{t+1} &= A^{t+1+\gamma}A^{t+\gamma}E^{t+\gamma},\\
E^t_{t+2} &= A^{t+2+\gamma}A^{t+1+\gamma}A^{t+\gamma}E^{t+\gamma},\\
&\ldots\\
E^t_{T-2} &= A^{T-2+\gamma}\ldots A^{t+2+\gamma}A^{t+1+\gamma}A^{t+\gamma}E^{t+\gamma},\\
E^t_{T-1} &= A^{T-1+\gamma}A^{T-2+\gamma}\ldots A^{t+2+\gamma}A^{t+1+\gamma}A^{t+\gamma}E^{t+\gamma}.
\end{aligned}$$

- Evidently, these matrices/operators may be computed by recursion.
- The same procedure may be applied for computing x^l_{T-1}.
- These formulas and problems (15.3) establish together an evident affinity of our method with the dynamic programming method.

15.3 Optimal Control of Linear Discrete-Time Processes with Periodic Echoes and Retroactive Future

Can the future influence the past? Generally, it can be appreciated as a question related to Ψ phenomena and it is rather to be a negative answer. But, our mathematical models give a positive answer and it has a simple explanation and practical application! If the final state (future) is fixed as x^{*T}, it influences precedent states, i.e. the intermediary states between x^0 and x^{*T}. If the initial state x^0 is not fixed, then it depends on the last state too.

Theorem 15.2 *Let problem (15.1) with the fixed last state x^{*T} be solvable. The control $(\bar{u}^1, \bar{u}^2, ..., \bar{u}^T)$ is optimal if and only if it is a solution of the linear programming problem*

$$
\begin{aligned}
\sum_{t=1}^{T}\left(\sum_{l=t}^{T} c^l\left(B_{l-1}^t + E_{l-1}^t\right) + d^t\right) u^t + \sum_{t=1}^{T} c^t x_{t-1}^0 \to \max, \\
\sum_{l=1}^{T}\left(B_{T-1}^l + E_{T-1}^l\right) u^l + x_{T-1}^0 &= x^{*T}, \\
G^1 u^1 &\leq g^1, \\
&\cdots \\
G^T u^T &\leq g^T.
\end{aligned}
\tag{15.7}
$$

Proof By applying formula (15.5), i.e.

$$
x_{T-1}^0 + \sum_{l=1}^{T}\left(B_{T-1}^l + E_{T-1}^l\right) u^l = x^{*T},
$$

to problem (15.1), which is an equivalent problem with (15.6), problem (15.1) becomes (15.7). □

Remark 15.5 If the last state is fixed, it influences the initial state. Generally, the optimal control in (15.7) depends both on initial and final states.

Let us consider the retroactive influence of the decision maker future decisions on the present state. In this context, we consider the following mathematical model

$$
\begin{aligned}
f(X, U) &= \sum_{t=1}^{T}(c^t x^t + d^t u^t) \to \max, \\
x^t &= A^{t-1} x^{t-1} + B^t u^t + E^t u^{t-\gamma} + \Psi^t u^{t+\gamma},\ t = 1, ..., T, \\
G^t u^t &\leq g^t,\ \ t = 1, ..., T,
\end{aligned}
\tag{15.8}
$$

where all the data and parameters are defined analogically as in (15.1)–(15.2), and $\Psi^t \in \mathbb{R}^{n \times m}$ is defined analogically with E^t and has appropriate interpretation, i.e. Ψ^t is null if t is not in the intervals where echoes and influences occur.

Theorem 15.3 *Let problem (15.8) be solvable. The control*

$$(\bar{u}^1, \bar{u}^2, ..., \bar{u}^T)$$

is optimal if and only if any its component $\bar{u}^t$ is a solution of a linear programming problem

$$\begin{aligned} \left(\sum_{l=t}^{T} c^l \left(B_{l-1}^t + E_{l-1}^t + \Psi_{l-1}^t \right) + d^t \right) u^t &\to \max, \\ G^1 u^1 &\le g^1, \\ &\dots \\ G^T u^T &\le g^T, \end{aligned} \tag{15.9}$$

where B_{l-1}^t, E_{l-1}^t are defined by (15.4) and

$$\Psi_l^t = \begin{cases} A^{l-\gamma} A^{l-1-\gamma} \dots A^{t-\gamma} \Psi^{t-\gamma}, \ l = T-1, T-2, \dots, t, \\ \Psi^{t-\gamma}, \ l = t-1. \end{cases} \tag{15.10}$$

Proof The theorem is provable by direct substitutions similar to that in the proof of Theorem 15.2, obtaining (15.9). □

Remark 15.6 We choose the same parameters for echoes and retroactive influences. Evidently, they may be different.

Remark 15.7 Theorem 15.2 may be simply extended to problem (15.8) maintaining simultaneously the same order of complexity.

15.4 Stackelberg Control of Linear Discrete-Time Processes with Periodic Echoes

Let us modify problem (15.1) by considering a Stackelberg (equilibrium type) control [1, 2, 6, 9], i.e. a Stackelberg game in which at each stage π ($\pi = 1, ..., T$) the player π selects his strategy-control u^π. Every player knows the strategies selected by the precedent players. After all strategy selections, the players compute their gains on a resulting profile. A Stackelberg decision process may be formalised as a mathematical problem

$$
\begin{aligned}
f_1(X, U) &= \sum_{t=1}^{T} (c^{1t} x^t + d^{1t} u^t) \underset{u^1}{\longrightarrow} \max, \\
f_2(X, U) &= \sum_{t=1}^{T} (c^{2t} x^t + d^{2t} u^t) \underset{u^2}{\longrightarrow} \max, \\
&\dots \\
f_T(X, U) &= \sum_{t=1}^{T} (c^{Tt} x^t + d^{Tt} u^t) \underset{u^T}{\longrightarrow} \max, \\
x^t &= \begin{cases} A^{t-1} x^{t-1} + B^t u^t, & t \notin [k\tau, k\tau + \omega), \\ A^{t-1} x^{t-1} + B^t u^t + E^t u^{t-\gamma}, & t \in [k\tau, k\tau + \omega), \end{cases} \\
G^t u^t &\leq g^t, \\
& \quad t = 1, ..., T, \\
& \quad k = 1, 2, ..., \left\lfloor \tfrac{T}{\tau} \right\rfloor,
\end{aligned}
\tag{15.11}
$$

where $x^0, x^t \in \mathbb{R}^n$, $u^t \in \mathbb{R}^m$, $c^{\pi t} \in \mathbb{R}^n$, $d^{\pi t} \in \mathbb{R}^m$, $A^{t-1} \in \mathbb{R}^{n\times n}$, B^t, $E^t \in \mathbb{R}^{n\times m}$, $g^t \in \mathbb{R}^r$, $G^t \in \mathbb{R}^{r\times m}$, π, $t = 1, ..., T$, $r, n, m \in \mathbb{N}^*$. The parameters τ, γ, ω, have the same interpretation as in (15.1).

Remark 15.8 Further, the system of equations in (15.11) is substituted by its equivalent system (15.2).

Evidently, the strategy set of the player π, ($\pi = 1, 2, ..., T$), is defined by the admissible set of the linear programming problem

$$
\begin{aligned}
f_\pi(X, u^\pi \| u^{-\pi}) = \sum_{t=1}^{T} (c^{\pi t} x^t + d^{\pi t} u^t) \underset{u^\pi}{\longrightarrow} \max, \\
x^\pi = A^{\pi-1} x^{\pi-1} + B^\pi u^\pi + E^\pi u^{\pi-\gamma}, \\
G^\pi u^\pi \leq g^\pi.
\end{aligned}
\tag{15.12}
$$

Theorem 15.4 *Let problem (15.11) be solvable. The control*

$$(\bar{u}^1, \bar{u}^2, ..., \bar{u}^T)$$

is a Stackelberg equilibrium control if and only if the component $\bar{u}^\pi$ is an optimal solution to the linear programming problem

$$
\begin{aligned}
\left(\sum_{l=\pi}^{T} c^{\pi l} \left(B^\pi_{l-1} + E^\pi_{l-1} \right) + d^{\pi\pi} \right) u^\pi \underset{u^\pi}{\longrightarrow} \max, \\
G^\pi u^\pi \leq g^\pi,
\end{aligned}
\tag{15.13}
$$

for $\pi = 1, ..., T$, where B^π_{l-1}, $E^\pi_{l-1} \in \mathbb{R}^{n\times m}$, $l = 1, \ldots, T$, are defined by (15.4).

Proof The decision problem of the player π ($\pi = 1, 2, ..., T$) is defined by linear programming problem (15.12). By performing direct substitutions as in the proof of theorem 15.1, problem (15.12) transforms problem (15.11) into

$$
\begin{aligned}
& f_\pi(u^\pi||u^{-\pi}) = \\
& = \sum_{t=1}^{T}(c^{\pi t}x^t + d^{\pi t}u^t) = \sum_{t=1}^{T} c^{\pi t}x^0_{t-1} \\
& +(c^{\pi T}B^1_{T-1} + c^{\pi T-1}B^1_{T-2} + \cdots + c^{\pi 2}B^1_1 + c^{\pi 1}B^1_0 \\
& +c^{\pi T}E^1_{T-1} + c^{\pi T-1}E^1_{T-2} + \cdots + c^{\pi 2}E^1_1 + c^{\pi 1}E^1_0 + d^{\pi 1})u^1 \\
& +(c^{\pi T}B^2_{T-1} + c^{\pi T-1}B^2_{T-2} + \cdots + c^{\pi 3}B^2_2 + c^{\pi 2}B^2_1 \\
& +c^{\pi T}E^2_{T-1} + c^{\pi T-1}E^2_{T-2} + \cdots + c^{\pi 3}E^2_2 + c^{\pi 2}E^2_1 + d^{\pi 2})u^2 \\
& +\cdots+ \\
& +(c^{\pi T}B^T_{T-1} + c^{\pi T}E^T_{T-1} + d^{\pi T})u^T \\
& = \sum_{t=1}^{T} c^{\pi t}x^0_{t-1} \\
& +\sum_{t=1}^{T}\left(\sum_{l=t}^{T} c^{\pi l}\left(B^t_{l-1} + E^t_{l-1}\right) + d^{\pi t}\right)u^t \underset{u^\pi}{\rightarrow} \max, \\
& \pi = 1, ..., T, \\
& G^\pi u^\pi \le g^\pi, \pi = 1, ..., T.
\end{aligned}
$$

The decision variables are separable in the objective function and system of linear inequalities of the last problem. So, problem (15.11) is equivalent to (15.13). □

Theorem 15.4 establishes a principle for solving (15.11).

15.5 Stackelberg Control of Linear Discrete-Time Processes with Periodic Echoes and Retroactive Future

Can the future influence the past in the Stackelberg decision process? The mathematical model permit to construct simply a solution in the case of future decisions. Unfortunately, the problem remains unsolved in the case of the fixed last state x^{*T}.

Let us consider the retroactive influence of the players future decisions on the present state

$$
\begin{aligned}
f_1(X,U) &= \sum_{t=1}^{T}(c^{1t}x^t + d^{1t}u^t) \underset{u^1}{\longrightarrow} \max, \\
f_2(X,U) &= \sum_{t=1}^{T}(c^{2t}x^t + d^{2t}u^t) \underset{u^2}{\longrightarrow} \max, \\
&\dots \\
f_T(X,U) &= \sum_{t=1}^{T}(c^{Tt}x^t + d^{Tt}u^t) \underset{u^T}{\longrightarrow} \max, \\
x^\pi &= A^{\pi-1}x^{\pi-1} + B^\pi u^\pi + E^\pi u^{\pi-\gamma} + \Psi^\pi u^{\pi+\gamma}, \\
& \quad\quad \pi = 1, \dots, T, \\
G^\pi u^\pi &\le g^\pi, \quad \pi = 1, \dots, T,
\end{aligned}
\tag{15.14}
$$

where all the data and parameters are defined analogically as in (15.8) and (15.11).

Theorem 15.5 *Let problem (15.14) be solvable. The control*

$$(\bar{u}^1, \bar{u}^2, \dots, \bar{u}^T)$$

is a Stackelberg equilibrium control if and only if any its component $\bar{u}^\pi$ is a solution of the linear programming problem

$$
\begin{aligned}
\left(\sum_{l=t}^{T} c^{\pi l}\left(B^\pi_{l-1} + E^\pi_{l-1} + \Psi^\pi_{l-1}\right) + d^{\pi\pi}\right) u^\pi &\underset{u^\pi}{\longrightarrow} \max, \\
G^\pi u^\pi &\le g^\pi,
\end{aligned}
\tag{15.15}
$$

where $B^t_{l-1}, E^t_{l-1}, \Psi^t_l$, are defined by (15.4) and (15.10).

Proof The theorem is proved by providing direct substitutions similar to that in the proves of Theorems 15.1 and 15.4. □

Remark 15.9 Evidently, the parameters for echoes and retroactive influences may be different not only in Theorem 15.5, but in general mathematical model.

15.6 Pareto-Stackelberg Control of Discrete-Time Linear Processes with Periodic Echoes and Retroactive Future

Let us consider a control of the Pareto-Stackelberg type in problem (15.14). At each stage a single player makes decision, selects his strategy by considering all own criteria and communicates his choice and precedent players choices to the following player. At the last stage, after the all stage strategy selections, the players compute their gains.

The decision process is formalized mathematically as follows:

$$\begin{aligned} f_1(X, U) &= \sum_{t=1}^{T}(c^{1t}x^t + d^{1t}u^t) \underset{u^1}{\rightarrow} \max, \\ f_2(X, U) &= \sum_{t=1}^{T}(c^{2t}x^t + d^{2t}u^t) \underset{u^2}{\rightarrow} \max, \\ &\dots \\ f_T(X, U) &= \sum_{t=1}^{T}(c^{Tt}x^t + d^{Tt}u^t) \underset{u^T}{\rightarrow} \max, \\ x^\pi &= A^{\pi-1}x^{\pi-1} + B^\pi u^\pi + E^\pi u^{\pi-\gamma} + \Psi^\pi u^{\pi+\gamma}, \\ &\quad \pi = 1, ..., T, \\ G^\pi u^\pi &\leq g^\pi, \quad \pi = 1, ..., T, \end{aligned} \tag{15.16}$$

where $c^{\pi t} \in \mathbb{R}^{k_\pi \times n}$, $d^{\pi t} \in \mathbb{R}^{k_\pi \times m}$, and the rest data and parameters are defined analogically as in (15.8), (15.11), and (15.14).

Remark 15.10 Problem 15.16 is a multi-criteria maximization problem.

The strategy set of the player π ($\pi = 1, 2, ..., T$) is defined formally by the problem

$$\begin{aligned} f_\pi(X, u^\pi || u^{-\pi}) &= \sum_{t=1}^{T}(c^{\pi t}x^t + d^{\pi t}u^t) \underset{u^\pi}{\rightarrow} \max, \\ x^\pi &= A^{\pi-1}x^{\pi-1} + B^\pi u^\pi + E^\pi u^{\pi-\gamma} + \Psi^\pi u^{\pi+\gamma}, \ \pi = 1, ..., T, \\ G^\pi u^\pi &\leq g^\pi, \quad \pi = 1, ..., T, \end{aligned}$$

Theorem 15.6 *Let problem (15.16) be solvable. The control*

$$(\bar{u}^1, \bar{u}^2, ..., \bar{u}^T)$$

forms a Pareto-Stackelberg equilibrium control if and only if the component $\bar{u}^\pi$ is an efficient solution of the multi-criteria linear programming problem

$$\begin{aligned} \left(\sum_{l=t}^{T} c^{\pi l}\left(B^t_{l-1} + E^t_{l-1} + \Psi^t_{l-1}\right) + d^{\pi\pi}\right) u^\pi &\underset{u^\pi}{\rightarrow} \max, \\ G^\pi u^\pi &\leq g^\pi, \end{aligned} \tag{15.17}$$

for $\pi = 1, ..., T$, where $B^t_{l-1}, E^t_{l-1}, \Psi^t_l$, are defined by (15.4) and (15.10).

Proof By performing direct transformations as above, problem (15.16) is transformed into the multi-criteria linear programming problem

$$
\begin{aligned}
f_\pi(u_\pi||u_{-\pi}) = \sum_{t=1}^{T} c^{\pi t} x^0_{t-1} + \\
+ \sum_{t=1}^{T} \left(\sum_{l=t}^{T} c^{\pi l} \left(B^t_{l-1} + E^t_{l-1} + \Psi^t_{l-1} \right) + d^{\pi t} \right) u^\pi \underset{u^\pi}{\longrightarrow} \max, \\
\pi = 1, \ldots, T, \\
G^1 u^1 \leq g^1, \\
G^2 u^2 \leq g^2, \\
\ldots \\
G^T u^T \leq g^T.
\end{aligned}
\tag{15.18}
$$

For every player π, the payoff function and constraints are additively separable with respect to u_π in (15.18). So, every player π selects his optimal strategy as an efficient solution of multi-criteria linear programming problem (15.17). □

Remark 15.11 It must be mentioned once again that the method of Pareto-Stackelberg control determining established by Theorem 15.6 needs to solve T multi-criteria linear programming problems.

15.7 Nash-Stackelberg Control of Linear Discrete-Time Processes with Echoes and Retroactive Future

Let us modify problem (15.14) by considering a control of Nash-Stackelberg type with T stages and $\nu_1 + \nu_2 + \ldots + \nu_T$ players, where $\nu_1, \nu_2, \ldots, \nu_T$ are the numbers of players at the stages $1, 2, \ldots, T$. Every player is identified by two numbers (indices) (s, π), where s is the number of the stage on which the player selects his strategy and $\pi \in \{1, 2, \ldots, \nu_s\}$ is his number at the stage s. In a such game, at each stage s players $1, 2, \ldots, \nu_s$ play a Nash game by selecting simultaneously their strategies and communicating their and the strategies of all the precedent players to the following $s+1$ stage players. After the all stage strategy selections, the players compute their gains on the resulting profile.

The mathematical model of the Nash-Stackelberg decision process is formalized as follows

$$
\begin{aligned}
& f_{s\pi}(X, u^{s\pi}||u^{-s\pi}) = \sum_{t=1}^{T} \left(c^{s\pi t} x^t + \sum_{\mu=1}^{\nu_t} d^{s\pi t\mu} u^{t\mu} \right) \underset{u^{s\pi}}{\longrightarrow} \max, \\
& \qquad s = 1, \ldots, T, \ \pi = 1, \ldots, \nu_s, \\
& x^t = A^{t-1} x^{t-1} + \sum_{\mu=1}^{\nu_t} B^{t\mu} u^{t\mu} + \sum_{\mu=1}^{\nu_t} E^{t\mu} u^{t-\gamma\mu} + \sum_{\mu=1}^{\nu_t} \Psi^{t\mu} u^{t+\gamma\mu}, \\
& \qquad t = 1, \ldots, T, \\
& G^{s\pi} u^{s\pi} \leq g^{s\pi}, \quad s = 1, \ldots, T, \ \pi = 1, \ldots, \nu_s,
\end{aligned}
\tag{15.19}
$$

where $x^0, x^t, c^{s\pi t} \in \mathbb{R}^n$, $u^{s\pi}, d^{s\pi t\mu} \in \mathbb{R}^m$, $A^{t-1} \in \mathbb{R}^{n\times n}$, $B^{t\mu} \in \mathbb{R}^{n\times m}$, $g^{s\pi} \in \mathbb{R}^r$, $G^{s\pi} \in \mathbb{R}^{r\times m}$, $E^{t\mu}, \Psi^{t\mu} \in \mathbb{R}^{n\times m}$, $t,\ s = 1, ..., T, \pi = 1, ..., \nu_s, \mu = 1, ..., \nu_t$.

Remark 15.12 Let us recall once again that

- $E^{t\mu} = \Psi^{t\mu} = 0 \in \mathbb{R}^{n\times m}, t \notin [k\tau, k\tau + \omega), k = 1, 2, ..., \lfloor \frac{T}{\tau} \rfloor$,
- $E^{t\mu} = \Psi^{t\mu} = 0$, for $t > T$ and $t < 1$,
- $A^t = I$, for $t \geq T$, where I is the identity matrix.

Theorem 15.7 *Let problem (15.19) be solvable. The sequence*

$$(\bar{u}^{11}, \bar{u}^{12}, ..., \bar{u}^{T\nu_T})$$

forms a Nash-Stackelberg equilibrium control if and only if any its component $\bar{u}^{t\pi}$ is an optimal solution of the linear programming problem

$$\left(\sum_{l=t}^{T} c^{t\pi l}\left(B^{t\pi}_{l-1} + E^{t\pi}_{l-1} + \Psi^{t\pi}_{l-1}\right) + d^{t\pi t\pi}\right) u^{t\pi} \to \max, \qquad (15.20)$$
$$G^{t\pi} u^{t\pi} \leq g^{t\pi},$$

where

$$B^{t\pi}_{l-1},\ E^{t\pi}_{l-1},\ \Psi^{t\pi}_{l-1} \in \mathbb{R}^{n\times m},$$

and

$$B^{t\pi}_l = \begin{cases} A^l A^{l-1} \dots A^t B^{t\pi},\ l = T-1, T-2, \dots, t, \\ B^{t\pi},\ l = t-1, \end{cases}$$
$$E^{t\pi}_l = \begin{cases} A^{l+\gamma} A^{l-1+\gamma} \dots A^{t+\gamma} E^{t+\gamma\pi},\ l = T-1, \dots, t, \\ E^{t+\gamma\pi},\ l = t-1. \end{cases} \qquad (15.21)$$
$$\Psi^{t\pi}_l = \begin{cases} A^{l-\gamma} A^{l-1-\gamma} \dots A^{t-\gamma} \Psi^{t-\gamma\pi},\ l = T-1, \dots, t, \\ \Psi^{t-\gamma\pi},\ l = t-1. \end{cases}$$

for $t = 1, ..., T, \pi = 1, ..., \nu_t$.

Proof Let us perform direct transformations:

$$
\begin{aligned}
x^1 = &\ A^0x^0 + \sum_{\mu=1}^{\nu_1} B^{1\mu}u^{1\mu},\\
x^2 = &\ A^1x^1 + \sum_{\pi=1}^{\nu_2} B^{2\mu}u^{2\mu} =\\
= &\ A^1\left(A^0x^0 + \sum_{\mu=1}^{\nu_1} B^{1\mu}u^{1\mu}\right) + \sum_{\mu=1}^{\nu_2} B^{2\mu}u^{2\mu} =\\
= &\ A^1A^0x^0 + A^1\sum_{\mu=1}^{\nu_1} B^{1\mu}u^{1\mu} + \sum_{\mu=1}^{\nu_2} B^{2\mu}u^{2\mu},\\
x^3 = &\ A^2x^2 + \sum_{\mu=1}^{\nu_3} B^{3\mu}u^{3\mu} =\\
= &\ A^2\left(A^1A^0x^0 + A^1\sum_{\mu=1}^{\nu_1} B^{1\mu}u^{1\mu} + \sum_{\mu=1}^{\nu_2} B^{2\mu}u^{2\mu}\right) +\\
&+\sum_{\mu=1}^{\nu_3} B^{3\mu}u^{3\mu} = A^2A^1A^0x^0 +\\
&+A^2A^1\sum_{\mu=1}^{\nu_1} B^{1\mu}u^{1\mu} + A^2\sum_{\mu=1}^{\nu_2} B^{2\mu}u^{2\mu} + \sum_{\mu=1}^{\nu_3} B^{3\mu}u^{3\mu},\\
\ldots&\\
x^t = &\ A^{t-1}x^{t-1} + \sum_{\mu=1}^{\nu_t} B^{T\mu}u^{t\mu} =\\
= &\ \prod_{l=t-1}^{0} A^l x^0 + \prod_{l=t-1}^{1} A^l \sum_{\mu=1}^{\nu_1} B^{1\mu}u^{1\mu} +\\
&+ \prod_{l=t-1}^{2} A^l \sum_{\mu=1}^{\nu_2} B^{2\mu}u^{2\mu} + \ldots +\\
&+A^{t-1}\sum_{\mu=1}^{\nu_{t-1}} B^{t-1\mu}u^{t-1\mu} + \sum_{\mu=1}^{\nu_t} B^{t\mu}u^{t\mu} =
\end{aligned}
$$

$$
\begin{aligned}
&= x^0_{t-1} + \sum_{\mu=1}^{\nu_1} B^{1\mu}_{t-1}u^{1\mu} + \sum_{\mu=1}^{\nu_2} B^{2\mu}_{t-1}u^{2\mu} + \ldots +\\
&\quad + \sum_{\mu=1}^{\nu_{t-1}} B^{t-1\mu}_{t-1}u^{t-1\mu} + \sum_{\mu=1}^{\nu_t} B^{t\mu}_{t-1}u^{t\mu} =\\
&= x^0_{t-1} + \sum_{l=1}^{t}\sum_{\mu=1}^{\nu_l} B^{l\mu}_{t-1}u^{l\mu}.
\end{aligned}
$$

The last formula is true for $t < \tau$. If $\tau \leq t \leq \tau + \omega$, the formula must include echoes and retroactive influence of the future

$$x^t = x^0_{t-1} + \sum_{l=1}^{t} \sum_{\mu=1}^{\nu_l} \left(B^{l\mu}_{t-1} + E^{l\mu}_{t-1} + \Psi^{l\mu}_{t-1} \right) u^{l\mu}. \tag{15.22}$$

By considering the values of $E^{l\mu}_{t-1}$ and $\Psi^{l\mu}_{t-1}$, we can conclude that formula (15.22) is valid for all the values of t.

By subsequent substitution of formula (15.22) in the objective functions, problem (15.19) is reduced to

$$\begin{aligned}
f_{s\pi}(u^{s\pi} || u^{-s\pi}) = & \sum_{t=1}^{T} c^{s\pi t} x^0_{t-1} + \\
& + \sum_{t=1}^{T} c^{s\pi t} \sum_{l=1}^{t} \sum_{\mu=1}^{\nu_l} \left(B^{l\mu}_{t-1} + E^{l\mu}_{t-1} + \Psi^{l\mu}_{t-1} \right) u^{l\mu} + \\
& + \sum_{t=1}^{T} \sum_{\mu=1}^{\nu_t} d^{s\pi t\mu} u^{t\mu} \underset{u^{s\pi}}{\longrightarrow} \max, \\
& s = 1, ..., T, \pi = 1, ..., \nu_s, \\
G^{s\pi} u^{s\pi} \leq & \, g^{s\pi}, s = 1, ..., T, \pi = 1, ..., \nu_s.
\end{aligned}$$

Evidently, the last problem defines a strategic game for which Nash-Stackelberg equilibria are Nash equilibria too. By observing that the objective functions and the system of inequalities are additively separable with respect to $u^{s\pi}$, we can conclude that the Nash-Stackelberg equilibria are simply computed as the solutions of linear programming problems (15.20), where, for simplicity's sake and convenience, the notation s is substituted by t.

So, the theorem is proved. □

15.8 Pareto-Nash-Stackelberg Control of Discrete-Time Linear Processes with Echoes and Retroactive Future

Let us unite models (15.16) and (15.19) by considering a Pareto-Nash-Stackelberg control with T stages and $\nu_1 + \nu_2 + ... + \nu_T$ players, where $\nu_1, \nu_2, ..., \nu_T$ are the player numbers on the stages $1, 2, ..., T$. Any player is identified with two numbers as above in the Nash-Stackelberg control: s (or t) is the stage on which the player selects his strategy and π (or μ) is player's number at the stage s. In a such game, at the each stage s the players $1, 2, ..., \nu_s$, play a Pareto-Nash game by selecting simultaneously their strategies accordingly to their criteria ($k_{s1}, k_{s2}, ..., k_{s\nu_s}$ are the numbers of criteria of the respective players) and by communicating their and all previous selected strategies to the following $s + 1$ stage players. After the all stage strategy selections, all the players compute their gains on the resulting profile. This type of control is called a Pareto-Nash-Stackelberg control.

A mathematical model of the Pareto-Nash-Stackelberg control processes may be established as

$$\begin{aligned} & f_{s\pi}(X, u^{s\pi} \| u^{-s\pi}) = \sum_{t=1}^{T} \left(c^{s\pi t} x^t + \sum_{\mu=1}^{\nu_t} d^{s\pi t\mu} u^{t\mu} \right) \underset{u^{s\pi}}{\longrightarrow} \max, \\ & \qquad s = 1, ..., T, \ \pi = 1, ..., \nu_s, \\ & x^t = A^{t-1} x^{t-1} + \sum_{\mu=1}^{\nu_t} B^{t\mu} u^{t\mu} + \sum_{\mu=1}^{\nu_t} E^{t\mu} u^{t-\gamma\mu} + \sum_{\mu=1}^{\nu_t} \Psi^{t\mu} u^{t+\gamma\mu}, \\ & \qquad t = 1, ..., T, \\ & G^{s\pi} u^{s\pi} \le g^{s\pi}, \ s = 1, ..., T, \ \pi = 1, ..., \nu_t, \end{aligned} \tag{15.23}$$

where x^0, $x^t \in \mathbb{R}^n$, $c^{s\pi t} \in \mathbb{R}^{k_{s\pi} \times n}$, $u^{s\pi} \in \mathbb{R}^m$, $d^{s\pi t\mu} \in \mathbb{R}^{k_{s\pi} \times m}$, $A^{t-1} \in \mathbb{R}^{n \times n}$, $B^{t\mu}$, $E^{t\mu}, \Psi^{t\mu} \in \mathbb{R}^{n \times m}$, $g^{s\pi} \in \mathbb{R}^r$, $G^{s\pi} \in \mathbb{R}^{r \times m}$, t, $s = 1, ..., T$, $\pi = 1, ..., \nu_s$, $\mu = 1, ..., \nu_t$, the parameters τ, γ, ω, are defined as above for the precedent controls and the values of the matrices $E^{t\mu}$ and $\Psi^{t\mu}$ are supposed to be attributed accordingly.

Theorem 15.8 *Let problem (15.23) be solvable. The control*

$$\left(\bar{u}^{11}, \bar{u}^{12}, ..., \bar{u}^{\nu_T T}\right)$$

forms a Pareto-Nash-Stackelberg equilibrium control if and only if every component $\bar{u}^{t\pi}$ *is an efficient solution of the multi-criteria linear programming problem*

$$\begin{aligned} \left(\sum_{l=t}^{T} c^{t\pi l} \left(B_{l-1}^{t\pi} + E_{l-1}^{t\pi} + \Psi_{l-1}^{t\pi} \right) + d^{t\pi t\pi} \right) u^{t\pi} & \underset{u^{t\pi}}{\longrightarrow} \max, \\ G^{t\pi} u^{t\pi} & \le g^{t\pi}, \end{aligned} \tag{15.24}$$

where $B_{l-1}^{t\pi}$, $E_{l-1}^{t\pi}$, $\Psi_l^{t\pi} \in \mathbb{R}^{n \times m}$, *are defined by (15.21).*

Proof Technically, the proof is analogical to the proves of Theorems 15.6 and 15.7. The particularity of problem (15.23) is taken into account by connecting to problem (15.20) the multi-criteria essence of the objective functions in (15.23), and obtaining on this basis (15.24). □

15.9 Concluding Remarks

Real dynamic processes may be very different. As a rule, the problem of their control is very difficult. Different types of control may be highlighted: optimal control, Stackelberg control, Pareto-Stackelberg control, Nash-Stackelberg control, Pareto-

Nash-Stackelberg control, etc. These types of control where studied earlier, e.g. in [1–3]. But, real processes may include various and specific parameters of secondary phenomena such as echoes and retroactive future. For such type of dynamic processes we exposed above the mathematical models of Pareto-Nash-Stackelberg control and established the principles/methods for a control determining by a special type decomposition of the initial problems. It is important to observe that we established pseudo-polynomial complexity of the afferent problems.

The presented mathematical models may be seen as a starting point to considering other interesting decision and control processes, both discrete and continuous. The variety of control problems that arises is impressively large. They indubitably are an interesting subject for the future investigations.

References

1. Ungureanu V. 2013. Linear discrete-time Pareto-Nash-Stackelberg control problem and principles for its solving. *Computer Science Journal of Moldova*, 21, No. 1:(61), 65–85.
2. Ungureanu, V. 2008. Solution principles for simultaneous and sequential games mixture. *ROMAI Journal* 4 (1): 225–242.
3. Ungureanu, V., and V. Lozan. 2013. Linear discrete-time set-valued Pareto-Nash-Stackelberg control processes and their principles. *ROMAI Journal* 9 (1): 185–198.
4. Pontryagin, L.S., Boltyanskii, V.G., Gamkrelidze, R.V., and Mishchenko, E.F. 1961. *Mathematical theory of optimal processes*. Moscow: Nauka, 1961, 393 pp. (in Russian).
5. Bellman, R. 1957. *Dynamic Programming*, 365. New Jersey: Princeton University Press.
6. Von Stackelberg, H. 1934. *Marktform und Gleichgewicht (Market Structure and Equilibrium)*. Vienna: Springer Verlag, XIV+134 pp. (in German).
7. Nash, J. 1951. Noncooperative games. *Annals of Mathematics* 54 (2): 280–295.
8. Von Neumann, J., and Morgenstern, O. 1944. *Theory of Games and Economic Behavior*. Princeton, New Jersey: Annals Princeton University Press, (2nd edn., 1947), 674 pp.
9. Leitmann, G. 1978. On Generalized Stackelberg Strategies. *Journal of Optimization Theory and Applications* 26: 637–648.
10. Nahorski, Z., H.F. Ravn, and R.V.V. Vidal. 1983. *Optimization of Discrete Time System: the Upper Boundary Approach*, vol. 51, 139., Lecture Notes in Control and Information Sciences Berlin: Springer.
11. Ter-Krikorov, A.M. 1977. *Optimal Control ans Mathematical Economy*. Moscow: Nauka, 217 pp. (in Russian).
12. Anderson, B.D.O., and J.B. Moore. 1971. *Linear Optimal Control*, 413. New Jersey: Prentice-Hall Inc, Englewood Cliffs.
13. Ashmanov, S.A., and A.V. Timohov. 1991. *The optimization theory in problems and exercises*, 142–143. Moscow: Nauka.

Index

V. Ungureanu, *Pareto-Nash-Stackelberg Game and Control Theory*,
Smart Innovation, Systems and Technologies 89,
https://doi.org/10.1007/978-3-319-75151-1

L

M

N

O

S

The manufacturer's authorised representative in the EU is Springer Nature Customer Service Centre GmbH, Europaplatz 3, 69115 Heidelberg, Germany. If you have any concerns regarding our products, please contact ProductSafety@springernature.com

Printed and bound by CPI Group (UK) Ltd, Croydon, CR0 4YY
31/07/2026
02182922-0001